Lecture Notes in Mathematics

Edited by A. Dold and B. Eckmann

T0184413

445

Model Theory and Topoi

A Collection of Lectures by Various Authors

Edited by F. W. Lawvere, C. Maurer and G. C. Wraith

Springer-Verlag
Berlin · Heidelberg · New York 1975

Editors
Dr. F. William Lawvere
Department of Mathematics
State University of New York
at Buffalo
Amherst, N. Y. 14226/USA

Dr. Christian Maurer
Freie Universität Berlin
Institut für Mathematik II
D–1 Berlin 33
Königin-Luise-Str. 24–26

Dr. Gavin C. Wraith
School of Mathematical
and Physical Sciences
University of Sussex
Brighton BN1 9QH/England

Library of Congress Cataloging in Publication Data

Model theory and topoi.

(Lecture notes in mathematics ; 445)
Bibliography: p.
Includes index.
1. Model theory. 2. Toposes. I. Lawvere, F. W.
II. Maurer, Christian, 1945- III. Wraith, Gavin C.,
1939- IV. Series.
QA3.L28 no. 445 [QA9.7] 510'.8s [511'.8] 75-20007

AMS Subject Classifications (1970): 02B15, 02C15, 02G20, 02H10, 02J05, 02J15, 02J99, 02K10, 06A20, 14A20, 14A99, 18A05, 18A15, 18A25, 18A30, 18B05, 18B99, 18C05, 18C10, 18D15, 18E15, 18E99, 18F10, 18F20, 18F99

ISBN 3-540-07164-4 Springer-Verlag Berlin · Heidelberg · New York
ISBN 0-387-07164-4 Springer-Verlag New York · Heidelberg · Berlin

This work is subject to copyright. All rights are reserved, whether the whole or part of the material is concerned, specifically those of translation, reprinting, re-use of illustrations, broadcasting, reproduction by photocopying machine or similar means, and storage in data banks.

Under § 54 of the German Copyright Law where copies are made for other than private use, a fee is payable to the publisher, the amount of the fee to be determined by agreement with the publisher.

© by Springer-Verlag Berlin · Heidelberg 1975
Printed in Germany
Offsetdruck: Julius Beltz, Hemsbach/Bergstr.

TABLE OF CONTENTS

PART I

(Manuscripts received by the editors in October 1973)

Introduction to Part I

F. William Lawvere

Part I of this volume consists of three of the first papers on functorial model theory, developing concretely the approach to algebraic logic according to which a "theory" (understood in a sense invariant with respect to various "presentations" by means of particular atomic formulas and particular axioms) is actually a category T having certain properties P and a model of T is any set-valued P - preserving functor. As a rough general principle, one could choose for P any collection of categorical properties which the category of sets satisfies, the choice then determining the "doctrine" of theories of kind P, which is thus a (non-full) subcategory of the category of small categories. For example, the doctrine of universal algebra thus springs from the fact that the category of sets has the property P of having finite cartesian products, while the doctrine of higher-order logic springs from the property of being a topos. The much-researched intermediate doctrine of (classical) first-order logic corresponds to the fact P that the category of sets has finite limits, complements of subsets, and images of mappings (related by the condition of being a "regular" category, which is essentially the logical rule $\exists x[A \wedge B(x)] \equiv A \wedge \exists x\, B(x)$ for A independent of x). The usual syntactical preoccupations of logic appear in the following way: once the logical operations and rules of inference are fixed (by the choice of P) the question arises of investigating free objects and hence presentation of arbitrary objects in the category of all P-categories T. But the often encountered suggestion that "syntax comes first" is refuted: the essential role of theories is to describe their models, and the same applies also to presentations of theories when the latter are needed for calculation. We often encounter and deal with groups for which we do not know or do not use any presentation: the same is true of theories.

Of course, for an arbitrary given P there is no guarantee of "completeness" in the usual sense, i.e. an arbitrary P-category T may fail to have enough models in the originally-envisioned category $\mathcal{S}$ of sets, sometimes paradoxically due to the fact that abstract sets are too "constant"; on the other hand it has become clear in the past decade that we are for reasons of geometry and analysis in fact interested in models in more general categories of variable sets such as sheaves over a topological space, Boolean-valued sets, algebraic spaces, permutation representations of a group, etc. - it is because of that that the interaction between the geometrical and logical aspects of general topoi has become an object of investigation, for example in the Bangor and Berlin parts of this volume.

Since a variable set may be partly empty and partly non-empty, the traditional model-theoretic banishment of empty models cannot be maintained, bringing to light a certain difficulty which the banishment obscured. Some claim that this difficulty is the "fact" that "entailment is not transitive", contrary to mathematical experience. However, the actual "difficulty" is that the traditional logical way of dealing with variables is inappropriate and hence should be abandoned. This traditional method (which by the way is probably one of the reasons why most mathematicians feel that a logical presentation of a theory is an absurd machine strangely unrelated to the theory or its subject matter) consists of declaring that there is one set I of variables on which all finitary relations depend, albeit vacuously on most of them; e.g. a binary relation on X is interpreted as $X^I \to 2$ depending vacuously on all but two of the variables in I. This is of course not totally absurd, since in the case of non-empty single-sorted structures, such an interpretation can be associated (in an infinite number of different but equivalent ways) to a correct interpretation. However, the fact that 2^{X^I} is a single Boolean algebra (claimed sometimes to be a "convenience") implies that propositional operators such as $\wedge, \vee, \Rightarrow$, applied indiscriminately to finitary relations, can be given a "meaning", a highly dubious "gain in generality", especially when, as noted above, the useful generalization to many sorts and/or partly empty domains is made.

Actually the (binary) propositional operators can only meaningfully be applied to (pairs of) relations having the <u>same</u> free variables. This may seem to prohibit such combinations as

$$(*) \qquad A(x,y) \wedge A(y,z) \Rightarrow A(x,z)$$

but consider the actual meaning: A denotes some subobject of the square X^2 of some sort X, and $(*)$ denotes a certain subobject of the cube X^3. The three projection maps $X^3 \rightrightarrows X^2$ induce three different substitution operators which to a binary relation A associate three different ternary relations $\sigma_{12}A$, $\sigma_{23}A$, $\sigma_{13}A$. Since conjunction and implication can meaningfully be applied to ternary relations, there is a ternary relation

$(\sigma_{12}A) \wedge (\sigma_{23}A) \Rightarrow \sigma_{13}A$ of which (*) is an abbreviation. Thus a syntax for presenting theories can be given in which propositional operators operate only among formulas with each fixed finite set of free variables, while substitution operators on an equal footing with quantifiers operate to change the set of free variables of a formula. These substitution operators have the structure (not of a monoid but) of a category with finite cartesian products; they need not consist only of tuples of projections, diagonal maps, etc. for if the presentation contemplates also function symbols, any m-tuple of terms in n free variables denotes a map $X^n \xrightarrow{f} X^m$ and hence induces a substitution f^* from m-ary relations to n-ary relations. If several basic sorts are considered, it is reasonable to consider that X^n, X^m are themselves further sorts V, W and that the m-tuple f of terms just referred to is simply another kind of term $V \xrightarrow{f} W$; it is then sensible to regard quantifications $\exists_f, \forall_f$ along an arbitrary such f , not only quantifications $\exists x, \forall x$ along projection maps $W \times X \to W$. The meaning of $\exists_f$, applied to a relation A of sort (or type) V is simply the relation $\exists_f A$ of type W which is the <u>image</u> of the composite map $A \rightarrowtail V \xrightarrow{f} W$; for any relation B of type W ,

$$\exists_f A \vdash_W B \quad \text{iff} \quad A \vdash_V f^* B$$

$$B \vdash_W \forall_f A \quad \text{iff} \quad f^* B \vdash_V A$$

are the rules of inference which characterize the two quantifiers as being respectively left and right adjoint to substitution. The subscripts V,W indicate that also entailments are only meaningful if both hypothesis and conclusion have the same set of free variables; the semantical meaning of entailment is inclusion between subjects of V (respectively of W).

It may be objected that in the above description of doctrines of theories the primacy of syntax has not been overturned since the determining property P must presumably be written in some language of categories. Since a general investigation of something like a "category of doctrines" has so far not seemed useful, the possible productive consequences of this contradiction, if any, are not known. However, one striking fact should be pointed out: While classes of theories with complicated definitions have been investigated in particular, the distinctive <u>general</u> classes which have actually been of interest, namely universal algebra, positive first-order logic, first-order logic, weak second order logic (= the "arithmetic universes" of Joyal), higher-order logic, etc, are all definable within an <u>equational</u> metatheory. More precisely the definition of such a doctrine amounts itself to a <u>cartesian</u> category (= category with finite limits) obtained by adjoining to the universal Horn theory of categories certain additional operators (usually denoting functors or natural transformations) whose domain is defined by equations, and imposing certain equations (which may hold only on equationally defined subvarieties) - usually in fact these equations express adjointness or distributivity of limits. Thus no disjunctions or existential quantifiers, nor any genuine occurence of universal quantifiers or implication, are involved in the definition of these doctrines. Here by a genuine occurence of a universal quantifier I mean something like the definition of a generator G

$$\forall x[G \xrightarrow{x} X \Rightarrow fx = gx] \vdash f = g$$

but not a universal Horn sentence

$$\forall x[A(x) \Rightarrow B(x)]$$

which can be replaced by a (free variable) inclusion of subobjects of X

$$A \vdash_X B$$

Even the "strong" conditions which distinguish a topos of "constant" sets from a general topos of variable sets,

(Axiom of Choice) For $X \xrightarrow{f} Y$,

if $1_Y \vdash \exists_f(1_X)$ then there exists $Y \xrightarrow{x} X$ with $f \circ x = 1_Y$

(Two-valuedness) For $1 \underset{\psi}{\overset{\varphi}{\rightrightarrows}} 1 + 1$

if $1 \vdash \varphi \vee \psi$ then $1 \vdash \varphi$ or $1 \vdash \psi$

do not involve genuine occurrences of universal quantification or implication, but do involve there exists and or on the right-hand side of an inference; hence, while not expressible in a cartesian (= Horn) metatheory, they are expressible in a pretopos metatheory so that the full algebraico-geometric method of coherent classifying topoi is applicable to them.

The paper by Orville Kean (his 1971 U. of Penn. dissertation) considers the case of theories which can be presented by axioms having the form of universal Horn sentences, i.e. the extension of "equational" universal algebra to the case in which some of the postulated identities between operations may hold only on "algebraic varieties" defined by equations between some other operations. Were one to consider an arbitrary set of "sorts", varying from theory to theory, rather than limiting oneself to the "one base set" for an algebra as is customary in universal algebra, and were one to allow further the possibility of partial operations whose domains of definition were such "algebraic varieties", then the appropriate condition on a category T would simply be: T is any small category with finite inverse limits (i.e. terminal object and pullbacks, hence finite products and equalizers, exist in T). Kean however takes care to analyze the further conditions on T corresponding to the restriction to <u>one</u> base sort on which all operations are defined. With or without these further conditions, the correct notion of model is simply any functor T → S which preserves finite limits (i.e. which is "left exact") and the category of models is the category Lex(T,S) of all such functors and <u>all</u> natural transformations between them. These categories of models retain the features from the equational universal algebra of being complete and having a set of generators which are "finitely-presented" objects in a categorically invariant sense, but in general <u>fail</u> to satisfy the two further properties characteristic of <u>equational</u> universal algebra that these generators can be taken as projective objects and that equivalence relations are effective (= "precongruences are congruences" in the terminology of my 1963 articles). The precise definition of "finitely-presented objects" can be found in Gabriel & Ulmer's Springer Lecture Notes volume 221, which also (implicitly) shows that "the functor Semantics has a functor Structure adjoint to it", but does not take any account of the relation with the logical concept of universal Horn axioms as Kean does. Another important feature of equational universal algebra which remains valid is the existence of left adjoints to the "algebraic" (syntactically induced) functors; i.e. if T' → T is any functor preserving finite limits between small categories having them, then the induced "forgetful" functor $\text{Lex}(T,\mathbb{S}) \to \text{Lex}(T',\mathbb{S})$ has a left adjoint. Here, since preferred "sorts" have less invariant significance in this doctrine, there is less motivation for requiring $T' \to T$ to preserve them even if they are there; this has of course the effect that such "forgetful" functors need not be faithful, but the added generality is mathematically very natural. For example, the functor SO(2) from the category of commutative rings to the category of abelian groups is induced by a functor $T' \to T$ which does not preserve the base sort, since the base sort of the Horn theory of abelian groups is mapped to the subobject $\{\langle x,y\rangle | x^2 + y^2 = 1\}$ of the square of the base sort of the theory of commutative rings, but it is clear that this latter functor should be considered as an interpretation of the theory of abelian groups into the theory of commutative rings, indeed an interpretation "definable" within the doctrine of Horn theories.

The completeness of the category of models and the existence of left adjoints for induced functors are properties which in general do not carry over to theories more complicated than Horn theories, though it now seems that the adjoints may be recovered by allowing the "set-theory" $\mathbb{S}$ to vary along with the models (see remarks below).

The first detailed development of a purely categorical concept corresponding to full first-order theories was in the 1971 Dalhousie dissertation of Volger, on which the second article in this volume is based. The various sets of conditions on a category T which are considered in this article are corrections and improvements of a set conjectured earlier by me which exploited special properties of the Boolean case and coded formulas as morphisms into an object Ω which in various cases may be interpreted roughly as the object of sentences or the truth-value object. Volger considers throughout an arbitrary set of sorts, both because it is no more difficult and because various results, in particular his completeness theorem, then apply without change to type theory, which, whatever the exact notion of first-order theory T, means one which as a category is cartesian closed. Another feature which has remained invariant through the various experimentation which has gone on is the interpretation of quantifiers as functors adjoint to substitutions. Volger also outlines a modification of the completeness proof due to Andre Joyal which has played a role in the further unpublished development of the subject which has taken place since these papers were written.

These early calculations in categorical logic played a role in the development of the elementary theory of topoi (see, in addition to the present volume, SLN 274 and articles by Barr, Johnstone, W. Mitchell, Osius, and Paré in the Journal of Pure and Applied Algebra and the Bulletin of the AMS, Freyd's article in the Bulletin of the Australian Math Soc., for some of these developments) which in turn has affected the recent work in functorial model theory. In particular, using topoi, Kock and Mikkelsen (in the Victoria Symposium, SLN 369) generalized and clarified some basic constructions of non-standard analysis, which was one of the spurs to the further simplifications and application contained in Volger's second paper (1972) in this volume.

In the remainder of this introduction I sketch briefly some more recent developments in geometric logic wherein theories are modelled functorially in general topoi or in other words continuously variable models are studied. In this the doctrine of positive logic, i.e. $\exists, \wedge, \vee$, but no special attention to $\forall, \Rightarrow$, necessarily plays a distinguished role, since it is just this logic which is preserved under arbitrary continuous change of parameter space (the $\vee$ may be allowed to be infinitary) and also because an arbitrary Grothendieck topos can be viewed as the "classifying topos" for such a theory. However, full first-order logic can also be handled using the method due to Kripke and refined by Joyal and Freyd. More details can be found in my forthcoming paper in the Proceedings of the 1973 Bristol Logic meeting and in papers of Freyd, Johnstone, Joyal, Reyes and Wraith and by Benabou and his students.

In fact, important in algebraic geometry, that a sheaf of local rings is just a "local ring object" in the category of set-valued sheaves, remains valid when the theory of local rings is replaced by any many sorted theory in which only the logical operations $\wedge \vee \exists$ are considered and when sheaves are taken to mean objects in any topos. Here the truth of an existential statement or disjunction in the intrinsic logic of the topos is found by the adjointness rules of inference to mean locally, existence or locally, disjunction. The discrepancy between true (globally) and globally true (which is due to the fact that epimorphisms need not have sections and which gives rise to cohomology) may be exemplified by the fact that sheaf theoretically complex exponentiation is an epimorphism and hence the statement that the logarithm exists is true globally, but the actual existence takes place on a covering only. Intuitionistically, the same sort of relation between local and global holds even for a cubic. This class of theories may be considered to include any <u>classical</u> theory, since the negations of formulas may be considered as further atomic formulas and the axioms of negation considered as particular axioms rather than general axioms.

But the doctrine is basically intuitionistic, as is the intrinsic logic of the topoi where models are to be valued. The geometrically invariant condition on T to be a theory according to this doctrine is precisely that it should be a pretopos in the sense of Grothendieck-Verdier Exposé VI in Springer Lecture Notes Volume 270. The finite-covering topology on T leads to a topos T which, as pointed out by Reyes, has the property that for any topos X the category of continuous maps $X \to T$ is equivalent to the category of models in the "set theory" X of the theory T. The topos T is coherent in the sense of SLN 270 and all such arise from such theories; one may consider T as $\mathcal{S}[U]$, the "set theory" obtained by freely adjoining to the category of sets an indeterminate model U of T. Even for the theory T of equality, this construction is instructive; T in that case is the functor category $\mathcal{S}^{S_0}$ (where S_0 is the category of finite sets) which is a non-trivial topos whose category of points is equivalent to the category of sets, and we have that for any topos X, the sheaves on X are just the continuous functions from X into the (generalized) space T of sets.

The theorem of Deligne that every coherent topos has enough (set-valued) points is seen from the above discussion to be equivalent with the fact that every many-sorted intuitionistic theory taking account only of $\wedge, \vee, \exists$ has enough set-valued models. Further, the Kripke completeness theorem (preserving also $\forall, \Rightarrow$ when they exist) has been elegantly proved by Joyal in the invariant setting. The Kripke-Joyal Theorem constructs a model $\mathcal{S}^{\mathbb{D}} \to T$ in a functor category rather than in sets $\mathcal{S}$; while the model itself preserves $\forall, \Rightarrow$ the "models" in $\mathcal{S}$ derived by evaluating at a given "stage of knowledge" $D \in \mathbb{D}$ usually do not.

Varying the topos in which we take models is quite essential for certain universal problems. For example consider the interpretation $T \to \overline{T}$ of the theory of commutative rings into the theory of local rings and consider any given ring A. The problem of finding a local ring $\overline{A}$ universal among all those to which A maps has no solution if we consider only one topos, but on the other hand if we allow the set theory to spread out, there is such a universal local ring in the topos called spec (A); thus the universal problem involves finding the natural domain of variation for the quantities in A, which will usually not be only the single point which corresponds to the topos of constant sets. When the topos of departure does not satisfy the axiom of choice, spec (A) does not have enough internal points (contrary to the incorrect statement in my paper for the 1970 International Congress) but Joyal has given a very simple internal construction of it using the notion of distributive lattice object. Since spec (A) is coherent* if $A \in \mathcal{S}$, Deligne's theorem yields enough external points for it when $\mathcal{S}$ does satisfy the axiom of choice. When the base topos of departure does not satisfy the axiom of choice, i.e. when it consists of variable sets varying in an organic fashion, a suitable formulation along these lines of a general completeness theorem for first-order theories in it has still to be found; such a formulation would presumably partly reflect the fact that in the real world consistency of a theory is not sufficient for the existence of models.

---

*To prevent a possible delay in understanding the important exposé VI (SLN 270) of Grothendieck-Verdier cited above, it should be pointed out that their statement to the effect that separated coherent spaces are finite is incorrect; in fact these spaces are just the Stone spaces of arbitrary Boolean algebras, while arbitrary coherent topoi which are generated by their open sets are just "Stone spaces" of arbitrary distributive lattices. This is also a good place to point out that my statement in Springer Lecture Notes 274 that universal quantification in a topos leads to a triple is also incorrect; what was intended there is simply that universal quantification and infinite internal intersection satisfy the reasonable formal laws.

Abstract Horn Theories

Orville Keane

Introduction

In this paper we apply a technique similar to functorial semantics [6] to obtain a characterization of categories whose classes of objects are models for universal Horn theories and whose maps are homomorphisms (i.e. maps which preserve atomic formulas) between the models. A universal Horn theory is a formal first-order theory whose axioms are all of the form (1) A or (2) $(A_1 \wedge \ldots \wedge A_n) \longrightarrow B$ where A, B, A_i, $i = 1,\ldots,n$ are all atomic formulas. Partially ordered sets, torsion free groups and (equational) algebraic theories are all examples of universal Horn theories.

The categorical counterpart of a universal Horn theory is called an Abstract Horn Theory and is defined in Chapter 1 as a small, skeletal, finitely complete category with a cogenerator M such that: (1) every object can be embedded in a finite power of M so that M looks injective with respect to the embedding, and (2) the maps which make M look injective are closed under the formation of pullbacks and products. If $\mathcal{J}$ is an Abstract Horn Theory then we denote the category whose objects are the finitely continuous functors from $\mathcal{J}$ to the category of sets S, and whose maps are the natural transformations between the functors by $S^{(\mathcal{J})}$. Given an abstract Horn theory $\mathcal{J}$, there exists an associated universal Horn theory denoted $H_{\mathcal{J}}$ such that $S^{(\mathcal{J})}$ is equivalent to the category whose objects are the models for $H_{\mathcal{J}}$ and whose maps are homomorphisms between the models (Proposition 1.4.1). Conversely given a universal Horn theory H we derive an associated abstract Horn theory in the following way. Let C_H denote the category whose objects are models for H and whose maps are homomorphisms and let A be an n-ary formula in $L(H)$ which is a conjunction of atomic formulas. Then the functor:

$$U_A\colon\ C_H \longrightarrow S$$

such that $U_A(N) = \{\langle a_1,\dots,a_n\rangle \mid N \models A(a_1,\dots,a_n)\}$ is representable (Corollary 1.5.2). Let $\mathcal{A}$ be a full sub-category of C_H whose objects are the models of H which represent the U_A's. Then $\mathcal{A}^{op}$ is an abstract Horn theory (Proposition 1.7.2) . Furthermore, $S^{(\mathcal{A}^{op})} \simeq C_H$ (Corollary 2.2.3).

Chapter 3 is the Characterization Theorem. Fittler introduced the notion of Lowenheim-Skolem Dense (L.S.D.) in [2]. We use this notion in our characterization which states that a category $\mathcal{A}$ is equivalent to a category of models of a universal Horn theory iff it contains a small full subcategory $\mathcal{B}$ such that (1) $\mathcal{B}$ is L.S.D. in $\mathcal{A}$ and (2) $\mathcal{B}^{op}$ is an abstract Horn theory.

In Chapter 4 we discuss maps between abstract Horn theories. A Horn theory map is defined as a finitely continuous functor between two abstract Horn theories which preserve the cogenerator and the maps which make the cogenerator look injective. Horn theory maps induce maps (going in the opposite direction) between the corresponding categories of models. The induced map always has an adjoint (Proposition 3.1.4). Theorem 3.2.5 states that a functor from one category of models of a universal Horn theory to another which preserves underlying sets is induced by a Horn theory map iff it preserves submodels, products and direct limits. A corollary states roughly that a functor between categories of models of two universal Horn theories is induced by a map on the language level iff the conditions stated in the theorem are satisfied.

Chapter I

Abstract Horn Theories

1.1 Abstract Horn Theories

By an abstract Horn theory we mean a small skeletal finitely complete category $\mathcal{T}$ with a cogenerator M such that:

(1) Every object in $\mathcal{T}$ can be embedded in a finite power of M so that M looks injective with respect to the embedding.

(2) The maps which make M look injective are closed under pullbacks and products.

If $\mathcal{T}$ is an abstract Horn theory then we shall use the term monic to refer to the maps which make the co-generator look injective. Note that

$$E \xrightarrow{p} X \underset{y}{\overset{x}{\rightrightarrows}} Y$$

is an equalizer diagram iff

$$\begin{array}{ccc} E & \xrightarrow{p} & X \\ {\scriptstyle p}\downarrow & & \downarrow{\scriptstyle (1,y)} \\ X & \xrightarrow[(1,x)]{} & X \times Y \end{array}$$

is a pullback, hence condition (2) in the definition above implies that equalizers are monics.

The category of finite cardinal numbers with 2 as a cogenerator is an example of an abstract Horn theory. Also, every algebraic theory as defined by Lawvere in [6] is an abstract Horn theory.

Let S be the category of sets. If $\mathcal{T}$ is an abstract Horn theory, then by $S^{(\mathcal{T})}$ we mean the category whose class of objects are the finitely continuous functors from $\mathcal{T}$ to $\mathcal{S}$ and whose maps are the natural transformations between the functors. Suppose $T \in S^{(\mathcal{T})}$ and $X \xrightarrow{p} M^n$ is monic in $\mathcal{T}$. Then

$$\begin{array}{ccc} X & \xrightarrow{1} & X \\ {\scriptstyle 1}\downarrow & & \downarrow{\scriptstyle p} \\ X & \xrightarrow[p]{} & M^n \end{array}$$

is a pullback hence $T(p) : T(X) \longrightarrow \{T(M)\}^n$ is a monomorphism. Thus in a sense $T(p)$ defines an n-ary predicate on $T(M)$.

1.2 The Associated Universal Horn Theory

A universal Horn theory is a formal first-order theory H whose axioms are all of the form:

(1) A where A is an atomic formula

(2) $(A_1 \wedge \ldots \wedge A_n) \longrightarrow B$ where $A_1,\ldots,A_n,B$ are all atomic formulas.

Examples of universal Horn theories are partially ordered sets, torsion free groups and any algebraic theory which can be defined equationally.

For every abstract Horn theory $\mathcal{T}$ there exists an associated universal Horn theory, which we shall denote by $H_{\mathcal{T}}$. We construct $H_{\mathcal{T}}$ as follows:

1. The language $L(H_{\mathcal{T}})$

   a) f is an n-ary function symbol in $L(H_{\mathcal{T}})$ iff $M^n \xrightarrow{f} M$ is in $\mathcal{T}$.

   b) p is an n-ary predicate symbol in $L(H_{\mathcal{T}})$ iff $X \xrightarrow{p} M^n$ is monic in $\mathcal{T}$.

2. The axioms of $H_{\mathcal{J}}$.

a) If $M^{\ell} \xrightarrow{f} M^m \xrightarrow{g} M^n = M^{\ell} \xrightarrow{h} M^n$ then

i) If $m \neq 0$, $n \neq 0$, $\bigwedge_{i=1}^{n} (h_i(t_1,\ldots,t_{\ell}) = g_i(f_1(t_1,\ldots,t_{\ell}),\ldots,f_n(t_1,\ldots,t_{\ell})))$ is an axiom

ii) If $m = 0$, $n \neq 0$, then $\bigwedge_{i=1}^{n} (h_i(t_1,\ldots,t_{\ell}) = g_i)$ is an axiom

b) If $X \xrightarrow{f} Y$ in $\mathcal{J}$ and $X \xrightarrow{p} M^m$, $Y \xrightarrow{q} M^n$ are monic, then

i) If $n \neq 0$, then for each $M^m \xrightarrow{g} M^n$ such that $pg = fq$, (there must exist at least one such map), $p(t_1,\ldots,t_m) \longrightarrow g(g_1(t_1,\ldots,t_n),\ldots,g_n(t_1,\ldots,t_m))$ is an axiom.

ii) If $n = 0$ then $p(t_1,\ldots t_m) \longrightarrow q$ is an axiom.

c) If $p = 1_{M^n}$, then $p(t_1,\ldots,t_n)$ is an axiom.

d) If $E \xrightarrow{p} M^m \underset{g}{\overset{f}{\rightrightarrows}} M^n$ is an equalizer diagram then $p(t_1,\ldots,t_m) \longleftrightarrow \left[\bigwedge_{i=1}^{n} (f_i(t_1,\ldots,t_m) = g_i(t_1,\ldots,t_m))\right]$ is an axiom.

e) If $X \xrightarrow{p} M^n$ and $Y \xrightarrow{q} M^n$ are monic and

$$\begin{array}{ccc} Z & \xrightarrow{z} & Y \\ {\scriptstyle w}\downarrow & & \downarrow{\scriptstyle q} \\ X & \xrightarrow[p]{} & M^n \end{array}$$

is a pullback diagram then

$r(t_1,\ldots,t_n) \longleftrightarrow (p(t_1,\ldots,t_n) \wedge q(t_1,\ldots,t_n))$

is an axiom, where $r = wp$.

f) If $X \xrightarrow{p} M^m$ and $Y \xrightarrow{q} M^n$ are monic and $r = p \times q$ then $r(t_1,\ldots,t_m,s_1,\ldots,s_n) \longleftrightarrow (p(t_1,\ldots,t_m) \wedge q(s_1,\ldots,s_n))$ is an axiom.

Whereas some of the above axioms are not in the form of universal Horn formulas, each is easily seen to be logically equivalent to a conjunction of universal Horn formulas. Thus $H_{\mathcal{T}}$ is logically equivalent to a universal Horn theory.

1.3 Categories of Models of Horn Theories

If H is a universal Horn theory then by C_H we mean the category whose class of objects are the (normal) models of H and whose maps are the homomorphisms between the models, (i.e. the maps which preserve the atomic formulas). We shall use U to denote the forgetful functor from C_H to the category of sets S. If A is an n-ary atomic formula in $L(H)$ then we shall use U_A to denote the functor from C_H to S such that

(1) $U_A(N) = \{\langle a_1,\ldots,a_n\rangle \in U^n(N) \mid N \models A(a1,\ldots,a_n)\}$

(2) If $N_1 \xrightarrow{f} N_2$ in C_H then $U_A(f)$ is the unique map from $U_A(N_1)$ to $U_A(N_2)$ such that the following diagram commutes:

$$\begin{array}{ccc} U_A(N_1) & \xrightarrow{U_A(f)} & U_A(N_2) \\ \big\downarrow & & \big\downarrow \\ U^n(N_1) & \xrightarrow[U^n(f)]{} & U^n(N_2) \end{array}$$

It is understood that if A is 0-ary, then $U_A(N) = 1$ if $N \models A$ and $U_A(N) = \emptyset$ if $N \models \neg A$.

1.4 The Equivalence of $S^{(\mathcal{J})}$ and $C_{H_\mathcal{J}}$

If $\mathcal{J}$ is an abstract Horn theory and $T \in S^{(\mathcal{J})}$, then there is an obvious way in which T can be made into a model for $H_\mathcal{J}$. That is to say, there is a rather obvious functor V from $S^{(\mathcal{J})}$ to $C_{H_\mathcal{J}}$. V is constructed as follows:

If $T \in S^{(\mathcal{J})}$ then $V(T)$ must be a model for $H_\mathcal{J}$, hence we have to define an $L(H_\mathcal{J})$-structure on $V(T)$. This is done as follows:

(1) $U[V(T)] = T(M)$, where M is the cogenerator in $\mathcal{J}$.

(2) If p is an n-ary predicate in $H_\mathcal{J}$, then $X \xrightarrow{p} M^n$ is monic in $\mathcal{J}$. We define:

$$U_p(V(T)) = Im(T(p)) .$$

(3) If f is an n-ary function symbol in $H_\mathcal{J}$ then $\bar{f} = T(f) : U^n(V(I)) \longrightarrow U(V(T))$.

If $n: T_1 \longrightarrow T_2$ is a map in $S^{(\mathcal{J})}$ then $V(n) = n_M: U(V(T_1)) \longrightarrow U(V(T_2))$ is a homomorphism.

The proof that V is a well defined functor is straightforward (though tedious). One checks the axioms. In fact V is an equivalence. Thus we have the following proposition.

Proposition 1.4.1: If $\mathcal{J}$ is an abstract Horn theory then $S^{(\mathcal{J})} \approx C_{H_\mathcal{J}}$.

We omit the proof as it is very long (at least ten pages), but completely straightforward. The functor W from $C_{H_\mathcal{J}}$ to $S^{(\mathcal{J})}$ uses the Axiom of Choice and is constructed in the following manner:

Let $\mathcal{O}$ be a well-ordering of the monics in $\mathcal{J}$ with range M^n for $n = 0,1,2,\ldots$. For each X in $Ob(\mathcal{J})$ we define P_X as follows:

$$P_X = \begin{cases} 1_{M^n} & \text{if } X = M^n \\ \text{The first map in } \mathcal{G} \text{ with domain } X & \\ \text{if } X \neq M^n,\ n = 0,1,2,\ldots & \end{cases}$$

If $N \in C_{H_{\mathcal{T}}}$, then $W(N)$ is a functor from $\mathcal{T}$ to S. We define $W(N)$ on objects and maps in $\mathcal{T}$ as follows:

(1) If $X \in \mathrm{Ob}(\mathcal{T})$, then $[W(N)](X) = U_{P_X}(N)$.

(2) If $X \xrightarrow{f} Y$ is a map in $\mathcal{T}$ then the definition of an abstract Horn theory implies the existence of at least one map g such that the following diagram commutes:

$$\begin{array}{ccc} X & \xrightarrow{f} & Y \\ {\scriptstyle P_X}\downarrow & & \downarrow{\scriptstyle P_Y} \\ M^m & \xrightarrow[g]{} & M^n \end{array}$$

We define $[W(N)](f)$ to be the unique map such that the following diagram commutes for all g which makes the above diagram commute:

$$\begin{array}{ccccccc} [W(N)](X) & = & U_{P_X}(N) & \xrightarrow{[W(N)](f)} & U_{P_Y}(N) & = & [W(N)](Y) \\ & & \downarrow & & \downarrow & & \\ [W(N)](M^n) & = & U^m(N) & \xrightarrow[\bar{g}]{} & U^n(N) & = & [W(N)](M^n) \end{array}$$

The uniqueness of definition of $[W(N)](f)$ follows from the fact that if P_X equalizes g and h then there exists an axiom in $H_{\mathcal{T}}$ which forces $U_{P_X} \hookrightarrow U^m(N)$ to equalize $\bar{g}$ and $\bar{h}$.

1.5 Standard Complete Categories of Models

Let T be a first-order theory, F a set of formulas in the language $L(T)$. We shall use $C_{T,F}$ to denote the category whose objects are the (normal) models for T and whose maps are the maps between the models which preserve the formulas in F [3].

If A is an n-ary formula which is a conjunction of formulas in F, then we shall us U_A to denote the functor from $C_{H,F}$ to S such that

(1) $U_A(N) = \{\langle a_1,\ldots,a_n\rangle \in U^n(N) \mid N \models A(a_1,\ldots,a_n)\}$

(2) If $N_1 \xrightarrow{\theta} N_2$ is a map in $C_{T,F}$, then $U_A(\theta)$ is the unique map from $U_A(N_1)$ to $U_A(N_2)$ such that the following diagram commutes:

$$\begin{array}{ccc} U_A(N_1) & \xrightarrow{U_A(\theta)} & U_A(N_2) \\ \downarrow & & \downarrow \\ U^n(N_1) & \longrightarrow & U^n(N_2) \end{array}$$

Notice that for $n \geq 1$, $U^n = U_{A_n}$ where $A_n \equiv \left(\bigwedge_{i=1}^{n} (x_i = x_i)\right)$.

We say that a limit (colimit) in a category of models $C_{T,F}$ is standard if it is preserved by U_A for every $A \in F$. If $C_{T,F}$ is complete category and it has standard limits then we say that it is standard complete.

Proposition 1.5.1: Let $C_{T,F}$ be a standard complete category. Let A be a conjunction of formulas in F. Then there exists a left-adjoint R_A to U_A. In particular there exists a model F_A in $C_{T,F}$ such that F_A represents U_A.

Proof: As in Freyd [3], $C_{T,F}$ is well-powered and the solution set condition follows from the Löwenheim-Skolem Theorem. Thus if $A \in F$, as U_A is continuous, the proposition follows immediately. If $A \notin F$ then U_A is a limit of the following type where U_{A_i}, $i = 1,\ldots,k$, is in F.

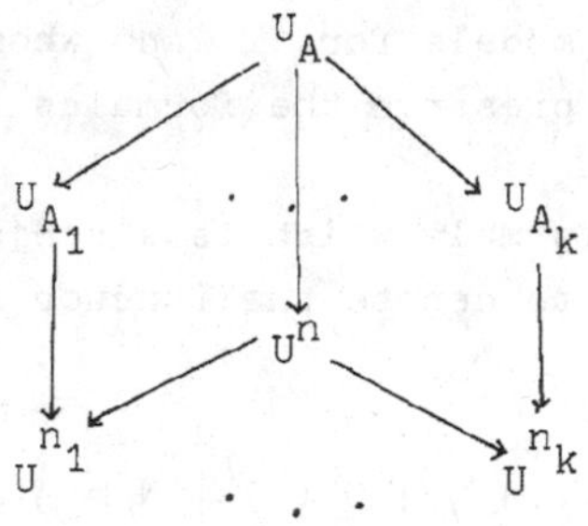

Thus U_A is continuous and has a left adjoint R_A. Now
$U_A(N) \simeq (1, U_A(N))_S \simeq (R_A(1), N)_{C_{T,F}}$. Thus $R_A(1)$ is a model in
$C_{T,F}$ which represents U_A.

Q.E.D.

If H is a universal Horn theory then it is well known that C_H is
complete. A terminal object in C_H is a one elm model in which
every predicate is true. Terminal objects are standard.

If $N = \prod_{\alpha \in \beta} N_\alpha$ in C_H and A is either a predicate in $L(H)$ or
$A \equiv (t(x_1, \ldots, x_k) = s(x_1, \ldots, x_k))$ then $U_A(N) = \prod_{\alpha \in \beta} U_A(N_\alpha)$

If $A = P(t_1(x_1, \ldots, x_k), \ldots, t_n(x_1, \ldots, x_k))$ then the following
diagram is a pullback where $\overline{t} = (\overline{t}_1, \ldots, \overline{t}_k)$.

$$
\begin{array}{ccc}
U_A & \longrightarrow & U_p \\
\downarrow & & \downarrow \\
U^R & \xrightarrow{\ \overline{t}\ } & U^n
\end{array}
$$

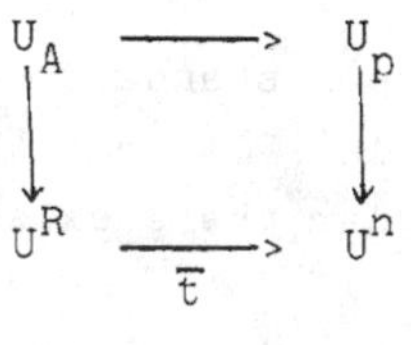

Therefore U_A preserves products.

If $N_1 \underset{\theta}{\overset{\phi}{\rightrightarrows}} N_2$ are two homomorphisms between models in C_H, then
there exists a substructure N of N_1 such that $a \in U(N)$ iff

$\phi(a) = \theta(a)$. As H is a universal theory it follows that N is a model for H and that $N \hookrightarrow N_1$ is a standard equalizer for Φ and θ. Thus C_H is standard complete. We have the following corollary to Proposition 1.5.1.

Corollary 1.5.2: If H is a universal Horn theory and A is a conjunction of atomic formulas, then the functor:

$$U_A: C_H \longrightarrow S$$

is representable.

From this point on we shall use H to denote a universal Horn theory.

1.6 Construction of F_n and F_B

We shall use the notation F_n to denote an nth free model, $n = 0,1,\ldots,$ and the notation F_B to denote a model which represents U_B where B is an appropriate n-ary formula.

If H is a universal Horn theory we may assume that there exist variables $x_1, x_2, \ldots$ in L(H) such that $i \neq j$ implies $x_i \neq x_j$.

Let

$$G_n = \begin{cases} \{t \mid t \text{ is a term in } L(H) \text{ with variables in a subset of the set } \{x_1,\ldots,x_n\}\} & \text{for } n = 1,2,\ldots \\ \{t \mid t \text{ is a variable free term in } L(H)\} & \text{if } n = 0 \end{cases}$$

For each $t \in G_n$ define $\overline{t}$ as the set $\{t_\alpha \in G_n \mid H \vdash t_\alpha = t\}$. Let $F_n = \{\overline{t} \mid t \in G_n\}$. We define the following L(H)-structure on F_n. If $s_1 \in \overline{t}_1, \ldots, s_m \in \overline{t}_m$; f an m-ary function symbol and p an m-ary predicate symbol in L(H), $m = 1,2,\ldots,$

then we define

(1) $\overline{f}(\overline{t}_1,\ldots,\overline{t}_m) = \overline{f(s_1,\ldots,s_m)}$

(2) $F_n \models p(\overline{t}_1,\ldots,\overline{t}_m)$ iff $H \vdash p(s_1,\ldots,s_m)$

If P is a 0-ary predicate symbol then $F_n \models P$ iff $H \vdash P$. That F_n as constructed is an nth free model for C_H is straightforward.

If A is a formula in $L(H)$ with n free variables which is a conjunction of atomic formulas, then for each $t \in G_n$ defined above we define:

$$\tilde{t} = \{t_\alpha \in G_n \mid H \vdash A \longrightarrow (t_\alpha = t)\}$$

Let $F_A = \{\tilde{t} \mid t \in G_n\}$. We define an $L(H)$-structure on F_A as follows. If $s_1 \in \tilde{t}_1, \ldots, s_m \in \tilde{t}_m$, f an m-ary function symbol and P an m-ary predicate symbol in $L(H)$, $m = 1,2,\ldots$, then we define

(1) $f(\tilde{t}_1,\ldots,\tilde{t}_m) = \widetilde{f(s_1,\ldots,s_m)}$

(2) $F_A \models P(\tilde{t}_1,\ldots,\tilde{t}_m)$ iff
$H \vdash A(x_1,\ldots,x_n) \longrightarrow P(s_1,\ldots,s_m)$.

If P is 0-ary then $F_A \models P$ iff $H \vdash A(x_1,\ldots,x_n) \longrightarrow P$. In particular $F_A \models A(\tilde{x}_1,\ldots,\tilde{x}_n)$.
F_A with the $L(H)$-structure defined above is a model for H and represents the functor $U_A\colon C_H \longrightarrow S$.

The map $\Phi_A\colon F_n \longrightarrow F_A$ via $\Phi_A\colon \overline{t} \longmapsto \tilde{t}$ is clearly an onto map and will be referred to as the canonical map.

Note than any 0th-free model is an initial object in C_H.

1.7 Special Subcategories of C_H

A subcategory $\mathcal{A}$ of C_H is said to be an RAF ("represents atomic formulas") subcategory of C_H if $\mathcal{A}$ is a full skeletal subcategory whose class of objects are models which represent U_A for every A which is a conjunction of atomic formulas in $L(H)$. We assume that a model F_0 which represents U^0 is also in $\mathcal{A}$.

Lemma 1.7.1: If $\mathcal{A}$ is an RAF subcategory of C_H, then $\mathcal{A}$ is fintely cocomplete and the inclusion functor $I: A \hookrightarrow C_H$ is finitely cocontinuous.

Proof: Suppose N_1 and N_2 are both in $Ob(\mathcal{A})$. Then there exist formulas A_1 and A_2 such that N_1 represents U_{A_1} and N_2 represents U_{A_2}. We may assume that A_1 and A_2 have no variables in common. Let $A \equiv A_1 \wedge A_2$. Then there exists an $N \in Ob(\mathcal{A})$ such that N represents U_A. It is easy to see that $N = N_1 + N_2$ in C_H, hence also in $\mathcal{A}$.

Let $N_1 \overset{\Phi}{\underset{\theta}{\rightrightarrows}} N_2$ be a pair of maps in $\mathcal{A}$. We may assume that N_1 represents U_{A_1} and N_2 represents U_{A_2} where A_1 and A_2 are formulas which have m and n free variables respectively. Let $\{\tilde{s}_1,\ldots,\tilde{s}_m\}$ generate N_1 and $\{\tilde{t}_1,\ldots,\tilde{t}_n\}$ generate N_2. Then there exist f_i, g_i, $i = 1,\ldots,m$ in $L(H)$ such that

$$\Phi(\tilde{s}_i) = f_i(\tilde{t}_1,\ldots,\tilde{t}_n)$$

and

$$\theta(\tilde{s}_i) = g_i(\tilde{t}_1,\ldots,\tilde{t}_n) \quad ;$$

$i = 1,\ldots,m$. Let

$A(u_1,\ldots,u_n) \equiv$

$$\left\{A_2(u_1,\ldots,u_n) \wedge \left[\bigwedge_{i=1}^{m} (f_i(u_1,\ldots,u_n) = g_i(u_1,\ldots,u_n))\right]\right\}.$$

then there exists an $N \in \mathrm{Ob}(\mathcal{C})$ such that N represents U_A. If $\{\bar{u}_1,\ldots,\bar{u}_n\}$ generates N and $\varphi\colon N_2 \longrightarrow N$ is the map which sends $\tilde{t}_j$ to $\bar{u}_j$, $j = 1,\ldots,n$, then φ coequalizes Φ and θ in C_H. Hence it coequalizes them in $\mathcal{C}$. The model in $\mathcal{C}$ which represents U^0 is the initial object. Thus $\mathcal{C}$ is finitely cocomplete and I is finitely cocontinuous.

Q.E.D.

If $\theta\colon N_1 \longrightarrow N_2$ is a homomorphism between two models of H, then the (set) image of θ is a substructure, hence a submodel of N_2. Thus the category C_H has standard images. It follows that coequalizers are onto maps and pushouts of onto maps are onto maps in C_H.

Proposition 1.7.2: If $\mathcal{C}$ is an RAF subcategory of C_H, then $\mathcal{C}^{op}$ is an abstract Horn theory.

Proof: We know that $\mathcal{C}^{op}$ is a small skeletal finitely continuous category. The cogenerator is F_1 and the monic maps are the maps which are onto maps in $\mathcal{C}$. The rest follows immediately.

Q.E.D.

Chapter II

The Characterization Theorem

2.1 Bicompleteness, Direct Limits and LSD Subcategories

As C_H is complete, it is cocomplete, hence bicomplete by a theorem of Freyd's.(cf. [4]). If D is a functor from a directed category $\mathcal{D}$ into C_H, then

$U^n(\underset{\rightarrow}{\lim}\, D(\alpha)) = \underset{\rightarrow}{\lim}\, (U^n(D(\alpha)))$. Also if A is either a predicate symbol or an equation in L(H) then

$U_A(\underset{\rightarrow}{\lim}\, D(\alpha)) = \underset{\rightarrow}{\lim}\, (U_A(D(\alpha)))$ (cf: [2]).

Lemma 2.1.1: C_H has standard direct limits.

Proof: Let $A \equiv P(f_1(x_1,\ldots,x_n),\ldots,f_m(x_1,\ldots,x_n))$ where P is an n-ary predicate symbol in L(H). Let F_m, F_p, F_n, F_A with generators $\langle\bar{y}_i\rangle_{i=1}^m$, $\langle\tilde{y}\rangle_{i=1}^m$, $\langle\bar{x}_j\rangle_{j=1}^n$, $\langle\tilde{x}_j\rangle_{j=1}^n$ represent U^m, F^P, U^n and F_A respectively. Then the following diagram is a pushout where:

$f\colon \bar{y}_i \longmapsto \bar{f}_i(\bar{x}_1,\ldots,\bar{x}_n)$ and $f'\colon \tilde{y}_i \longmapsto \bar{f}_i(\tilde{x}_1,\ldots,\tilde{x}_n)$,

$i = 1,\ldots,n$.

$$\begin{array}{ccc} F_m & \xrightarrow{\ f\ } & F_n \\ {\scriptstyle \varphi_P}\downarrow & & \downarrow{\scriptstyle \varphi_A} \\ F_P & \xrightarrow[\ f'\ ]{} & F_A \end{array}$$

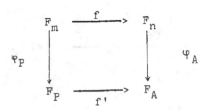

Suppose $N = \varinjlim D(\alpha)$ where D is a functor from a directed category $\mathfrak{D}$ to C_H. If $d\colon F_A \longrightarrow N$, then since U_P and U^n preserve direct limits there exists an $\alpha \in Ob(\mathfrak{D})$ and maps $g\colon F_n \longrightarrow D(\alpha)$, $h\colon F_P \longrightarrow D(\alpha)$ such that

$$F_n \xrightarrow{\varphi_A} F_A \xrightarrow{d} N = F_n \xrightarrow{g} D(\alpha) \xrightarrow{i_\alpha} N$$

and

$$F_P \xrightarrow{f'} F_A \xrightarrow{d} N = F_P \xrightarrow{h} D(\alpha) \xrightarrow{i_\alpha} N .$$

So

$$F_m \xrightarrow{\varphi_P g} D(\alpha) \xrightarrow{i_\alpha} N = F_m \xrightarrow{fh} D(\alpha) \xrightarrow{i_\alpha} N .$$

As $U(N) = \varinjlim (U(D(\alpha)))$ and F_m is finitely generated it follows that there exists a $\beta \in Ob(\mathfrak{D})$ such that

$$D(\alpha) \xrightarrow{i_\alpha^\beta} D(\beta) \xrightarrow{i_\beta} N = D(\alpha) \xrightarrow{i_\alpha} N$$

and

$$F_m \xrightarrow{\varphi_P g i_\alpha^\beta} D(\beta) = F_m \xrightarrow{fh i_\alpha^\beta} D(\beta)$$

Hence there exists a unique $d'\colon F_A \longrightarrow D(\beta)$ such that

$$F_m \xrightarrow{fh i_\alpha^\beta} D(\beta) = F_m \xrightarrow{f\varphi_A} F_A \xrightarrow{d'} D(\beta) \quad .$$

But

$$F_n \xrightarrow{\varphi_A} F_A \xrightarrow{d'} D(\beta) \xrightarrow{i_\beta} N = F_n \xrightarrow{\varphi_A} F_A \xrightarrow{d} N$$

Since φ_A is onto it follows that $d'i_\beta = d$. Hence

$U_A(\varinjlim D(\alpha)) \simeq (F_A, \varinjlim D(\alpha))_{C_H} \simeq \varinjlim (F_A, D(\alpha))_{C_H} \simeq \varinjlim (U_A(D(\alpha)))$.

Suppose $A \equiv A_1 \wedge A_2$ where A_1 and A_2 are atomic formulas. Let F_{A_1}, F_{A_2} and F_A be models in C_H which represent U_{A_1}, U_{A_2} and U_A respectively. Then there exists a pushout diagram in C_H of the following form:

$$\begin{array}{ccc} F_m & \longrightarrow & F_{A_1} \\ \downarrow & & \downarrow \\ F_{A_2} & \longrightarrow & F_A \end{array}$$

Using the same technique as above it can be shown that

$(F_A, \varinjlim D(\alpha))_{C_H} \simeq \varinjlim (F_A, D(\alpha))_{C_H}$. The rest of the proof follows from finite induction.

Q.E.D.

Lemma 2.1.2: If N is a model in C_H and $\mathcal{A}$ is an RAF subcategory of C_H, then there is a directed category $\mathfrak{D}$ and a functor

$$D: \mathfrak{D} \longrightarrow C_H$$

such that D factors through the inclusion functor $I: \mathcal{A} \hookrightarrow C_H$ and $N = \varinjlim D(\alpha)$.

Proof: Let N be a model in C_H and let $\mathcal{A} \hookrightarrow C_H$ be an RAF subcategory of C_H. We construct a directed category $\mathfrak{D}$ as follows:

The objects of $\mathfrak{D}$ consist of triples

$\langle\langle a_1,\ldots,a_n\rangle,\Phi,F_A\rangle$ where:

(1) $\langle a_1,\ldots,a_n\rangle \in N^n$

(2) A is an n-ary formula which is a conjunction of atomic formulas and $F_A \in \mathrm{Ob}(\mathcal{C})$ represents U_A.

(3) $\Phi\colon F_A \longrightarrow N$ with $\Phi\colon \tilde{x}_i \longrightarrow a_i$, where $\langle\tilde{x}_i\rangle_{i=1}^n$ generate F_A.

A map from $\langle\langle a_1,\ldots,a_m\rangle,\Phi_1,F_A\rangle$ to $\langle\langle b_1,\ldots,b_n\rangle,\Phi_2,F_B\rangle$ exists only if $m < n$. Such a map is a map $\gamma\colon F_A \longrightarrow F_B$ such that $\Phi_1 = \gamma^{\Phi}{}_2$ and $Y\colon \tilde{x}_i \longrightarrow \tilde{y}_{\pi(i)}$ where $\langle\tilde{x}_i\rangle_{i=1}^m$ and $\langle\tilde{y}_j\rangle_{j=1}^n$ generate F_A and F_B respectively, and $\pi \in S_n$, the permutation group on n elements. If there is more than one map between F_A and F_B which satisfy the criteria stated above, use the Axiom of Choice to select the one which will be in $\mathcal{D}$.

If $X = \langle\langle a_1,\ldots,a_n\rangle,\Phi_1,F_A\rangle$ and $Y = \langle\langle b_1,\ldots,b_m\rangle,\Phi_2,F_B\rangle$ then it is easy to see that both X and Y enjoy a map into:

$$Z = \left\langle\langle a_1,\ldots,a_n,b_1,\ldots,b_m\rangle, \binom{\Phi_1}{\Phi_2}, F_A + F_B\right\rangle.$$

Therefore $\mathcal{D}$ is a directed category. Define the functor:

$$D\colon \mathcal{D} \longrightarrow C_H$$

as follows:

$$D\colon \langle\langle a_1,\ldots,a_n\rangle,\Phi,F_A\rangle \longmapsto F_A$$

$$D\colon \gamma \longmapsto \gamma$$

Then $N = \varinjlim D$. The rest is a standard exercise in diagram chasing.

Q.E.D.

Let $\mathcal{A}$ be a small full subcategory of a category $\mathcal{B}$. Let K be the class of all functors from α-directed categories into $\mathcal{B}$ which factor through the inclusion functor:

$$I: \mathcal{A} \hookrightarrow \mathcal{B}.$$

$\mathcal{A}$ is said to be LSK ("Löwenheim-Skolem-Dense") of type K in $\mathcal{B}$ (Fitter [2]) if:

(1) $\varinjlim D$ exist in $\mathcal{B}$, for all $D \in K$

(2) If $N \in Ob(\mathcal{B})$ then there is a $D \in K$ such that $N = \varinjlim D(\alpha)$

(3) For all $N \in Ob(\mathcal{A})$, for all $D \in K$: $\varinjlim_{\alpha \in \mathcal{D}} (N, D(\alpha)) \approx (N, \varinjlim_{\alpha \in \mathcal{D}} D(\alpha))$.

Proposition 2.1.3: If $\mathcal{A} \hookrightarrow C_H$ is an RAF subcategory, then $\mathcal{A}$ is LSD of type K in C_H where K is the classof all functors from directed categories into C_H which factor through $\mathcal{A} \hookrightarrow C_H$.

Proof: $\mathcal{A}$ is obviously a small full subcategory of C_H. (1) is true since C_H is cocomplete. (2) and (3) follow from the two previous lemmas.

Q.E.D.

2.2. The Characterization Theorem

Lemma 2.2.1: If $\mathcal{J}$ is an abstract Horn theory then $\mathcal{J}^{op}$ is equivalent to an RAF subcategory of $C_{H_{\mathcal{J}}}$.

Proof: We shall use the notation of section 1.4 as well as the functors V and W defined there in the proof.

$\mathcal{J}^{op}$ is equivalent to the category whose objects are the representable functors from $\mathcal{J}$ to S and whose maps are the natural transformations between the functors. Hence $\mathcal{J}^{op}$ can be embedded in $S^{(\mathcal{J})}$ in a natural way.

Since $V: S^{(\mathcal{J})} \longrightarrow C_{H_{\mathcal{J}}}$ is an equivalence, it is a full embedding. Hence it suffices to show that for every A which is a conjunction of atomic formulas in $L(H_{\mathcal{J}})$ there is an $X \in Ob(\mathcal{J})$ such that $V(H^X)$ represents U_A. We will give the X for every such A. The verification that $V(H^X)$ represents U_A is straightforward and will be left to the reader. It follows from the definition of $H_{\mathcal{J}}$ and the fact that V and W are both equivalences.

(1) If P is a predicate in $H_{\mathcal{J}}$, then P is a monic map in $\mathcal{J}$. Let P have domain X. Then $V(H^X)$ represents U_P.

(2) If $A \doteq (f(x_1,\ldots,x_\ell,y_1,\ldots,y_m) = g(x_1,\ldots,x_\ell,z_1,\ldots,z_n))$ then

$M^{\ell+m} \xrightarrow{f} M$ and $M^{\ell+n} \xrightarrow{g} M$.

are both in J. Let X be the limit of the following diagram:

$$\begin{array}{ccc} M^{\ell+m+n} & \xrightarrow{P_{\ell+n}} & M^{\ell+n} \\ {\scriptstyle P_{\ell+m}}\downarrow & & \downarrow{\scriptstyle g} \\ M^{\ell+m} & \xrightarrow[f]{} & M \end{array}$$

Then $V(H^X)$ represents U_A.

(3) If $A \equiv (A_1(x_1,\ldots,x_\ell,y_1,\ldots,y_m)$
$\wedge\ A_2(x_1,\ldots,x_\ell,z_1,\ldots,z_n))$
where X_1 and X_2 represent U_{A_1} and U_{A_2} respectively

and

$$X_1 \xrightarrow{a_1} M^{\ell+m} \quad \text{and} \quad X_2 \xrightarrow{a_2} M^{\ell+n}$$

are monic maps in $\mathcal{J}$. Then let the following diagram be a pullback:

$$\begin{array}{ccccc}
X & & \longrightarrow & & X_2 \\
\downarrow & & & & \downarrow{\scriptstyle a_2} \\
 & & & & M^{\ell+n} \\
 & & & & \downarrow{\scriptstyle P_\ell} \\
X_1 & \xrightarrow{a_1} & M^{\ell+m} & \xrightarrow{P_\ell} & M^{\ell}
\end{array}$$

Then $V(H^X)$ represents U_A.

(4) If $A \equiv P(f_1(x_1,\ldots,x_n),\ldots,f_m(x_1,\ldots,x_n))$

where P is an m-ary predicate in $H_{\mathcal{J}}$, then

$H_{\mathcal{J}} \vdash A \longleftrightarrow B$ where:

$$B \equiv \left(P(y_1,\ldots,y_m) \wedge \left(\bigwedge_{i=1}^{m} (y_i = f_i(x_1,\ldots,x_n))\right)\right).$$

Hence $U_A \approx U_B$ and (3) and (4) can be used to find an X such that $V(H^X)$ represents U_A.

Q.E.D.

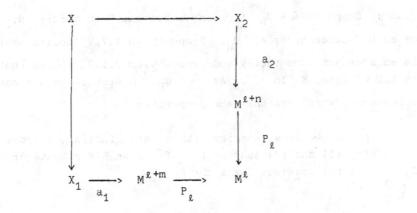

Theorem 2.2.2 (Characterization): A category $\mathcal{A}$ is equivalent to a category whose class of objects are the (normal) models for a universal Horn theory and whose maps are homomorphisms between the models iff it contains a small full subcategory $\mathcal{B}$ such that:

(1) $\mathcal{B}^{op}$ is an abstract Horn theory

(2) $\mathcal{B}$ is LSD of type K in $\mathcal{A}$ where K is the class of all functors from directed categories into $\mathcal{A}$ which factor through the inclusion functor I: $\hookrightarrow \mathcal{A}$.

Proof: Suppose $\mathcal{A} \simeq \mathcal{C}_H$ for some universal Horn theory H. Let C be an RAF subcategory of $\mathcal{C}_H$. Proposition 1.7.2 implies that $\mathcal{C}^{op}$ is an abstract Horn theory and proposition 2.1.3 implies that $\mathcal{C}$ is LSD of type K in $\mathcal{C}_H$. As $\mathcal{A} \simeq \mathcal{C}_H$, $\mathcal{A}$ must contain a small full subcategory which has the same properties.

Suppose $\mathcal{A}$ is a category with a small full subcategory satisfying (1) and (2) above. Let $\mathcal{E}$ be an RAF subcategory of $C_{H_{\mathcal{B}^{op}}}$. By the previous lemma $\mathcal{B} \simeq \mathcal{E}$.

Now Fittler [2] has shown that $\mathcal{X}$ is LSD of type K in $\mathcal{Y}$ iff $\mathcal{Y} \simeq K(\mathcal{X},S)$ where the objects of $K(\mathcal{X},S)$ are functors from $\mathcal{X}$ to S of the form $\varinjlim_{\alpha \in \mathcal{D}} H_{D(\alpha)}$, $D \in K$ and whose maps are natural transformations between the functors. As $\mathcal{B} \simeq \mathcal{E}$, it follows that:

$$\mathcal{A} \simeq K(\mathcal{B},S) \simeq K(\mathcal{E},S) \simeq C_{H_{\mathcal{B}^{op}}} .$$

Q.E.D.

Corollary 2.2.3: Iff $\mathcal{A}$ is an RAF subcategory of C_H, then

$$C_H \simeq S^{(\mathcal{A}^{op})} .$$

Chapter III

Horn Theory Maps

3.1. Horn Theory Maps

Let $\mathcal{J}_1$ and $\mathcal{J}_2$ be abstract Horn theories. By a Horn theory map $T: \mathcal{J}_2 \longrightarrow \mathcal{J}_1$ we mean a finitely continuous functor which preserves both the cogenerator and the monic maps.

If $N \in S^{(\mathcal{J}_1)}$ and $T: \mathcal{J}_2 \longrightarrow \mathcal{J}_1$ is a Horn theory map, then $TN \in S^{(\mathcal{J}_2)}$. Also if $\eta: N_1 \longrightarrow N_2$ in $S^{(\mathcal{J}_1)}$, then there exists a $\varphi: TM_1 \longrightarrow TM_2$ in $S^{(\mathcal{J}_2)}$ such that:

$$\varphi_X = \eta_{T(X)} \quad \text{for all} \quad X \in Ob(\mathcal{J}_2) .$$

Therefore T induces a functor from $S^{(\mathcal{J}_1)}$ to $S^{(\mathcal{J}_2)}$. We shall denote this functor by S^T.

Let $V: S^{(\mathcal{J})} \longrightarrow C_{H_{\mathcal{J}}}$ and $W: C_{H_{\mathcal{J}}} \longrightarrow S^{(\mathcal{J})}$ be the functors defined in §1.5. Note that the definitions of V and W imply that:

$$C_{H_{\mathcal{J}}} \xrightarrow{W} S^{(\mathcal{J})} \xrightarrow{V} C_{H_{\mathcal{J}}} = 1.$$

If $T: \mathcal{J}_2 \longrightarrow \mathcal{J}_1$ is a Horn theory map, we shall use the notation S^T to denote the composition:

$$C_{H_{\mathcal{J}_1}} \xrightarrow{W_1} S^{(\mathcal{J}_1)} \xrightarrow{S^T} \dot{S}^{(\mathcal{J}_2)} \xrightarrow{V_2} C_{H_{\mathcal{J}_2}} .$$

If $N \in C_{H_{\mathcal{J}_1}}$ and P is a predicate in $L(H_{\mathcal{J}_2})$, then:

$$\begin{aligned} U_P(S^{\dot{T}}(N)) &= \operatorname{Im}\{(S^T[W_1(N)])(P)\} \\ &= \operatorname{Im}\{[W_1(N)](T(P))\} \\ &= U_{T(P)}[V_1W_1(N)] \\ &= U_{T(P)}(N) \end{aligned}$$

In particular if $N = M_1$, the cogenerator for $\mathcal{J}_1$, then we have $U(S^{\dot{T}}(N)) = U(N)$. Hence $S^{\dot{T}}$ preserves underlying sets.

Note that if A is an atomic formula in $L(H_{\mathcal{J}})$ then there is a predicate $P \in L(H_{\mathcal{J}})$ such that $U_A = U_P$. Hence to check that two models are the same it suffices to show that they agree on U_P for all predicates $P \in L(H_{\mathcal{J}})$.

Lemma 3.1.1: If $T\colon \mathcal{J}_1 \longrightarrow \mathcal{J}_2$ is a Horn theory map, then $S^{\dot{T}}$ preserves terminal objects and submodels.

Proof: Let N be a terminal object in $S^{(\mathcal{J}_1)}$. We may assume that $U_{P'}(N) = 1$ for all predicate symbols $P' \in L(H_{\mathcal{J}_1})$. Then for all predicate symbols $P' \in L(H_{\mathcal{J}_2})$ we have:

$$U_P(S^{\dot{T}}(N)) = U_{T(P)}(N) = 1 .$$

Hence N is a terminal object in $S^{(\mathcal{J}_2)}$.

Suppose $N_1 \hookrightarrow N_2$ in $C_{H_{\mathcal{J}_1}}$ and P is an n-ary predicate in

$L(H_{\mathcal{J}_2})$, then

$$U_P(S^{\dot{T}}(N_1)) = U_{T(P)}(N_1) = [U_{T(P)}(N_2)] \cap [U^n(N_1)]$$

$$= [U_P(S^{\dot{T}}(N_2))] \cap [U^n(N_1)]$$

Therefore, $S^{\dot{T}}(N_1)$ is a submodel of $S^{\dot{T}}(N_2)$.

Q.E.D.

Corollary 3.1.2: $S^T: S^{(\mathcal{J}_1)} \longrightarrow S^{(\mathcal{J}_2)}$ preserves terminal objects and subfunctors.

Lemma 3.1.3: S^T and $S^{\dot{T}}$ are both continuous functors.

Proof: Let $N = \prod_{\alpha \in \beta} N_\alpha$ in $C_{H_{\mathcal{J}_1}}$. Then for every predicate symbol in

$L(H_{\mathcal{J}_2})$:

$$U_P(S^{\dot{T}}(\prod_{\alpha \in \beta} M_\alpha)) = U_{T(P)}(\prod_{\alpha \in \beta} M_\alpha)$$

$$= \prod_{\alpha \in \beta} (U_{T(P)}(M_\alpha))$$

$$= \prod_{\alpha \in \beta} (U_P(S^{\dot{T}}(M_\alpha)))$$

$$= U_P(\prod_{\alpha \in \beta} (S^{\dot{T}}(M_\alpha)))$$

Therefore $S^{\dot{T}}$ preserves products.

Let $N_1 \xrightarrow{\;f\;} N_2 \underset{h}{\overset{g}{\rightrightarrows}} N_3$ be an equalizer diagram in $C_{H_{\mathcal{J}_1}}$. As

equalizers are standard in $C_{H_{\mathcal{J}_1}}$ and $S^{\dot{T}}$ preserves underlying sets

it follows that

$$U(S^{\dot{T}}(N_1)) \xrightarrow{\;U(S^{\dot{T}}(f))\;} U(S^{\dot{T}}(N_2)) \underset{U(S^{\dot{T}}(h))}{\overset{U(S^{\dot{T}}(g))}{\rightrightarrows}} U(S^{\dot{T}}(N_3))$$

is an equalizer diagram in S. Since $N_1 \xrightarrow{\;f\;} N_2$ is an embedding (in
the model theory sense of the word) it follows that

$$S^{\dot{T}}(f): S^{\dot{T}}(N_1) \longrightarrow S^{\dot{T}}(N_2)$$

is also an embedding. Hence $S^{\dot{T}}(f)$ equalizes $S^{\dot{T}}(g)$ and $S^{\dot{T}}(h)$.

Since $S^{\dot{T}}$ also preserves the terminal object it must be continuous.

Q.E.D.

Proposition 3.1.4: If $T: \mathcal{J}_1 \longrightarrow \mathcal{J}_2$ is a Horn theory map then:

$$S^T: S^{(\mathcal{J}_1)} \longrightarrow S^{(\mathcal{J}_2)} \quad \text{and} \quad S^{\dot{T}}: C_{H_{\mathcal{J}_1}} \longrightarrow C_{H_{\mathcal{J}_2}}$$

both have left adjoints.

Proof: $C_{H_{\mathcal{J}_1}}$ is complete and well powered. $S^{\dot{T}}$ is continuous. Hence
it suffices to show that the solution-set condition is satisfied.

For each model $N \in C_{H_{\mathcal{J}_2}}$ let η_N be a set which contains exactly
one model from each isomorphism class with models of cardinality
$\leq \#(U(N) \overset{\cdot}{\cup} L(H_{\mathcal{J}_1}))$ in $C_{H_{\mathcal{J}_1}}$.

If $N \xrightarrow{f} \dot{S}^T(L)$ in $C_{H_{\mathcal{T}_2}}$ then there is a submodel L' of L of cardinality $\leq \#(U(N) \dot{\cup} L(H_{\mathcal{T}_1}))$ such that $U(\text{Im } f) \subseteq U(L') = U(\dot{S}^T(L'))$.

Without loss of generality we may assume that $L' \in \eta_N$. Let

$$i: L' \hookrightarrow L$$

be the canonical inclusion map. Then

$$\dot{S}^T(i): \dot{S}^T(L') \longrightarrow \dot{S}^T(L)$$

is also the canonical inclusion map. As $\text{Im}(f) \subset \text{Im}(\dot{S}^T(i))$ it follows that f must factor through $\dot{S}^T(i)$.

Q.E.D.

3.2 Lawvere Functors

A Lawvere functor is a functor $T: C_{H_{\mathcal{T}_1}} \longrightarrow C_{H_{\mathcal{T}_2}}$ which preserves underlying sets. We are interested in determining the conditions under which Lawvre functors are induced by Horn theory maps.

Let $\mathcal{A}$ be a category which has direct limits. An object X in $\mathcal{A}$ is said to be small if

$$(X, \varinjlim D(\alpha))_{\mathcal{A}} \approx \varinjlim (X, D(\alpha))_{\mathcal{A}} \text{ for every direct limit in } \mathcal{A}.$$

Lemma 3.2.1: A model N in C_H is small iff there exists a $B \in L(H)$ such that B is a conjunction of atomic formulas and N represents U_B.

Proof: Lemma 2.2.1 is one-half of the proof.

Suppose N is a model in C_H which is not finitely generated. Then there is a functor D from a directed category $\mathcal{D}$ into C_H such that the values of D are the finitely generated submodels of N and such that $N = \varinjlim_{\alpha} D(\alpha)$. 1_N does not factor through any $i_\alpha : D(\alpha) \longrightarrow N$. Hence N is not small.

Therefore, if N is small we may assume that N is finitely generated. Let n, a finite ordinal, be a minimal generating set for N. Then there is a functor $D: \mathcal{D} \longrightarrow C_H$ where $\mathcal{D}$ is a directed category, such that

(1) $N = \varinjlim D(\alpha)$

(2) For all $\alpha \in Ob(\mathcal{D})$, $D(\alpha)$ represents U_{B_α} where B_α is m-ary, a conjunction of atomic formulas, and $m \leqq n$ generated $D(\alpha)$

(3) If $i_\alpha : D(\alpha) \longrightarrow N$ is the canonical map, then $x \in m$ implies $i_\alpha(x) \in n$.

Now suppose there exists an $\alpha \in \mathcal{D}$ and a map $\theta: N \longrightarrow D(\alpha)$ such that $N \xrightarrow{\theta} D(\alpha) \xrightarrow{i_\alpha} N = 1_N$. Then we have the following diagram where $D(\alpha) \xrightarrow{e} E$ is the coequalizer of $1_{D(\alpha)}$ and $i_\alpha \theta$.

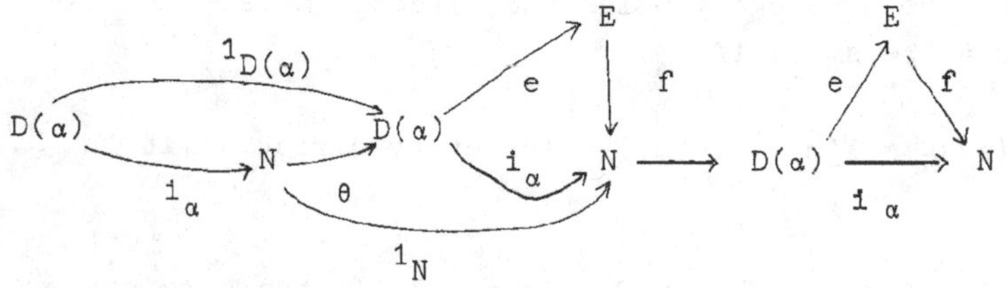

Since E is the coequalizer of two maps between $D(\alpha)$ it follows that there is a formula A which is a conjunction of atomic formulas such that E represents U_A.

$$D(\alpha) \xrightarrow{i_\alpha} N \xrightarrow{\theta e} E = D(\alpha) \xrightarrow{e} E$$

is onto. Hence $N \xrightarrow{\theta e} E$ is onto.

$$N \xrightarrow{\theta e} E \xrightarrow{f} N = 1_N \quad .$$

Also

$$N \xrightarrow{\theta e} E \xrightarrow{f} N \xrightarrow{\theta e} E = N \xrightarrow{\theta e} E \xrightarrow{1_E} E$$

As θe is onto this implies that $E \xrightarrow{f} N \xrightarrow{\theta e} E = 1_E$. Therefore $N \approx E$.

Q.E.D.

Corollary 3.2.2: $M \in S^{(J)}$ is small iff $M \approx H^X$ for some $X \in Ob(J)$.

Lemma 3.2.3: S^T and S^T both preserve direct limits.

Proof: Let $D: \mathfrak{D} \longrightarrow C_{H_J}$ be a functor from a direct category $\mathfrak{D}$ into C_{H_J}. Since direct limits are standard in C_{H_J} it follows that for every predicate symbol $P \in L(H_J)$ it is the case that:

$$\begin{aligned} U_P(S^{\dot{T}}(\varinjlim D(\alpha)) &= U_{T(P)}(\varinjlim D(\alpha)) \\ &= \varinjlim (U_{T(P)}(D(\alpha))) \\ &= \varinjlim (U_P(S^{\dot{T}}(D(\alpha)))) \\ &= U_P(\varinjlim (S^{\dot{T}}(D(\alpha)))) \end{aligned}$$

Q.E.D.

Lemma 3.2.4: If $T: C_{H_1} \longrightarrow C_{H_2}$ is a Lawvere functor which has a left adjoint R, then R preserves onto maps.

Proof: If R is aleft-adjoint for T then for each $A \in Ob(C_{H_2})$ there is a map $\eta_A: A \longrightarrow TR(A)$ in C_{H_2} which is functorial. Suppose $f: A_1 \longrightarrow A_2$ is an onto map in C_{H_2}. Let:

$$R(A_1) \xrightarrow{x} B \xrightarrow{y} R(A_2) = R(A_1) \xrightarrow{T(f)} R(A_2)$$

where B is the standard image of $T(f)$. Then we have the following commutative diagram in C_{H_2}.

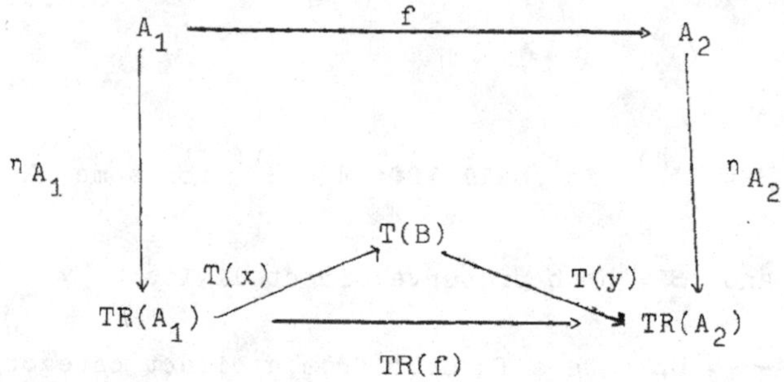

Since T preserves underlying sets it also preserves standard images. Hence as f is onto it follows that:

$$Im(\eta_{A_2}) \subset Im\ TR(f) = T(B) .$$

Let $w: A_2 \longrightarrow T(B)$ be the map such that

$$A_2 \xrightarrow{w} T(B) \xrightarrow{T(y)} TR(A_2) = A_2 \xrightarrow{\eta_{A_2}} TR(A_2)$$

Then there must exist a unique map $z: R(A_2) \longrightarrow B$ in C_H such that

$$A_2 \xrightarrow{w} T(B) = A_2 \xrightarrow{\eta_{A_2}} TR(A_2) \xrightarrow{T(z)} T(B).$$

Thus we have

$$A_2 \xrightarrow{\eta_{A_2}} TR(A_2) \xrightarrow{T(zy)} TR(A_2)$$

$$= A_2 \xrightarrow{\eta_{A_2}} TR(A_2) \xrightarrow{T(1_{A_2})} TR(A_2) .$$

It follows that $zy = 1_{A_2}$. Hence y is onto. Therefore R(f) is onto.

Q.E.D.

Theorem 3.2.5: Let $Q: C_{H_{\mathcal{J}_1}} \longrightarrow C_{H_{\mathcal{J}_2}}$ be a Lawvre functor. Then there is a Horn theory map $T:\mathcal{J}_2 \longrightarrow \mathcal{J}_1$ such that $Q \approx S^{\dot{T}}$ iff Q preserves: (1) products, (2) submodels, (3) direct limits.

Proof: We have already shown that $S^{\dot{T}}$ satisfies (1), (2) and (3).

Let Q be a functor from $C_{H_{\mathcal{J}_1}}$ to $C_{H_{\mathcal{J}_2}}$ which satisfies the four conditions stated in the theorem. Then Q is continuous and as the solution-sets condition is satisfied it follows that Q has a left adjoint R.

Let:

$$Q' = (S^{(\mathcal{J}_1)} \xrightarrow{V_1} C_{H_{\mathcal{J}_1}} \xrightarrow{Q} C_{H_{\mathcal{J}_2}} \xrightarrow{W_2} S^{(\mathcal{J}_2)})$$

and

$$R' = (S^{(\mathcal{J}_2)} \xrightarrow{V_2} C_{H_{\mathcal{J}_2}} \xrightarrow{R} C_{H_{\mathcal{J}_1}} \xrightarrow{W_1} C^{(\mathcal{J}_1)})$$

R' is a left adjoint for Q'. Also Q' satisfies the four conditions stated in the theorem.

If $Y \in Ob(\mathcal{J}_2)$, then H^Y is small in $S^{\mathcal{J}_2}$. Hence for every direct limit in $S^{(\mathcal{J}_2)}$ we have:

$$
\begin{aligned}
(R'(H^Y), \varinjlim N_\alpha)_{S^{(\mathcal{J}_1)}} &\approx (H^Y, Q'(\varinjlim N_\alpha))_{S^{(\mathcal{J}_2)}} \\
&\approx (H^Y, \varinjlim Q'N_\alpha)_{S^{(\mathcal{J}_2)}} \\
&\approx \varinjlim (H^Y, Q'N_\alpha)_{S^{(\mathcal{J}_2)}} \\
&\approx \varinjlim (R'(H^Y), N_\alpha)_{S^{(\mathcal{J}_1)}}
\end{aligned}
$$

Therefore $R'(H^Y)$ is small. Corollary 3.2.3 and the fact that $\mathcal{J}_1$ is skeletal implies that there exists a unique X $Ob(\mathcal{J}_1)$ such that

$$R'(H^Y) \approx H^X$$

We denote such an X by $T(Y)$.

Also if $y: Y_1 \longrightarrow Y_2$ in $\mathcal{J}_2$ then there exists a unique $x \in \mathcal{J}_1$ such that the following diagram commutes in $S^{(\mathcal{J}_1)}$

$$
\begin{array}{ccc}
R(H^{Y_2}) & \xrightarrow{R(H^y)} & R(H^{Y_1}) \\
\wr\downarrow & & \wr\downarrow \\
H^{R'(Y_2)} & \xrightarrow[H^x]{} & H^{R'(Y_1)}
\end{array}
$$

We denote such an x by $T(y)$.

T defines a functor from J_2 to J_1. As $\mathcal{J}_2^{op} \hookrightarrow S^{(\mathcal{J}_2)}$ is finitely cocontinuous and R' is cocontinuous it follows that T is finitely continuous. Also

$$R'(H^{M_2}) \simeq H^{M_1}$$

Therefore $T(M_2) = M_1$. If $y \in \mathcal{J}_2$ is monic, then H^y is an onto map in $S^{(\mathcal{J}_2)}$. Lemma 3.2.4 implies that $R'(H^y)$ is onto. Hence $T(y)$ is monic in $\mathcal{J}_1$.

Therefore $T\colon \mathcal{J}_2 \longrightarrow \mathcal{J}_1$ is a Horn theory map.

To show that $S^T \simeq Q'$ it suffices to show that they are equivalent in $\mathcal{J}_1^{op}$. This follows from the fact that $\mathcal{J}_1^{op}$ is L.S.D. with respect to direct limits in $S^{(\mathcal{J}_1)}$ and both functors preserve direct limits.

If $X \in Ob(\mathcal{J}_1)$ then:

$$\begin{aligned} S^T(H^X) &\simeq (X, T(\_\_))_{\mathcal{J}_1} \\ &\simeq (R'(\_\_), H^X)_{S^{(\mathcal{J}_1)}} \Big|_{\mathcal{J}_2^{op}} \\ &\simeq (\_\_, Q'(H^X))_{S^{(\mathcal{J}_2)}} \Big|_{\mathcal{J}_2^{op}} \\ &\simeq Q'(H^X)\ , \end{aligned}$$

since $\mathcal{J}_2^{op}$ is L.S.D. with respect to direct limits in $S^{(\mathcal{J}_2)}$.

Similarly one can show that if $x\colon X_2 \longrightarrow X_1$ in $\mathcal{J}_1$ then the following diagram commutes.

$$\begin{array}{ccc} S^T(H^{X_1}) & \xrightarrow{S^T(H^x)} & S^T(H^{X_2}) \\ \downarrow\wr & & \downarrow\wr \\ Q'(H^{X_1}) & \xrightarrow[Q'(H^x)]{} & Q'(H^{X_2}) \end{array}$$

Therefore $S^T \approx Q'$. So $S^T \overset{\cdot}{\approx} Q$.

Q.E.D.

Corollary 3.2.6: If $\mathcal{J}_1$ and $\mathcal{J}_2$ are abstract Horn theories then a functor $Q: S^{(\mathcal{J}_1)} \longrightarrow S^{(\mathcal{J}_2)}$ is equivalent to one induced by a Horn theory map from $\mathcal{J}_2$ to $\mathcal{J}_1$ iff the composition

$$C_{H_{\mathcal{J}_1}} \xrightarrow{W_1} S^{(\mathcal{J}_1)} \xrightarrow{Q} S^{(\mathcal{J}_2)} \xrightarrow{V_2} C_{H_{\mathcal{J}_2}}$$

is a Lawvre functor which satisfies the four conditions stated in Theorem 3.2.5.

If T_1 and T_2 are two first order theories, then by a theory map F from T_2 to T_1 we mean a map:

$$F: L(T_2) \longrightarrow L(T_1)$$

which preserves equality, negation, conjunction, arity, and such that:

$$T_2 \vdash A \text{ implies } T_1 \vdash F(A)$$

If $F: H_2 \longrightarrow H_1$ is a theory map then there is an induced functor

$$C^F: C_{H_1} \longrightarrow C_{H_2}$$

such that $U_A(C^F(N)) = U_{F(A)}(N)$ for every A which is a conjunction of atomic formulas in H_2.

Corollary 3.2.7: Let Q be a functor from C_{H_1} to C_{H_2}. Then there is a theory map $F: H_2 \longrightarrow H_1$ such that $Q = C^F$ iff Q is a Lawvre functor which satisfies the four conditions stated in Theorem 3.2.5.

BIBLIOGRAPHY

[1] Cohn, P.M., Universal Algebra. Harper & Row, New York, 1965.

[2] Fittler, R., Direct Limits of Models, to appear in Zeitschrift für Mathematische Logik und Grundlagen der Mathematik.

[3] Freyd, P.J., Abelian Categories. Harper & Row, New York, 1964.

[4] Freyd, P.J., The Theories of Functors and Models, Symposium on the Theory of Models, North Holland Pub. Co., Amsterdam, 1965, 107-120.

[5] Freyd, P.J., Algebra Valued Functors in General and Tensor Products in Particular, Colloq. Math. vol. 14, 1966, 89-106.

[6] Lawvere, F.W., Functorial Semantics of Algebraic Theories, Proc. Nat. Acad. Sci. 50 (5), 1963 , 869-872.

[7] Schoenfield, J.R., Mathematical Logic. Addison-Wesley, Reading, 1967.

Completeness theorem for logical categories *)**)

Hugo Volger

Introduction:

In [12] Lawvere introduced the method of functorial semantics in order to study categories of algebras. For this purpose he developed the concept of an algebraic theory. An algebraic theory is a small category $\underline{T}$ with products such that product-preserving functors from $\underline{T}$ into the category of sets correspond to algebras of a certain similarity type. Moreover, if two morphisms in $\underline{T}$ are identified by all product-preserving functors, then they have to be equal. This ensures that the class of algebras is defined by equations. This categorical concept has proven to be very useful in universal algebra.

In [13] Lawvere proposed a definition of elementary theories for model theory. An elementary theory should be a small category $\underline{T}$ such that structure-preserving functors from $\underline{T}$ into the category of sets correspond to relational structures of a certain similarity type. Moreover, if two morphisms in $\underline{T}$ are identified by all structure preserving functors, they should be equal. This ensures that the class of relational structures is defined by first order formulas. This condition corresponds to the completeness theorem of first-order logic.

This concept of an elementary theory may also be viewed as an algebraization of first-order logic by categorical means in the following sense. The elementary theory and the structure-preserving functors between them correspond to polyadic algebras and homomorphisms between them. A model of an elementary theory is a

---

* During the preparation of this article the author was supported by an NRC post-graduate and an NRC post-doctoral fellowship.

** Most of the results contained in this paper are part of the thesis of the author.

structure-preserving functor into a full subcategory of the category of sets which is an elementary theory, whereas a model of a polyadic algebra is a homomorphism into a functional two-valued polyadic algebra (cf. Halmos [7]). In this context the above condition corresponds to the representation theorem for polyadic algebras. The connections between elementary theories and polyadic algebras have been studied by Daigneault in [5].

In this paper we will prove the completeness theorem for elementary theories, suggested by Lawvere in [14]. The proof is categorical, but it can be said that it follows, in a sense, the lines of the completeness proof in Henkin [8]. We will use the slightly more general notion of a logical category. Aside from having some technical advantages, this permits an extension of the results to higher order logic. Thus we obtain an equivalent to Henkin's completeness theorem for higher logic in [9]. It should be remarked that our proof of the completeness theorem requires the addition of two new conditions to the original definition of elementary theories in [13]. They are concerned with certain pullbacks involving quantification and substitution. Two similar conditions occur already in a different context in Lawvere [15].

In the first chapter the basic definitions will be given. The second chapter contains the proof of the completeness theorem for logical categories and a criterion for the consistency of pushouts in the category of logical categories. This shows that the interpolation theorem of Craig [3], the consistency lemma of Robinson [18], as well as the amalgamation theorem of Daigneault [4] are equivalent. In the third chapter the completeness theorem will be extended to logical categories with exponentiation i.e. to higher order logic. In the last chapter we will introduce the notion of a semantical category i.e. a category of set-like objects in which quantification is replaced by the notion of direct image. The main result can be stated as follows. For every logical category $\underline{C}$ one can construct the free semantical category $S(\underline{C})$ which contains $\underline{C}$ as a subcategory. Hence every logical functor from $\underline{C}$ into a semantical category can be extended uniquely to $S(\underline{C})$. In particular every model of $\underline{C}$ can be extended to $\underline{C}$.

With regard to the necessary background from category theory and logic the reader is referred to Mitchell [16], chapters 1, 2 and 5, and to Shoenfield [21], chapters 1-5.

1. Basic definitions:

The completeness theorem states that the concept of an elementary theory is the abstraction of its models. Thus the definition may be developed by an analysis of the notion of a relational structure.

Therefore let P be a non-empty set and let $\underline{T}_P$ be the following subcategory of the category of sets. The objects are a two-element set 2 and the finite powers P^n of the set P. The morphisms of $\underline{T}_P$ are either of the form $P^n \to 2$ or $P^n \to P^m$. Thus $\underline{T}_P$ contains n-ary relations on P and m-tuples of n-ary operations on P for every n, $m \in N$.

The set 2 is a boolean algebra, whose operations are defined by the usual truth-tables. This implies that $\underline{T}_P(P^n, 2)$ is a boolean algebra for every $P^n \in ob(\underline{T}_P)$ and the substitution $\underline{T}_P(f, 2): \underline{T}_P(P^m, 2) \to \underline{T}_P(P^n, 2)$ is a boolean homomorphism for every $f: P^n \to P^m \in \underline{T}_P$. This determines the propositional structure on $\underline{T}_P$.

Let us denote the subset corresponding to a morphism $\varphi: P^n \to 2$ by $\varphi^{\#}$. Then the substitution $\underline{T}_P(f, 2): \underline{T}_P(P^m, 2) \to \underline{T}_P(P^n, 2)$ for $f: P^n \to P^m$ corresponds to the inverse image under f i.e. $(\psi f)^{\#} = f^{-1}(\psi^{\#})$ for $\psi: P^m \to 2$. The direct image under f will be called existential quantification along f and is denoted by $\exists f[-]$. Hence we have $\exists f[\varphi]^{\#} = f(\varphi^{\#})$ for $\varphi: P^n \to 2$. The inverse and the direct image are related as follows: $\exists f[\varphi]^{\#} = f(\varphi^{\#}) \subseteq \psi^{\#}$ iff $\varphi^{\#} \subseteq f^{-1}(\psi^{\#}) = (\psi f)^{\#}$.

This generalized quantification reduces to the usual one for a projection $p: P^n \times P^m \to P^n$. In this case $\exists p[\varphi]$ is the existential quantification of the last m variables of the n+m-ary relation $\varphi^{\#}$. Another special case is $e_n = \exists \Delta_{P^n}[1_{P^n}]: P^n \times P^n \to 2$, which is the identity relation on P^n. The universal quantification $\forall f$ can be

These considerations motivate the definition of an elementary theory. If we replace the assumption that every object is a finite product of the two basis objects by the assumption that the category has finite products, then we obtain the more general notion of a logical category.

1.1. Definition: A category $\underline{C}$ is called logical, if it satisfies the following conditions:

(1) $\underline{C}$ has finite products and hence in particular a terminal object I. – The unique morphism from X to I is denoted by $!_X$.

(2) $\underline{C}$ has a specified object Ω which is a boolean algebra object i.e. there exist morphisms $0, 1 : I \to \Omega$, $\sim : \Omega \to \Omega$ and $\wedge : \Omega \times \Omega \to \Omega$ which satisfy the identities for a boolean algebra. – Hence $\underline{C}(X, \Omega)$ is a boolean algebra for every object X and $\underline{C}(f, \Omega)$ is a boolean homomorphism for every morphism f. Thus $\underline{C}(X, \Omega)$ has a category structure determined by the order relation and $\underline{C}(f, \Omega)$ is a functor, since it preserves the order.

(3) For every $f : X \to Y$ in $\underline{C}$ there exists a functor $\exists f : \underline{C}(X, \Omega) \to \underline{C}(Y, \Omega)$ which is left adjoint to $\underline{C}(f, \Omega) : \underline{C}(Y, \Omega) \to \underline{C}(X, \Omega)$ i.e. $\exists f[\varphi] \leq \psi$ iff $\varphi \leq \psi f$ for every φ and ψ in $\underline{C}$. – $\exists f$ is called the existential quantification along f.

(4) $\underline{C}$ satisfies the equation $\exists f_1[\varphi] f_2 = \exists g_2[\varphi g_1]$, if $(g_1, g_2) = pb(f_1, f_2)$ is pullback in $\underline{C}$ of one of the two following types:

(a) $\exists(X, f)[1_Y f] = \exists \Delta_Y [1_Y](f \times Y)$, where $f : X \to Y$ and $1_Y = 1 !_Y$

(b) $\exists q[\varphi]g = \exists q'[\varphi(X \times g)]$ for $g: Z \to Y$ and projections $q: X \times Y \to Y$, $q': X \times Z \to Z$.

$$\begin{array}{ccc} X & \xrightarrow{(X,f)} & X\times Y \\ {\scriptstyle f}\downarrow & (a) & \downarrow{\scriptstyle f\times Y} \\ Y & \xrightarrow{\Delta_Y} & Y\times Y \end{array} \qquad \begin{array}{ccc} X\times Z & \xrightarrow{q'} & Z \\ {\scriptstyle X\times g}\downarrow & (b) & \downarrow{\scriptstyle g} \\ X\times Y & \xrightarrow{q} & Y \end{array}$$

(5a) if $e_Y(f_1, f_2) = 1_X$ then $f_1 = f_2$, where e_Y, the <u>equality on Y</u>, is defined by $e_Y = \exists \Delta_Y[1_Y]$.

(5b) $e_\Omega = \Leftrightarrow$, where $\Leftrightarrow$ is the biimplication.

A functor $F: \underline{C} \to \underline{C}'$ between two logical categories $\underline{C}$ and $\underline{C}'$ is called <u>logical</u>, if F preserves finite products, the boolean algebra object Ω together with $0, 1, \sim, \wedge$ and if F preserves quantification. – F is called an <u>extension</u> if F is also bijective on objects.

1.2. <u>Definition</u>: A category $\underline{T}$ is called an <u>elementary theory</u> if $\underline{T}$ satisfies the following conditions:

(1) $\underline{T}$ has two basic objects A and Ω such that every object X different from Ω has a specified representation as a finite power A^n of A, and $\underline{T}(\Omega, X)$ is empty.

(2) $\underline{T}(A^n, \Omega)$ is a boolean algebra for every $A^n \in ob(\underline{T})$ and $\underline{T}(f, \Omega)$ is a boolean homomorphism for every $f: A^n \to A^m \in \underline{T}$. – Ω might be called an implicit boolean algebra object.

(3) For every $f: A^n \to A^m \in T$ there exists a quantification $\exists f: \underline{T}(A^n, \Omega) \to T(A^m, \Omega)$ which satisfies the conditions (3), (4), (5) of 1.1.

A functor $F: \underline{T} \to \underline{T}'$ between two elementary theories $\underline{T}$ and $\underline{T}'$ is called an <u>elementary functor</u>, if F preserves finite products, the basic objects A and Ω, the boolean structure of the set $T(A^n, \Omega)$ and

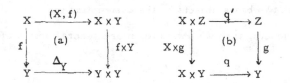

It should be noted that all the following results for logical categories are valid also for elementary theories with slight modifications.

In the following we will adopt the convention of writing binary propositional operations between the arguments.

1.3. Remarks to the previous definition:

(1) There are no explicit variables in $\underline{C}$. Their role is taken over by the objects of $\underline{C}$, which might be called the types of $\underline{C}$.

(2) It should be remarked that Lawvere used in [13] the following stronger condition for the object Ω: Ω is the coproduct $I+I$ with the injections $0, 1: I \to \Omega$ and the functor $X \times (-): \underline{C} \to \underline{C}$ preserves this coproduct for every $X \in ob(\underline{C})$. This implies that Ω is a boolean algebra object, since the negation $\sim: \Omega \to \Omega$ and the conjunction $\wedge: \Omega \times \Omega \to \Omega$ can be defined by means of the coproduct property.

(3) The adjointness condition for the quantification is equivalent to the following two equations:

$$\varphi \wedge \exists f[\varphi]f = \varphi \tag{3.1}$$

$$\exists f[\varphi \wedge \psi f] = \exists f[\varphi] \wedge \psi \tag{3,2}$$

<u>The universal quantification along f</u> is defined by

$$\forall f[\varphi] = \neg \exists f[\neg \varphi] \tag{3.3}$$

Since negation is the dualization functor of a boolean algebra, $\forall f$ is right adjoint to $\underline{C}(f, \Omega)$ i.e. $\psi \leq \forall f[\varphi]$ iff $\psi f \leq \varphi$ for every φ and ψ.

(4) The formulas (4a) and (4b) are equivalent to the following three formulas:

(a) $\exists f[\varphi] = \exists q[\varphi p \wedge e_Y(fp, q)]$ for $f: X \to Y$ and projections $p: X \times Y \to X$, $q: X \times Y \to Y$

(b) $\exists q[\varphi]g = \exists q'[\varphi(X \times g)]$ for $g: Z \to Y$ and projections $q: X \times Y \to Y$, $q': X \times Z \to Z$

(c) $e_{X \times Y} = e_X(p_1, p_3) \wedge e_Y(p_2, p_4)$ where p_i are the projections of $X \times Y \times X \times Y$

(a) shows that quantification along an arbitrary morphism can be reduced to quantification along a projection and equality. The condition (b) states that the quantification along q can be interchanged with the substitution of g, if there is no collision of variables i.e. if g is substituted only for those variables which are not affected by the quantification along q. The condition (c) allows a reduction of the equality. (5a) This condition ensures that two morphisms which are equal in the sense of the equality predicate are equal. (5b) is a natural condition, since the biimplication acts as equality predicate in a boolean algebra.

Making use of the above definitions one can derive the familiar logical identities.

1.4. The following results will be used later on:

(1) $\psi \leq \exists f[\psi]f$, $\exists f[\psi f] \leq \psi$

(2) $\exists f[\psi f \wedge \lambda] = \psi \wedge \exists f[\lambda]$

(3) $\exists f[\psi_1 \vee \psi_2] = \exists f[\psi_1] \vee f[\psi_2]$

(4) $\exists f[\psi] = 0_Y$ iff $\psi = 0_X$

(5) $\exists gf[\psi] = \exists g[\exists f[\psi]]$

(7) if f is left invertible then $\exists f[\psi]f = \psi$

(8) if f is an isomorphism then $\exists f[\psi] = \psi f^{-1}$

(9) f is epic iff $\exists f[1_X] = 1_Y$

(10) $\exists X_1 \times f_2[\varphi(f_1 \times X_2)] = \exists Y_1 \times f_2[\varphi](f_1 \times Y_2)$

$$\begin{array}{ccc} X_1\times X_2 & \xrightarrow{X_1\times f_2} & X_1\times Y_2 \\ {\scriptstyle f_1\times X_2}\downarrow & (10) & \downarrow{\scriptstyle f_1\times Y_2} \\ Y_1\times X_2 & \xrightarrow[Y_1\times f_2]{} & Y_1\times Y_2 \end{array} \qquad \begin{array}{ccc} X\times Z & \xrightarrow[q]{} & Z \\ {\scriptstyle p}\downarrow & (11) & \downarrow{\scriptstyle g} \\ X & \xrightarrow[f]{} & Y \end{array}$$

(11) $\exists f[\varphi]g = \exists q[\varphi_p \wedge e_Y(fp, gq)]$

(12) $e_Y(f, f) = 1_X$

(13) $e_Y(f, g) = e_Y(g, f)$

(14) $e_Y(f, g) \le e_Z(hf, hg)$

(15) $e_{Y\times Y'}(f, f', g, g') = e_Y(f, g) \wedge e_{Y'}(f', g')$

(16) $e_Y(f, g) \wedge \psi f \le \psi g$

(17) $e_Y(f, g) \wedge e_Y(g, h) \le e_Y(f, h)$

(18) $\forall q[e_Y(f, f') \wedge e_\Omega(\varphi, \varphi')] \le e_\Omega(\exists f[\varphi], \exists f'[\varphi'])$

(19) $\forall !_X[e_Y(f, g)] = 1$ iff $f = g$

(20) if $\varphi \le e_Y(f_1, f_2)$ then $\exists f_1[\varphi] = \exists f_2[\varphi]$

Corresponding formulas for the universal quantification can be obtained by dualizing.

1.5. Example: $\underline{S}$, the category of sets, is a logical category.

The category $\underline{S}$ has finite limits and hence in particular finite products. The 2-element set 2 is a boolean algebra object in $\underline{S}$. Moreover, 2 classifies subobjects in $\underline{S}$, i.e. for every $\varphi: X \to 2$ there exists a unique subobject $\varphi^{\#}$ of X such that $\varphi^{\#} = eq(\varphi, 1_X)$. The substitution $\underline{S}(f, 2)$ for $f: X \to Y \in \underline{S}$ corresponds to the inverse image under f i.e. $(\psi f)^{\#} = f^{-1}(\psi^{\#})$. Since $\underline{S}$ has images, the quantification $\exists f$ can be defined by the direct image under f i.e. $\exists f[\varphi]^{\#} = f(\varphi^{\#})$. This implies the adjointness of the quantification, since we have $f(\varphi^{\#}) \subseteq \psi$ iff $\varphi^{\#} \subseteq f^{-1}(\psi^{\#})$.

Since images are preserved by pullbacks, we have $f_2^{-1} f_1(\varphi^{\#}) = g_2 g_1(\varphi^{\#})$ for a pullback $(g_1, g_2) = pb(f_1, f_2)$ and therefore $\exists f_1[\varphi] f_2 = \exists g_2[\varphi g_1]$. This gives in particular the conditions (4a) and (4b). Condition (5a) follows from $e_Y^{\#} = \Delta_Y(1_Y^{\#}) = \Delta_Y$ and $(e_Y(f_1, f_2))^{\#} = (f_1, f_2)^{-1}(\Delta_Y) = eq(f_1, f_2)$. (5b) follows from $e_\Omega^{\#} = \Delta_\Omega = eq(\Leftrightarrow, 1_{\Omega\times\Omega}) = \Leftrightarrow^{\#}$.

Later on we will need the following elementwise description of the quantification in $\underline{S}$. Since 1 is a projective generator in $\underline{S}$, we have

(1) $\exists f[\varphi]_y = 1$ for $y: I \to Y$ iff there exists $x: I \to X$ such that $y = fx$ and $\varphi x = 1$

(2) $e_Y(y, y') = 1$ iff $y = y'$ for $y, y': I \to Y$.

1.6. Example: Every first-order theory with equality determines an elementary theory.

Let Fm(n) be the set of formulas of the theory whose free variables have indices less than n, and let R_n be the relation of logical equivalence i.e. $(\varphi_1, \varphi_2) \in R_n$ iff $\varphi_1 \Leftrightarrow \varphi_2$ is a theorem. Then let $\underline{T}(A_n, \Omega)$ be the boolean algebra $Fm(n)/R_n$.

Similarly let Tm(n) be the set of terms of theory whose variables have indices less than n, and let S_n be the relation of provable equality i.e. $(f_1, f_2) \in S_n$ iff $f_1 \circleddash f_2$ is a theorem, where $\circleddash$ is the equality predicate. Then let $\underline{T}(A_n, A_1)$ be the quotient $Tm(n)/S_n$. - The equivalence class of $\psi \in Fm(n)$ resp. $f \in Tm(n)$ will be denoted by $\bar{\psi}$ resp. $\bar{f}$. Finally let $\underline{T}(A_n, A_m)$ be the set of m-tuples of elements of $\underline{T}(A_n, A_1)$ i.e. $\underline{T}(A_n, A_m) = \underline{T}(A_n, A_1)^m$.

The composition at the right with $\bar{f} \in \underline{T}(A_n, A_m)$ is given by the simultaneous substitution of the terms $f_o, f_1, .., f_{m-1}$. This is well-defined and makes T into a category with the objects Ω and A_n for $n \in N$. In particular, $T(\bar{f}, \Omega)$ is boolean homomorphism. Moreover, the object A_n is the n-th power of A_1 i.e. $A_n = (A_1)^n$. The projections are given by the variables $\bar{x}_i$ for $i<n$.

The quantification in $\underline{T}$ is defined as follows (cf.1.3.4a):

$$\exists(\bar{f}_o, .., \bar{f}_{n-1})[\bar{\psi}] = \overline{\exists x_n, .., x_{n+m-1}[\psi(x_i/x_{n+i}) \bigwedge_{k \neq o}^{n-1} x_k = f_k(x_i/x_{n+i})]}$$

The adjointness of quantification follows from inferences in first-order logic. Similarly the conditions (4)-(6) can be verified.

It should be remarked that $\underline{T}$ can be extended to a logical category $\underline{T}^+$ in which $X \times \Omega$ is the coproduct of X and X for every X in $\underline{T}^+$. The objects of $\underline{T}^+$ are the finite products $A^n \times \Omega^m$. Before building up $\underline{T}^+$ from $\underline{T}$ by taking product-maps and coproduct-maps, we have to enlarge $\underline{T}$ by conditional terms. These terms correspond to morphisms of the form $(f_o; f_1)(A^n, \mu): A^n \times \Omega \to A$ and are interpreted as "if μ then f_1 else f_o" in the category of sets(cf. [5]).

2. The completeness theorem for logical categories:

The notion of a model of a logical category is defined as follows:

2.0. Definition: Let $\underline{C}$ be a logical category. A logical functor M from $\underline{C}$ into the category of sets is called a $\underline{C}$-model. A natural transformation $\mu: M \to N$ between two $\underline{C}$-models M and N is called a $\underline{C}$-embedding.

Now the completeness theorem can be formulated:

2.1. Theorem: Let $\underline{C}$ be a small logical category which is consistent and nice i.e. $0_X \neq 1_X$ and $\exists!_X[1_X] = 1$ for every $X \in ob(\underline{C})$.

(1) For every pair of morphisms $f, g: X \to Y$ there exists a $\underline{C}$-model M such that $M(f) \neq M(g)$.

(2) There exists a $\underline{C}$ model M such that $card(M(X)) \leq card(\underline{C})$ for every $X \in ob(\underline{C})$.

Making use of the corresponding fact in the category of sets, we obtain the following immediate corollary:

Corollary: In every small logical category which is consistent and nice, the pullback-condition 1.1.4 $\exists f_1[\varphi]f_2 = \exists g_2[\varphi g_1]$ is satisfied for every pullback $(g_1, g_2) = pb(f_1, f_2)$, which is preserved by every product-preserving functor.

As in the well-known proof of the completeness of first-order logic in Henkin [8] we shall construct for the given logical category an extension $F:\underline{C}\to\underline{C}'$ for which $\underline{C}'(I,-):\underline{C}'\to \underline{\text{Sets}}$ is a model i.e. $\underline{C}'$ has a canonical model. Then the composition of $\underline{C}'(I,-)$ with the functor F gives the required $\underline{C}$-model.

The following proposition characterizes those logical categories which have canonical models.

2.2. Proposition: Let $\underline{C}$ be a logical category. Then $\underline{C}(I,-):\underline{C}\to\underline{S}$ is a $\underline{C}$-model iff

(1) $\underline{C}$ is maximally consistent: i.e. $C(I,\Omega) = \{0,1\}$

(2) $\underline{C}$ is rich i.e. for every $\varphi:X\to\Omega$ such that $\exists!_X[\varphi] = 1$ there exists $k:I\to X$ with $\varphi k = 1$.

It is obvious from 1.5 that these conditions are necessary. – However these conditions are also sufficient. The functor $\underline{C}(I,-)$ always preserves products. The condition $\underline{C}(I,\Omega) = \{0,1\}$ ensures that $\underline{C}(I,-)$ preserves Ω. Moreover, $\underline{C}(I,-)$ preserves equality, since $\underline{C}$ satisfies condition (5a) of 1.1. Because of the reduction formula (3a) of 1.3 it remains to be shown that $\underline{C}(I,-)$ preserves quantifications along projections. Hence we have to show that for $\exists q[\varphi]y = 1$, where $y\in\underline{C}(I,Y)$ and $q:X\times Y\to Y$ a projection, there exists $x\in\underline{C}(I,X)$ such that $\varphi(x,y) = 1$ (cf. 1.5.1). But we have

$\exists q[\varphi]y = \exists !_X[\varphi(Xxy]$ because of (4b) in 1.1. Now the required result follows from the fact that $\underline{C}$ is rich.

Now we have to construct an extension which is maximally consistent and rich. A maximally consistent extension of $\underline{C}$ can be obtained by constructing the quotient category with respect to an ultrafilter Δ in the set of sentences $\underline{C}(I,\Omega)$ of $\underline{C}$. This corresponds to an extension by a new set of axioms in first-order logic.

2.3. Proposition. Let $\underline{C}$ be a consistent logical category and let Δ be a filter in $\underline{C}(I,\Omega)$. Define a relation R on $\underline{C}$ by $(f,g) \in R$ iff $\forall !_X[e_Y(f,g)] \in \Delta$ for $f, g \in \underline{C}(X,Y)$. Then $\underline{C}/\Delta = \underline{C}/R$ is again a consistent logical category and the projection $P: \underline{C} \to \underline{C}/\Delta$ is an extension of $\underline{C}$. If Δ is an ultrafilter, then $\underline{C}/\Delta$ is maximally consistent.

It is sufficient to show that R is an equivalence relation on $\underline{C}$ which is compatible with composition, products and quantification. This is done in two steps. Making use of the properties of the universal quantification in 1.4, we can verify that $F(X) = \{\varphi \in \underline{C}(X,\Omega): \forall !_X[\varphi] \in \Delta\}$ is a set of filters which is closed under substitution and universal quantification. Then, making use of the properties of the equality in 1.4, it can be verified that $R(X,Y) = \{(f,g): e_Y(f,g) \in F(X)\}$ has the required properties. In particular, 1.4.18 implies that R is closed under quantification.

The following consequence of 1.4.19 characterizes those logical functors which are faithful. This corresponds to conservative extensions in the first-order logic.

2.4. Lemma. A logical functor $F: \underline{C} \to \underline{C}'$ is faithful iff $F(\varphi) = F(1)$ implies $\varphi = 1$ for every $\varphi \in \underline{C}(I,\Omega)$.

Since the construction of a rich extension involves a countable chain of logical categories, we need the following lemma on colimits of chains.

2.5. Lemma. Let $F_i:\underline{C}_i \to \underline{C}_{i+1}$ for $i \in N$ be a countable chain of extensions of logical categories. Then the direct limit $\underline{C}$ is again a logical category. If every $\underline{C}_i$ is maximally consistent resp. rich then $\underline{C}$ is again maximally consistent resp. rich.

Since the F_i are extensions (cf. 1.1) we may assume $ob(\underline{C}_0) = ob(\underline{C}_i)$. Define $\underline{C}$ by $ob(\underline{C}) = ob(\underline{C}_0)$ and $\underline{C}(X,Y) = \sum_{i\in N} \underline{C}_i(X,Y)/R$, where R is defined by $(f,f') \in R$ for $f \in \underline{C}_i(X,Y)$, $f' \in \underline{C}_{i'}(X,Y)$ iff there exists $j \geq i, i'$ such that $F_{j-1} \ldots F_i(f) = F_{j-1} \ldots F_{i'}(f')$. It is easy to show that $\underline{C}$ is a logical category, since $\underline{N}$ is a directed set and every condition in 1.1 involves only finitely many morphisms. Similarly it can be verified that the property of being maximally consistent resp. rich is inherited by $\underline{C}$.

As in first-order logic the construction of a rich extension of $\underline{C}$ involves an extension $\underline{C}[K]$ of $\underline{C}$ by a set of constants K i.e. morphisms of the form $I \to X$ with X in $\underline{C}$. Moreover, this extension should be conservative i.e. the functor $\underline{C} \to \underline{C}[K]$ has to be faithful.

The basic idea of the construction can be described as follows. Every morphism in $\underline{C}[K]$ is a morphism of $\underline{C}$ into which a finite sequence of constants from K has been substituted. In a first step we will define $\underline{K}^{\#}$, the category of finite sequences of K, together with a contravariant functor $A:\underline{K}^{\#} \to \underline{C}$.

2.6. Let $\underline{C}$ be a small logical category. A category $\underline{K}$ with $ob(\underline{K}) = ob(\underline{C})$ and $\underline{K}(X,Y) = \emptyset$ for $X \neq I$ is called a category of constants for $\underline{C}$.

Let $\underline{S}_0$ be the category of finite cardinals and arbitrary mappings. Then $\underline{K}^{\#}$ is defined by $\underline{K}^{\#} = \underline{S}_0/\underline{K}$, where $\underline{K}$ is viewed as a set. Thus the objects of $\underline{K}^{\#}$ are maps $c:n \to K$ with $n \in ob(\underline{S}_0)$, and $s:c \to c'$ is a morphism if $c's = c$ with $s:n \to n' \in \underline{S}_0$.

$\underline{K}^{\#}$ is a filtered category i.e. for $c, c' \in ob(\underline{K}^{\#})$ there exist $s:c \to c''$, $s':c' \to c''$ in $\underline{K}^{\#}$ and for $s, s':c \to c'$ in $\underline{K}^{\#}$ there exists $t:c' \to c''$ with $ts = ts'$. This follows from the fact that $\underline{S}_0$ has finite colimits.

Remembering the category structure of $\underline{K}$, we can define $A(c) = A(c(0)) \times \ldots \times A(c(n-1))$ for $c:n \to K$, where $A(c(i))$ is the codomain of $c(i)$ in K, which is also an object in $\underline{C}$. For $s:c \to c'$ in $\underline{K}^{\#}$ we define $A(s):A(c') \to A(c)$ by $p_k A(s) = q_{s(k)}$, where p_i resp. q_j are the projections of $A(c)$ resp. $A(c')$. This gives a contravariant functor $A:\underline{K}^{\#} \to \underline{C}$.

It can be verified easily that A is faithful and carries finite coproducts into products. Later on, we will need the following remark:

For $s, s':c \to d$ in $\underline{K}^{\#}$ with c monic and $c \neq \emptyset$ there exists $t:d \to c$ such that $ts = ts' = id_c$.

After these preparations we can construct the extension by constants.

2.7. Proposition. Let $\underline{C}$ be a nice logical category and let $\underline{K}$ be a category of constants for $\underline{C}$. Then there exists a faithful extension $\underline{C}[K]$ of $\underline{C}$ which contains $\underline{K}$ as a subcategory. Moreover, $\underline{C}[K]$ has the following universal property. Every logical functor $F:\underline{C} \to \underline{D}$ which coincides on objects with a functor $H:\underline{K} \to \underline{D}$ can be extended uniquely to $C[K]$.

Formalizing the idea mentioned above, we define $ob(\underline{C}[K]) = ob(\underline{C})$ and $\underline{C}[K](X, Y) = \{\langle f, c\rangle \mid f{:}A(c)\times X \to Y \in \underline{C}\}/R$, where we have $(\langle f, c\rangle, \langle f', c'\rangle) \in R$ iff there exist $s{:}c \to d,\ s'{:}c \to d \in K^{\#}$ such that $f(A(s)\times X) = f'(A(s')\times X)$. The definition is equivalent to $\underline{C}[K](X, Y) = \text{colimit}\ (\underline{C}(A(-)\times X, Y){:}\underline{K}^{\#} \to S)$, since the above description gives the construction of the colimit of this set-valued functor over a filtered category. The equivalence class of $\langle f, c\rangle$ will be denoted by $\langle f \mid c\rangle$, and $c^\frown d$ denotes the juxtaposition of c and d.

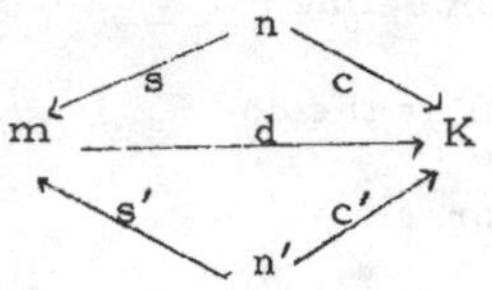

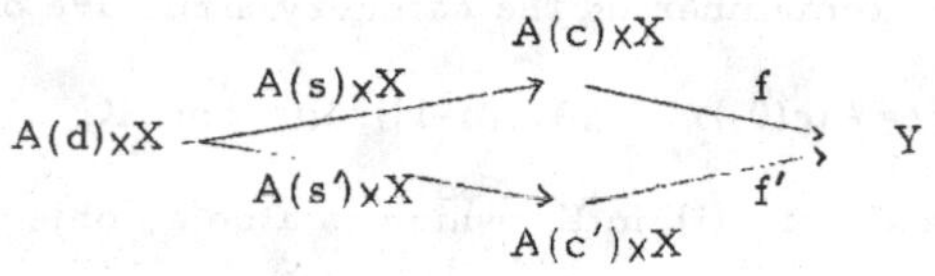

Using the above mentioned idea as guideline, we define:

(1) composition: $\langle f \mid c\rangle\langle g \mid d\rangle = \langle f(A(c)\times g) \mid c^\frown d\rangle$

(2) product-map: $(\langle f \mid c\rangle, \langle f' \mid c'\rangle) = \langle (fp, f'p') \mid c^\frown c'\rangle$

(3) $J_1{:}\ \underline{C} \to \underline{C}[K]{:}\ J_1(f) = \langle f \mid \emptyset\rangle$ with $\emptyset{:}\emptyset \to K$

(4) $J_2{:}\ \underline{K} \to \underline{C}[K]{:}\ J_2(k) = \langle X \mid k\rangle$ with $k{:}I \to X \in K$

(5) quantification: $\exists\langle f \mid c'\rangle[\langle \varphi \mid c\rangle] = \langle \exists A(c)\times(p, f)[\varphi(q\times X)] \mid c^\frown c'\rangle$

Now it can be verified that these definitions are well-defined and make $\underline{C}[K]$ into a logical category and $J_1{:}\underline{C} \to \underline{C}[K]$ into a logical functor. Except for the case of quantification all the computations are straightforward. For the products one might also use the fact that filtered colimits of set-valued functors commute with finite products. (cf. Schubert [20], Thm. 9.4.1)

In the case of quantification one verifies first that the definition satisfies the adjointness condition 1.1.3. Here we have to use the pullback conditions (4a) and (4b) in 1.1 in $\underline{C}$. Since the quantification is uniquely determined by the adjointness condition, this settles the question of being well-defined. Then one has to verify the conditions (4a) and (4b) of 1.1 for $\underline{C}[K]$, making use of them in $\underline{C}$. The remaining conditions in 1.1 follow from the fact that J_1 preserves quantification. In order to show that $J_1:\underline{C}\to\underline{C}[K]$ is faithful we have to use the assumption that $\underline{C}$ is nice.

The universal property of $\underline{C}[K]$ is verified as follows. The extension $F':\underline{C}[K]\to\underline{D}$ is defined by $F'(\langle f|c\rangle) = F(f)H(c)$ with $H(c) = (H(c(0), \ldots, H(c(n-1)))$.

Later on we will need the following statement about removing constants from an equation:

(6) if $\langle f|c\rangle = \langle f'|c'\rangle$ with c monic and $c \neq \emptyset$ then $f = f'$.

This is a consequence of the final remark in 2.6.

The rich extension of $\underline{C}$ is obtained by adding for every $\varphi: X\to\Omega$ with $\exists!_X[\varphi] = 1$ a constant together with a special axiom.

2.8. Proposition: Let $\underline{C}$ be a small logical category which is consistent and nice. Then there exist a rich logical category $\underline{C}'$ and a faithful extension $F:\underline{C}\to\underline{C}'$.

Let $\underline{C}[K]$ be the extension of $\underline{C}$ by constants in $\underline{K}$, where $\underline{K}$ is defined by $\underline{K}(I,X) = \{\varphi\in\underline{C}(X,\Omega):\exists!_X[\varphi] = 1\}$. Let Δ be the filter in $\underline{C}[K](I,\Omega)$ generated by the set of axioms $\{\varphi\varphi^* \mid \varphi\in\underline{C}(X,\Omega), \exists!_X[\varphi] = 1\}$, where φ^* is the constant corresponding to φ. Then $\underline{C}_1$ and $F_1:\underline{C}\to\underline{C}_1$ are defined by $\underline{C}_1 = \underline{C}[K]/\Delta$ and $F_1 = PJ_1:\underline{C}\to\underline{C}[K]\to\underline{C}[K]/\Delta$.

To show that F_1 is faithful it is sufficient — because of lemma 2.4 — to verify that $F_1(\psi) = 1$ implies $\psi = 1$ for $\psi\in\underline{C}(I,\Omega)$. $F_1(\psi) = 1$ implies $\psi\in\Delta$ and hence there exist $\varphi_i\in\underline{C}(X_i,\Omega)$ for $i = 1, \ldots, n$ such that $\bigwedge_{i=1}^{n} \varphi_i\varphi_i^* \leq \psi$ in $\underline{C}[K]$. This implies $\langle \bigwedge_{i=1}^{n} \varphi_i p_i | (\varphi_1^*, \ldots, \varphi_n^*)\rangle \leq \langle \psi!_X | (\varphi_1^*, \ldots, \varphi_n^*)\rangle$, where $X = X_1\times\ldots\times X_n$ with projections p_i. Removing the constants by means of 2.7.6 we obtain $\bigwedge_{i=1}^{n} \varphi_i p_i \le \psi !_X$. Making use of the adjointness of conjunction and implication, we obtain $1 = \psi_n$, where $\psi_k : Y_k \to \Omega$ for k=0, ..., n are defined by $\psi_0 = \psi$, $Y_0 = I$ and $\psi_k = \varphi_k r_k \Rightarrow \psi_{k-1} q_k$, $Y_k = X_k \times Y_{k-1}$ with projections $r_k : Y_k \to X_k$ and $q_k : Y_k \to Y_{k-1}$. In particular we have $(r_k, q_k) = pb(!_{X_k}, !_{Y_{k-1}})$. Now we can prove $\psi_0 = \psi = 1$ by descent. If $\psi_k = 1$ then $\varphi_k r_k \le \psi_{k-1} q_k$ and hence $1 = \exists !_{X_k} [\varphi_k] !_{Y_{k-1}} = \exists q_k [\varphi_k r_k] \le \exists q_k [\psi_{k-1} q_k] = \psi_{k-1}$ because of $\exists !_{X_k} [\varphi_k] = 1$, (4b) in 1.1 and 1.4.6. Here we have used the fact that the projection q_k is epic, since $\underline{C}$ is nice. This shows that F_1 is faithful.

Hence $\underline{C}_1$ is rich at least for morphisms of the form $F_1(\varphi)$. Iterating the above construction, we obtain a countable chain of faithful extensions. Then $\underline{C}'$, the colimit of this chain, is the desired rich extension of $\underline{C}$ and the injection $\underline{C} \to \underline{C}'$ is faithful.

Now we are in the position to prove the completeness theorem 2.1. As in the proof of Henkin the result is achieved by a rich extension followed by maximal consistent extension.

The assumption $f \ne g$ is equivalent to $\forall !_X [e_Y(f, g)] \ne 1$ because of 1.4.19 and 1.1.6. Now let $F: \underline{C} \to \underline{C}_1$ be the rich extension described in 2.8. Let Δ be an ultrafilter in $\underline{C}_1(I, \Omega)$ which does not contain $F(\forall !_X [e_Y(f, g)])$. Because of 2.3 we obtain a maximally consistent logical category $\underline{C}_2 = \underline{C}_1 / \Delta$ together with the canonical projection $P: \underline{C}_1 \to \underline{C}_2$. Moreover, $\underline{C}_2$ is still rich, since the functor P is full and surjective on objects. Hence $\underline{C}_2$ has a canonical model because of 2.2. In particular we have $PF(\forall !_X [e_Y(f, g)]) = 0$ and hence $PF(f) \ne PF(g)$. Now the composition $C_2(I, -)PF$ gives the required result, since $C_2(I, -)$ is faithful because of 2.4.

It should be remarked that until so far we had to use only the prime ideal theorem, which provided the existence of the ultrafilters involved. A careful inspection of the constructions which have been used yields the inequality $\mathrm{card}(\underline{C}(I,PF(X))) \leq \mathrm{card}(\underline{C})$ for every X in $\underline{C}$. This requires of course the full axiom of choice.

Making use of the methods developed above, we can prove the following criterion for the consistency of a pushout in Log, the category of logical categories. — It should be noted, that Log is complete and co-complete. The limits are the same as in Cat, whereas the colimits require a more elaborate construction.

2.9. Proposition: Let $H_1 : \underline{C}_0 \to \underline{C}_1$, $H_2 : \underline{C}_0 \to \underline{C}_2$ be logical functors, where $\underline{C}_0, \underline{C}_1$ and $\underline{C}_2$ are consistent. Let $(Q_1, Q_2) = P_0(H_1, H_2)$ be the pushout in the category of logical categories, then the following statements are equivalent:

$$\begin{array}{ccc} \underline{C}_0 & \xrightarrow{H_1} & \underline{C}_1 \\ {\scriptstyle H_2}\downarrow & & \downarrow{\scriptstyle Q_1} \\ \underline{C}_2 & \xrightarrow{Q_2} & \underline{C}_3 \end{array}$$

(1) If $Q_1(\varphi_1) \leq Q_2(\varphi_2)$ for $\varphi_i \in \underline{C}_i(I,\Omega)$, then there exists $\varphi \in \underline{C}_0(I,\Omega)$ such that $\varphi_1 \leq H_1(\varphi_0)$ and $H_2(\varphi_0) \leq \varphi_2$.

(2) If H_1 is faithful, then Q_2 is faithful.

(3) If H_1 and H_2 are faithful, then $\underline{C}_3$ is consistent.

(4) If $\underline{C}_3$ is not consistent, then there exists $\varphi \in \underline{C}_0(I,\Omega)$ such that $1 = H_1(\varphi)$ and $H_2(\varphi) = 0$.

(1) corresponds to the interpolation theorem of Craig (cf. [3]) for first-order logic. (3) corresponds to the amalgamation theorem of Daigneault (cf. [4], 2.6) for polyadic algebras. (4) corresponds to the consistency theorem of A. Robinson (cf. [18], 2.9) in model

theory. A related result concerning the equivalence of (1) and (3) can be found in Preller [17].

In the following we will use several times the lemma 2.4 which characterizes faithful logical functors.

(1) implies (2): Assume $Q_2(\varphi_2) = 1$. Since we have $Q_1(1) = 1 = Q_2(\varphi_2)$, there exists φ with $1 = H_1(\varphi)$ and $H_2(\varphi) \leq \varphi_2$. Since H_1 is faithful, $H_1(\varphi) = 1$ implies $\varphi = 1$ and hence $1 = H_2(\varphi) = \varphi_2$ as required.

(2) implies (3): Applying (2) twice we see that Q_1 and Q_2 have to be faithful. But now the consistency of $\underline{C}_1$ or $\underline{C}_2$ implies the consistency of $\underline{C}_3$.

(3) implies (4): The following argument is an adaptation of 4.5 in Daigneault [4]. Assume that there does not exist φ such that $1 = H_1(\varphi)$ and $H_2(\varphi) = 0$. Since the statement "(3) implies (4)" is true for boolean algebras (cf. Daigneault [4], 4.2), there exist ultrafilter Δ_i in $\underline{C}_i(I, \Omega)$ for $i=1,2$ such that $H_1^{-1}(\Delta_1) = H_2^{-1}(\Delta_2) = \Delta_0$. Because of 2.3 we can define $\bar{H}_1 : \underline{C}_0/\Delta_0 \to \underline{C}_1/\Delta_1$ and $\bar{H}_2 : \underline{C}_0/\Delta_0 \to \underline{C}_2/\Delta_2$ such that $\bar{H}_i P_0 = P_i H_i$ for $i=1,2$, where $P_i : \underline{C}_i \to \underline{C}_i/\Delta_i$ are the canonical projections. Since each Δ_i is an ultrafilter, each $\underline{C}_i/\Delta_i$ is maximally consistent and therefore $\bar{H}_1$ and $\bar{H}_2$ are automatically faithful. Now we can apply (3) to $\bar{H}_1$ and $\bar{H}_2$. This implies that $\bar{Q}_i : \underline{C}_i/\Delta_i \to \underline{C}_4$ for $i=1,2$ are faithful and $\underline{C}_4$ is consistent, where $\bar{Q}_1$ and $\bar{Q}_2$ are defined by $(\bar{Q}_1, \bar{Q}_2) = \mathrm{Po}(\bar{H}_1, \bar{H}_2)$. In particular we have $\bar{Q}_1 P_1 H_1 = \bar{Q}_1 \bar{H}_1 P_0 = \bar{Q}_2 \bar{H}_2 P_0 = \bar{Q}_2 P_1 H_2$. Hence there exists $H : \underline{C}_3 \to \underline{C}_4$ such that $HQ_i = \bar{Q}_i P_i$ for $i=1,2$. The existence of H and the consistency of

$\underline{C}_4$ imply the consistency of $\underline{C}_3$.

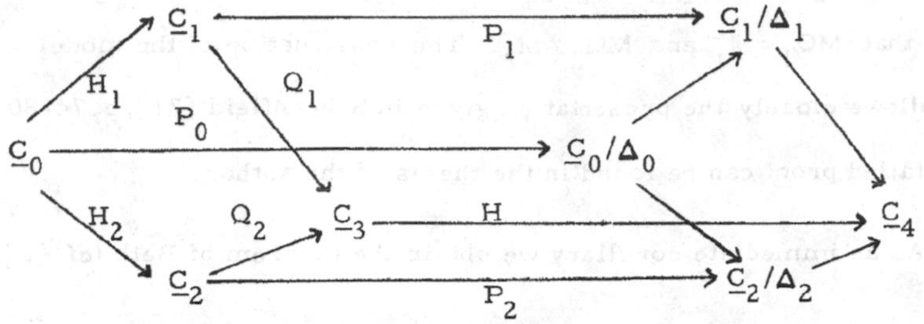

(4) implies (1): Assume $Q_1(\varphi_1) \leq Q_2(\varphi_2)$ with $\varphi_i \in \underline{C}_i(I, \Omega)$ for i=1, 2. Let Δ_1 resp. Δ_2 be the filters generated by φ_1 resp. φ_2 in $\underline{C}_1(I, \Omega)$ resp. $\underline{C}_2(I, \Omega)$. Then we can define $P_i : \underline{C}_i \to \underline{C}_i/\Delta_i$ for i=1, 2. If $\bar{Q}_1$ and $\bar{Q}_2$ are defined by $(\bar{Q}_1, \bar{Q}_2) = Po(P_1H_1, P_2H_2)$, then there exists a logical functor $Q : \underline{C}_3 \to \underline{C}_4$ with $\bar{Q}_iP_i = QQ_i$ for i=1, 2. Now we have $QQ_1(\varphi_1) \wedge QQ_2(\neg\varphi_2) = 0$ because of $Q_1(\varphi_1) \leq Q_2(\varphi_2)$ and on the other hand $\bar{Q}_1P_1(\varphi_1) \wedge \bar{Q}_2P_2(\neg\varphi_2) = 1$ because of the definition of Δ_1 and Δ_2. Hence $\underline{C}_4$ is inconsistent. Then (4) yields φ with $P_1(1) = P_1H_1(\varphi)$ and $P_2H_2(\varphi) = P_2(0)$; this implies $\varphi_1 \leq H_1(\varphi)$ and $H_2(\varphi) \leq \varphi_2$ because of the definition of Δ_1 and Δ_2 (cf. Shoenfield [21], p.80)

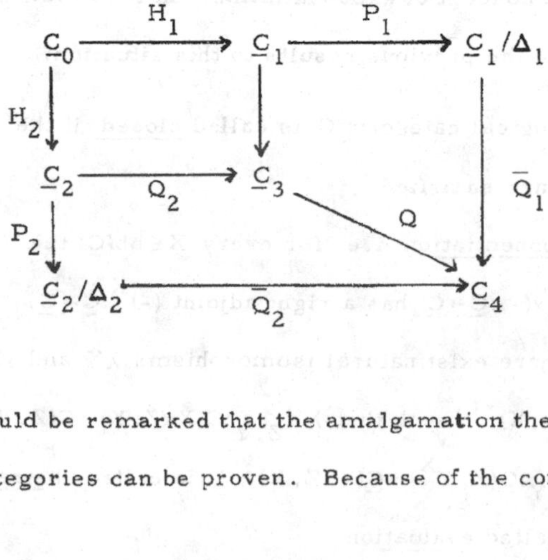

It should be remarked that the amalgamation theorem for nice logical categories can be proven. Because of the completeness theorem it is sufficient to prove that for two models M_1 and M_2 of $\underline{C}_1$ resp. $\underline{C}_2$ with $M_1H_1 = M_2H_2$ there exists a model M of $\underline{C}_3$ such that $MQ_1 = M_1$ and $MQ_2 = M_2$. The construction of the model M follows closely the presentation given in Schoenfield [21], p. 74-80. A detailed proof can be found in the thesis of the author.

As an immediate corollary we obtain the theorem of Beth (cf. [2]).

2.10. Proposition: Let $H:\underline{C}_0 \to \underline{C}_1$ be a logical functor, where $\underline{C}_0$ and $\underline{C}_1$ are nice, consistent logical categories. Let $(Q_1, Q_2) = P_0(H, H)$ be the pushout in the category of logical categories. Then the following two statements are equivalent:

(1) $\varphi \in \underline{C}_1(I, \Omega)$ is explicitly definable i.e. there exists $\psi \in \underline{C}_0(I, \Omega)$ such that $\varphi = H(\psi)$.

(2) $\varphi \in \underline{C}_1(I, \Omega)$ is implicitly definable i.e. $Q_1(\varphi) = Q_2(\varphi)$.

3. The completeness theorem for closed logical categories:

In order to take into account higher order logic with function types, we have to require that the set of morphisms from X to Y in a logical category $\underline{C}$ is represented by an actual object Y^X in $\underline{C}$. This leads to the concept of exponentiation. In the following we shall try to extend the previous results to this situation.

3.1. Definition: A logical category $\underline{C}$ is called closed if the following conditions are satisfied:

(1) $\underline{C}$ has exponentiation i.e. for every $X \in ob(\underline{C})$ the functor $X \times (-):\underline{C} \to \underline{C}$ has a right adjoint $(-)^X:\underline{C} \to \underline{C}$.
– Hence there exist natural isomorphisms λX and ϵX with $\lambda X = (\epsilon X)^{-1}$ such that $\lambda X_{Z,Y}:\underline{C}(X \times Z, Y) \to \underline{C}(Z, Y^X)$ and $\epsilon X_{Z,Y}:\underline{C}(Z, Y^X) \to \underline{C}(X \times Z, Y)$. λ is called abstraction and ϵ is called evaluation.

(2) $\underline{C}$ satisfies the <u>extensionality axiom</u> i.e. $e_Y X(\lambda X[f_1], \lambda X[f_2]) = \forall q[e_Y(f_1, f_2)]$ for $f_i: X \times Z \to Y$ and the projection $q: X \times Z \to Z$.

Having defined Y^h for $h: X' \to X$ by $\epsilon X'[Y^h] = \epsilon X[Y^X](h \times Y^X)$, we obtain the following two formulas:

(3) $g^X \lambda X[f] = \lambda X[gf]$ for $g: Y \to Y'$ and $f: X \to Y$

(4) $Y^h \lambda X[f] = \lambda X'[fh]$ for $h: X' \to X$ and $f: X \to Y$

Thus g^X resp. Y^h represents composition at the left with g resp. at the right with h.

In particular Ω^f represents the substitution $\underline{C}(f, \Omega)$. But also the quantification $\exists f$ is represented by a morphism $\overline{\exists f}: \Omega^X \to \Omega^Y$, where $\overline{\exists f}$ is defined by $\epsilon Y[\overline{\exists f}] = \exists f \times \Omega^X[\epsilon X[\Omega^X]]$ (cf. Scholz-Hasenjaeger [19], p.384). Thus the quantification $\exists f$ is represented by composition at the left with $\overline{\exists f}$:

(5) $\overline{\exists f} \lambda X[\varphi] = \lambda Y[\exists f[\varphi]]$ for $\varphi: X \to \Omega$.

The following special case of (2) shows that in a closed logical category quantification can be expressed by means of equality:

(6) $\forall q[\varphi] = e_{\Omega^X}(\lambda X[\varphi], \lambda X[1_{X \times Z}])$

for $\varphi: X \times Z \to \Omega$ and the projection $q: X \times Z \to Z$. – This corresponds to a result of Henkin about propositional types (cf. Henkin [10], 4.6).

Later on we will need the following consequence of (2):

(7) $e_Y X(f_1{}^Z, f_2{}^Z) = \forall q[e_Y(f_1, f_2) \in Z[X^Z]]$

with $f_i: X \to Y$ and the projection $q: Z \times X^Z \to X^Z$.

3.2. The completeness theorem for closed logical categories is not true with respect to models which preserve exponentiation. This follows from the incompleteness of Peano-arithmetic. However, the completeness theorem remains true if we admit non-standard models i.e. models M for which $M(X^Y)$ may be a subset of $M(X)^{M(Y)}$. This corresponds to the completeness theorem for higher order logic in Henkin [9].

This can be seen as follows. The propositions 2.3 - 2.8 remain true if logical categories and functors are replaced by closed logical categories and logical functors which preserve exponentiation. In 2.3 the extensionality axiom or rather its consequence 3.1.7 is used to show that the relation R is closed under exponentiation. The propositions 2.4-2.6 have obvious extensions, whereas in 2.7 we have to show that $\underline{C}[K]$ has exponentiation and that the inclusion of $\underline{C}$ preserves it. The exponentiation in $\underline{C}[K]$ is defined by $\langle f|c\rangle^Z = \langle f^Z(d_{A(c)} \times X^Z)|c\rangle$, where $d_Z = \lambda Z[q]:A(c) \to A(c)^Z$ is the diagonal morphism. Then in 2.8 we can obtain the rich extension as before.

For the reason stated above, the proposition 2.2 about canonical models cannot be strengthened in this way. However, if $\underline{C}$ is a closed logical category which satisfies the conditions (1) and (2) of 2.2, then the canonical morphism $\underline{C}(I, X^Y) \to \underline{C}(I, X)^{\underline{C}(I, Y)}$ is monic. Here we have to use the fact that 1 is a generator in the category of sets. — Combining these results, we obtain the completeness theorem stated above.

4. Semantical categories:

Following a suggestion of Lawvere, we will replace the concept of quantification by the concept of direct image. An analysis of the proof of 1.5 leads to the following concept of a semantical category.

4.1. <u>Definition</u>: A category $\underline{E}$ is called <u>semantical</u> if it satisfies the following conditions:

(1) $\underline{E}$ has finite limits.

(2) $\underline{E}$ has an object Ω which is a boolean algebra object.

(3) Ω classifies subobjects i.e. for every $\psi: X \to \Omega$ there exists a unique subobject $\psi^{\#}$ of X such that $\psi^{\#} = eq(\psi, 1_X)$ and every subobject is of this form.

(4) If $(f_1, f_2) = pb(g_2, g_1)$ is a pullback in $\underline{E}$ with g_2 epic then f_2 is epic.

(5) $\underline{E}$ has epimorphic images i.e. every morphism $f: X \to Y$ has a factorization $f = im(f)q$ with im(f) monic and q epic such that for any factorization $f = mh$ with m monic there exists a unique g such that $mg = im(f)$ and $h = gq$. - The subobject im(f) is called the <u>image</u> of f. Moreover, if n is a subobject of X, then im(fn) is called the direct image of n under f.

A functor $H: \underline{E} \to \underline{E}'$ between two semantical categories $\underline{E}$ and $\underline{E}'$ is called <u>semantical</u> if it preserves finite limits, epics and the boolean object Ω together with its operations $0, 1, \neg$ and $\wedge$. —Thus H preserves in particular monomorphisms and hence the image-factorization and direct image.

The proof of 1.5 shows now that the category of sets is a semantical category.

4.2. <u>Proposition</u>: Every semantical category resp. functor is logical. — In other words there is a forgetful functor from the category of semantical categories into the category of logical categories.

A review of the proof of 1.5 shows that we derived the fact that the category of sets is logical from the fact that it is semantical. An analogous proof shows that every semantical category is logical.

As we have remarked above, every semantical functor preserves direct images. Hence it preserves quantification which is given by direct images. This shows that every semantical functor is logical.

The main result for semantical categories is the following:

4.3. <u>Theorem:</u> For every logical category $\underline{C}$ one can construct a semantical category $S(\underline{C})$ together with a faithful logical functor $J:\underline{C} \to S(\underline{C})$ such that every logical functor from $\underline{C}$ into a semantical category $\underline{E}$ can be extended uniquely to a semantical functor from $S(\underline{C})$. If $\underline{C}$ is already semantical, then $S(\underline{C})$ is equivalent to $\underline{C}$.

In other words the category of semantical categories is a reflexive subcategory of the category of logical categories. The construction of $S(\underline{C})$ is divided in two parts. The first part involves the addition of subobjects for morphisms of the form $X \to \Omega$.

4.4. <u>Proposition</u>: For every logical category $\underline{C}$ one can construct a logical category $\underline{C}^*$ together with a full and faithful logical functor $K:\underline{C} \to \underline{C}^*$ such that every logical functor from $\underline{C}$ into a logical category with finite limits can be uniquely extended to a logical functor which preserves finite limits. Moreover, $\underline{C}^*$ satisfies the conditions (1), (2) and (4) of 4.1, whereas the conditions (3) and (5) are satisfied only up to morphisms which are monic and epic.

The category $\underline{C}^*$ is defined as follows. The objects are pairs of the form $(X|\varphi)$ with $X \in ob(\underline{C})$ and $\varphi:X \to \Omega \in \underline{C}$. A morphism from $(X|\varphi)$ to $(Y|\psi)$ should be the restriction of a morphism $f:X \to Y$ in $\underline{C}$ to the subobject $(X|\varphi)$. Hence the mprhisms are triples of the form $(\psi|f|\varphi)$ with $f:X \to Y$, $\varphi:X \to \Omega$, $\psi:Y \to \Omega$ in $\underline{C}$ and $\exists f[\varphi] \leq \psi$. However, two morphisms $(\psi|f_1|\varphi)$, $(\psi|f_2|\varphi)$ are considered to be equal if $\varphi \leq e_Y(f_1, f_2)$ i.e. if f_1 and f_2 agree on the subobject $(X|\varphi)$. The composition is defined by $(\lambda|gf|\varphi) = (\lambda|g|\psi)(\psi|f|\varphi)$. Making use of the properties of the equality, we can verify that we have obtained a category. Moreover, a functor $K:\underline{C}\to\underline{C}^*$ can be defined by $K(f) = (1_Y|f|1_X)$ for $f:X\to Y\in\underline{C}$. K is full and faithful. The latter makes use of condition (5a) in 1.1.

Using the concept of restrictions as guidelines, we make the following definitions:

(1) $(X_1|\varphi_1)\times(X_2|\varphi_2) = (X_1\times X_2|\varphi_1p_1\wedge\varphi_2p_2)$, where p_1, p_2 are the projections of $X_1\times X_2$ in $\underline{C}$.

(2) $eq((\psi|f_1|\varphi), (\psi|f_2|\varphi)) = (\varphi|X|\varphi\wedge e_Y(f_1,f_2))$ with $f_1, f_2:X\to Y$.

(3) $pb((\psi|f_1|\varphi_1), (\psi|f_2|\varphi_2)) = ((\varphi_1|p_1|\mu), (\varphi_2|p_2|\mu))$, where $f_i:X_i\to Y$ for $i=1,2$ and p_1, p_2 are the projections of $X_1\times X_2$ and μ is defined by $\mu=\varphi_1p_1\wedge\varphi_2p_2\wedge e_Y(f_1p_1, f_2p_2)$.

(4) $\exists(\psi|f|\varphi)[(1_\Omega|\lambda|\varphi)] = (1_\Omega|\exists f[\varphi\wedge\lambda]|\psi)$.

(5) $e_{(X|\varphi)} = (1_\Omega|e_X|\varphi p_1\wedge\varphi p_2)$ where p_1, p_2 are the projections of $X\times X$ in $\underline{C}$.

Making use of the corresponding properties of $\underline{C}$, we can verify now that $\underline{C}^*$ is a logical category with finite limits. For the adjointness of the quantification the following observation concerning the order relation is very useful: $(1_\Omega|\mu_1|\varphi)\leq(1_\Omega|\mu_2|\varphi)$ iff $\varphi\leq\mu_1\Rightarrow\mu_2$.

It should be remarked that $\underline{C}^*$ satisfies the pullback condition (4) of 1.1 for arbitrary pullbacks in $\underline{C}^*$. The proof makes use of (4a) and (4b) of 1.3 for $\underline{C}$.

It follows from the above definitions that $K:\underline{C}\to\underline{C}^*$ is a logical functor. Moreover, K is full and faithful. The latter is due to condition (5a) of 1.1 for $\underline{C}$.

We have shown so far that $\underline{C}^*$ satisfies the conditions (1) and (2) of 4.1. Since a morphism $f:X \to Y$ in a logical category is epic iff $\exists f[1_X] = 1_Y$ (cf. 1.4.9), the pullback-condition for arbitrary pullbacks implies that epimorphisms are stable under pullbacks. This gives condition (4).

Now let us consider condition (3). To every morphism $(1_\Omega|\lambda|\psi)$: $(Y|\psi) \to (\Omega|1_\Omega)$ we associate the subobject $(1_\Omega|\lambda|\psi)^\# = (\psi|Y|\psi\wedge\lambda) = eq((1_\Omega|\lambda|\psi),(1_\Omega|1_Y|\psi)):(Y|\psi\wedge\lambda) \rightarrowtail (Y|\psi)$. Conversely we associate to every subobject $(\psi|m|\varphi):(X|\varphi) \rightarrowtail (Y|\psi)$ the morphism $(\psi|m|\varphi)^b = (1_\Omega|\exists m[\varphi]|\psi)$. Then we have $(1_\Omega|\lambda|\psi)^{\# b} = (1_\Omega|\lambda|\psi)$ but $(\psi|m|\varphi) = (\psi|m|\varphi)^{b\#}(\exists m[\varphi]|m|\varphi)$. $(\exists m[\varphi]|m|\varphi)$ is monic, since $(\psi|m|\varphi)$ is monic. However, $(\exists m[\varphi]|m|\varphi)$ is also epic, since the epimorphisms in $\underline{C}^*$ have the form $(\exists f[\mu]|f|\mu)$ because of 1.4.9. This shows that $\underline{C}^*$ satisfies condition (3) of 4.1 up to a morphism which is monic and epic.

A morphism $(\psi|f|\varphi):(X|\varphi) \to (Y|\psi)$ can be factored as follows: $(\psi|f|\varphi) = (\psi|Y|\exists f[\varphi])(\exists f[\varphi]|f|\varphi)$ with $(\psi|Y|\exists f[\varphi])$ monic and $(\exists f[\varphi]|f|\varphi)$ epic. Let $(\psi|f|\varphi) = (\psi|m|\mu)(\mu|g|\varphi)$ be a factorization with $(\psi|m|\mu)$ monic. This implies in particular $\varphi \le e_Y(f, mg)$ and hence $\exists f[\varphi] = \exists mg[\varphi]$ because of 1.4.20. Thus the morphism $(\exists f[\varphi]|m|\exists g[\varphi])$ is not only monic but also epic. Now the following diagram shows that $\underline{C}^*$ satisfies condition (5) of 4.1 up to a morphism which is monic and epic.

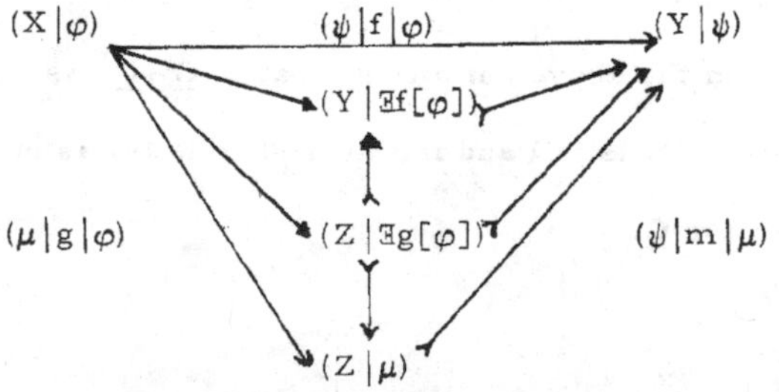

The extension $F^*: \underline{C}^* \to \underline{D}$ of a logical functor $F: \underline{C} \to \underline{D}$ into a logical category with finite limits can be defined as follows. $F^*((\psi|f|\varphi))$ is the unique map in $\underline{D}$ which satisfies the equation $eq(F(\psi), F(1_Y))F^*((\psi|f|\varphi)) = F(f)eq((F(\varphi), F(1_X)))$. It can be verified that F^* is a logical functor which preserves finite limits.

In the second step of the construction of $S(\underline{C})$ we have to invert the morphisms which are monic and epic, in order to make $\underline{C}^*$ into a semantical category. This can be done by means of a category of fractions(cf. Gabriel-Zisman [6]).

4.5. Proposition: Let $\underline{D}$ be a logical category which satisfies the conditions (1), (2), (4) of 4.1. If $\underline{D}$ satisfies the conditions (3) and (5) as in 4.4 up to morphisms from Σ, the set of morphisms which are monic and epic, then Σ is a calculus of right fractions, the category of fractions $\underline{D}\Sigma^{-1}$ is a semantical category and the canonical functor $P: \underline{D} \to \underline{D}\Sigma^{-1}$ is logical and faithful. Moreover, each finite limit preserving, logical functor $G: \underline{D} \to \underline{E}$ into a semantical category $\underline{E}$ can be extended uniquely to a semantical functor from $\underline{D}\Sigma^{-1}$. If every morphism in Σ is an isomorphism, then $\underline{D}\Sigma^{-1}$ is equivalent to $\underline{D}$.

It can be verified easily that Σ satisfies the following four conditions: (a) The identities are in Σ. (b) Σ is closed under composition. (c) Σ is closed under pullbacks. (d) If $sf_1 = sf_2$ with $s \in \Sigma$ then $f_1 = f_2$.

Now the category $\underline{D}\Sigma^{-1}$ together with the canonical functor $P: \underline{D} \to \underline{D}\Sigma^{-1}$ are defined as follows:

(1) $ob(\underline{D}\Sigma^{-1}) = ob(\underline{D})$

(2) $\underline{D}\Sigma^{-1}(X, Y) = \{(f, s) \mid f: Z \to Y,\ s: Z \to X,\ s \in \Sigma\}/\equiv$, where $(f_1, s_1) \equiv (f_2, s_2)$ iff there exist $t_1, t_2 \in \Sigma$ such that $f_1 t_1 = f_2 t_2$ and $s_1 t_1 = s_2 t_2$.

The equivalence class of (f, s) will be denoted by (f;s).

(3) $(f;s)(g;t) = (fg';ts')$ with $(s', g') = pb(g, s)$

(4) $P(f) = (f;X)$ for $f:X \to Y$

(5) $pb((f_1;s_1), (f_2;s_2)) = ((g_1, Q), (g_2, Q))$, where $(g_1, g_2) = pb(f_1, f_2)$ is a pullback in $\underline{D}$ and Q is the common domain of g_1, g_2.

Making use of these definitions, we can verify that $\underline{D}\Sigma^{-1}$ is a category with finite limits and that P is a functor which preserves finite limits. As a consequence we obtain that $P(\Omega) = \Omega$ is a boolean algebra object in $\underline{D}\Sigma^{-1}$. Moreover, P(s) is an isomorphism for $s \in \Sigma$. P is faithful, since the elements of Σ are epimorphisms. – It should be noted that our definition of the equivalence relation $\equiv$ differs from the one used in Gabriel-Zisman [6].

Until so far we have shown that $\underline{D}\Sigma^{-1}$ satisfies the conditions (1) and (2) of 4.1. For the following we will need the following observations:

(6) if f is epic then P(f) is epic

(7) if (f; s) is epic then f is epic

(8) if g is monic then P(g) is monic

(9) if (g;m) is monic then g is monic.

Thus the stability of epimorphisms under pullbacks in $\underline{D}$ implies the corresponding fact for $\underline{D}\Sigma^{-1}$ because of (5), (6), (7). This gives condition (4) of 4.1. Since $\underline{D}$ satisfies condition (3) of 4.1 up to morphisms in Σ, we can define $(\varphi;t)^{\#} = (t\varphi^{\#};dom(t))$ for a morphism $(\varphi;t):X \to \Omega$ in $\underline{D}\Sigma^{-1}$ and $(m;s)^b = (m^b;X)$ for a subobject of X in $\underline{D}\Sigma^{-1}$. An argument analogous to the one in 4.4 shows that $\underline{D}\Sigma^{-1}$ satisfies condition (3) of 4.1.

The condition (5) requires a more elaborate argument. Let $(f;s)=(m;t)(g;u)$ with $(m;t)$ monic be a factorization of $(f;s)$. This implies $(f;s)=(mg';ut')$ with $(t',g')=pb(g,t)$ and hence there exist $s',u'\in\Sigma$ such that $fs'=mg'u'$ and $ss'=ut'u'$ because of definition (2) and (3). Let $f=ip$ with i monic and p epic be the image-factorization of fs' since s' is epic. Since we have $fs'=mg'u'$ with m monic, there exists by assumption q, h and z with $z\in\Sigma$ such that $mh=iz$, $zq=ps'$ and $hq=g'u'$. However, this implies $(g;u)=(th;z)(p;s)$ and $(i;dom(i))=(m;t)(th;z)$, where $(th;z)$ is the required morphism. This completes the proof of the fact that $\underline{D}\Sigma^{-1}$ is a semantical category.

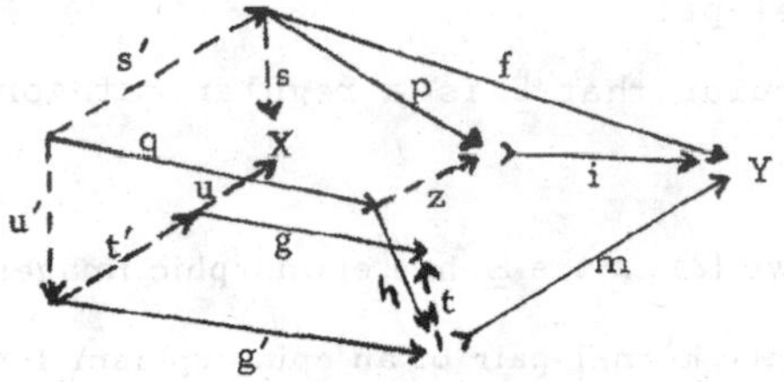

It remains to be shown that the functor P preserves quantification. The quantification in $\underline{D}\Sigma^{-1}$ is described by means of direct image. P preserves epimorphisms and monomorphisms because of (6) and (8). Hence we obtain $\exists(f;s)[(\psi;s)]^{\#}=\text{im}((f;s)(\psi;s)^{\#})=(\exists f[\psi];Y)^{\#}$. We can assume $s_1=s_2$ without loss of generality because of (c). This implies in particular that P preserves quantfication.

Now let $G:\underline{D}\to\underline{E}$ be a finite limit preserving logical functor into a semantical category $\underline{E}$. Since every morphism in $\underline{E}$ which is monic and epic is an isomorphism, we can define a semantical functor $\bar{G}:\underline{D}\Sigma^{-1}\to\underline{E}$ by $\bar{G}((f;s))=G(f)G(s)^{-1}$. - It should be remarked that $\underline{D}$ is equivalent to $\underline{D}\Sigma^{-1}$ if every morphism in $\underline{D}$ which is monic and epic is an isomorphism.

Combining 4.4 and 4.5 we obtain for every logical category $\underline{C}$ the desired free semantical category $S(\underline{C})=\underline{C}^*\Sigma^{-1}$. Moreover, since in a semantical category $k=eq(\psi,1_X)$ implies $\exists k[1_X] = \psi = k^b$, we obtain that $\underline{C}\Sigma^{-1}$ is equivalent to $\underline{C}$ in this case. This proves theorem 4.3 .

The following result was proven by Joyal (cf.|11|) in a slightly different context:

4.6. Proposition: In every semantical category $\underline{E}$ the following two statements are true:

(1) $\underline{E}$ has coequalizers of kernel-pairs.

(2) Every epimorphism in $\underline{E}$ is effective i.e. the coequalizer of its kernel-pair.

This implies in particular that $\underline{E}$ is a regular category in the sense of Barr (cf.[1]).

It is sufficient to prove (2), since $\underline{E}$ has epimorphic images. Let $(k_1, k_2) = kp(f) = pb(f, f)$ be the kernel-pair of an epimorphism $f:X\to Y$ and assume $gk_1 = gk_2$ for $g:X\to Z$. Instead of showing the existence of h with $g = hf$ directly, we will prove that a certain subobject of $Y\times Z$ is the graph of a morphism. Let jp with j monic and p epic be the image-factorization of $(f, g) = (f\times Z)(X, g):X\to Y\times Z$. Then we have $q_1jp = f$ and $q_2jp = g$, where q_1 and q_2 are the projections of $Y\times Z$. If q_1j is an isomorphism, then $h = q_2j(q_1j)^{-1}$ satisfies $g=hf$. Since f is epic, q_1j is epic, too. Since every morphism which is monic and epic is an isomorphism, it is sufficient to show that q_1j is monic. Making use of the fact that p is epic, we can verify that $kp(q_1jp) = kp(p)$ implies that q_1j is monic. But we have $kp(p) = kp(jp) = kp((f, g)) = kp(f)\cap kp(g)$, $kp(q_1jp) = kp(f)$ and $kp(f) = kp(f)\cap kp(g)$ because of $gk_1 = gk_2$.

4.7. Remark: It should be remarked, that one can prove now a completeness theorem for semantical categories. It implies the completeness theorem for logical categories because of 4.3. The proof follows the same pattern as the previous one. In the following we will give a sketch of this proof.

A semantical category $\underline{E}$ has a canonical model iff $\underline{E}$ is maximally consistent and the terminal object I is projective. Since $\underline{E}$ has pullbacks, I is projective iff every epimorphism into I is right invertible. This says that $\underline{E}$ is rich.

A maximal consistent extension can be obtained as before by means of an ultrafilter Δ in $\underline{E}(I, \Omega)$. However, in this context we have to use a calculus Σ of right fractions defined by $\Sigma = \{\varphi^{\#} : X_{\varphi} \rightarrowtail X \mid \forall !_X[\varphi] \in \Delta\}$ instead of a congruence relation.

In order to obtain an extension of $\underline{E}$ in which I is projective, we have to add right inverses for epimorphisms into I. This corresponds to the rich extension. Here we can use the method of A. Joyal, which he used in [11] in a similar context. We observe that the comma category $\underline{E}/X$ is again a semantical category and that the functor $!_X^* : \underline{E} \to \underline{E}/X$ is a faithful semantical functor if $!_X$ is epic. The functor $!_X^*$ consists of pulling back along $!_X$. Moreover, $!_X^*(!_X)$ has a right inverse, namely the diagonal.

Thus all the epimorphisms into I in $\underline{E}$ will become right invertible in $\underline{E}_1 = \mathrm{colim}(\underline{E}/\prod_{i=1}^{n} X_i \mid !_{X_i} : X_i \to I \text{ epic})$. Iterating this construction we obtain the required extension as a colimit of a countable chain.

Now the final result can be obtained as before by a maximally consistent extension preceded by an extension which makes I projective.

Notations:

$\underline{C}^{op}$ dual of $\underline{C}$

ob($\underline{C}$) objects of $\underline{C}$

$\underline{S}$ category of small sets, card(P) cardinality of P

$\underline{S}_0$ category of finite sets

$\underline{N}$ set of natural numbers

lim(F) limit of F

colim(F) colimit of F

dom(f) domain of f

cod(f) codomain of f

$(f_1, f_2): X \to X_1 \times X_2$ morphism into the product of X_1 and X_2

$\Delta_X: X \to X \times X$ diagonal of X

$(f_1; f_2): X_1 + X_2 \to X$ morphism from the coproduct of X_1 and X_2

$\nabla_X: X + X \to X$ codiagonal of X

$pb(f_1, f_2)$ pullback of f_1, f_2

kp(f) = pb(f, f) kernel pair of f

$po(f_1, f_2)$ pushout of f_1, f_2

$eq(f_1, f_2)$ equalizer of f_1, f_2

$coeq(f_1, f_2)$ coequalizer of f_1, f_2

im(f) image of f

$m: X \rightarrowtail Y$ monic, $q: X \twoheadrightarrow Y$ epic

$!_X: X \to I$ unique morphism into the terminal object I

$\prod_{k \in K} X_k$ product of X_k for $k \in K$

$\sim$ negation

$\wedge$ conjunction, $\vee$ disjunction

$\Rightarrow$ implication, $\Leftrightarrow$ biimplication

$\exists$ existential quantifier

$\forall$ universal quantifier

Bibliography:

[1] Barr, M., Grillet, P.A., van Osdol, D.H.: Exact categories and categories of sheaves, Springer Lecture Notes 236, 1971.

[2] Beth, E.W.: On Padoa's method in the theory of definitions, Indag. Math. 15 (1953), 330-339.

[3] Craig, W.: Three uses of the Herbrand-Gentzen theorem in relating model theory to proof theory, J. of Symb. Logic 22 (1957), 269-285.

[4] Daigneault, A.: Freedom in polyadic algebras and two theorems of Beth and Craig, Mich. Math. J. 11 (1964), 129-135.

[5] Daigneault, A.: Lawvere's elementary theories and polyadic and cylindric algebras, Fund. Math. 66, 3 (1970), 307-328.

[6] Gabriel, P., Zisman, M.: Calculus of fractions and homotopy theory, Ergebnisse der Mathematik und ihrer Grenzgebiete vol. 35, Springer Verlag, 1967.

[7] Halmos, P.: Algebraic logic, Chelsea Publ. Comp., 1962.

[8] Henkin, L.: The completeness of the first-order functional calculus, J. of Symb. Logic 14 (1949), 159-166.

[9] Henkin, L.: Completeness in the theory of types, J. of Symb. Logic 15 (1950), 81-91.

[10] Henkin, L.: A theory of propositional types, Fund. Math. 52 (1963), 323-344.

[11] Joyal, A.: Théorème de complétude pour les théories sémantiques, Talk at the University of Montréal in October 1971.

[12] Lawvere, F.W. Functorial semantics of algebraic theories, thesis, Columbia University, New York 1963.

[13] Lawvere, F.W.: Functorial semantics of elementary theories, J.of Symb.Logic 31 (1966), 294 (abstract).

[14] Lawvere, F.W.: Theories as categories and the completeness theorem, J.of Symb.Logic 32 (1967), 562 (abstract).

[15] Lawvere, F.W.: Equality in hyperdoctrines and comprehension schema as an adjoint functor, in Proc.of the AMS, Symp.Pure Math.17, Providence R.I. 1970, 1-14.

[16] Mitchell, B.: Theory of categories, Academic Press, 1965.

[17] Preller, A.: Interpolation et amalgamation, Publ.Dépt.Math. Lyon 6, 1 (1969), 49-65.

[18] Robinson, A.: A result on consistency and its application to the theory of definitions, Nederl.Akad.Wetensch.Proc.Ser. A59 (1956), 47-58.

[19] Scholz, H., Hasenjaeger, G.: Grundzuege der mathematischen Logik, Springer Verlag 1961.

[20] Schubert, H.: Kategorien I, II, Heidelberger Taschenbuecher, vol.65, 66, Springer Verlag 1970.

[21] Shoenfield, J.R.: Mathematical logic, Addison-Wesley, 1967.

[22] Volger, H.: Logical categories, thesis, Dalhousie University, Halifax, N.S., March 1971.

Logical categories, semantical categories and topoi

Hugo Volger

1. Introduction

In the paper "Completeness theorem for logical categories" [18] I introduced to concept of a semantical category. A semantical category is a logical category with a subobject classifier Ω i.e. with comprehension scheme in the sense of higher order logic resp. separation axiom in the sense of set theory. There I gave a construction of the free semantical category $Fr(\underline{C})$ over a given logical category $\underline{C}$. However, this construction could not be extended to logical categories with exponentiation.

Here I will give a new construction which works also for logical categories with exponentiation. Thus we obtain the free topos over a logical category with exponentiation, using the simplification of the axioms of an elementary topos by Mikkelsen [13]. This is done, using the category of functional relations of $\underline{C}$ rather than the category of restrictions of morphisms of $\underline{C}$ in which every epimonomorphism is inverted. The idea of this construction goes back to the observation of Lawvere that the invertibility of epimonomorphisms is closely related with the representability of functional relations by actual morphisms. Very helpful was also the remark of Kock, that it is sufficient to have exponents of Ω in order to have arbitrary exponentiation. Finally I found a variant of the construction which works even in the non-boolean case, because my friends in Aarhus insisted that the construction should work in this more general case.

As an application one obtains the result of Kock and Mikkelsen [4] on factorization of left exact functors between topoi, which generalizes the factorization of the ultra power functor used in non-standard analysis. Furthermore the above construction may be used to reduce the construction of the free topos over an arbitrary category to the construction of the free logical category with exponentiation over that category. More generally, certain problems for semantical categories can be transferred to logical categories and vice versa.

Perhaps it should be mentioned that beside logical and semantical categories an intermediate notion has been considered by Joyal |3| and Reyes [14] . They consider regular categories in the sense of Barr [1] with additional properties. Basically these are semantical categories without a subobject classifier.

2. Basic definitions

In order to obtain the theorem in full generality, we have to redefine the notions of a logical resp. semantical category which were introduced in |18|.

A category $\underline{C}$ is called <u>prelogical</u> resp. <u>logical</u> resp. <u>closed logical</u> if it satisfies the conditions (1) - (3) resp. (1) - (4) resp. (1) - (5).

(1) $\underline{C}$ has finite products.-Thus $\underline{C}$ has in particular a terminal object I. A projection onto X will be denoted by p_X.

(2) $\underline{C}$ has a Heyting semi-lattice object Ω i.e. there exist morphisms $\wedge:\Omega\times\Omega\to\Omega$, $\Rightarrow:\Omega\times\Omega\to\Omega$ and $1:I\to\Omega$ which satisfy the identities for a Heyting semi-lattice. - This makes $C(X,\Omega)$ into a Heyting semi-lattice homomorphism.

(3.1.)For every $f:X\to Y$ in $\underline{C}$ the orderpreserving map $\underline{C}(f,\Omega):\underline{C}(Y,\Omega)\to\underline{C}(X,\Omega)$ has a left adjoint $\exists f:C(X,\Omega)\to\underline{C}(Y,\Omega)$ i.e. $\exists f[\phi]\leq\psi$ iff $\phi\leq\psi f$ for all ϕ and ψ'. $\exists f$ is called the existential quantification along f. - In particular $\underline{C}$ has the equality on X θ_X for every X, which is defined by $\theta_X = \exists\Delta_X[1_X]$.

(3.2.)$\underline{C}$ satisfies the equation $\underline{C}(g_2,\Omega)\exists g_1 = \exists f_2\underline{C}(f_1,\Omega)$, where $(f_1,f_2) = pb(g_1,g_2)$ is a pullback of one of the following two types:

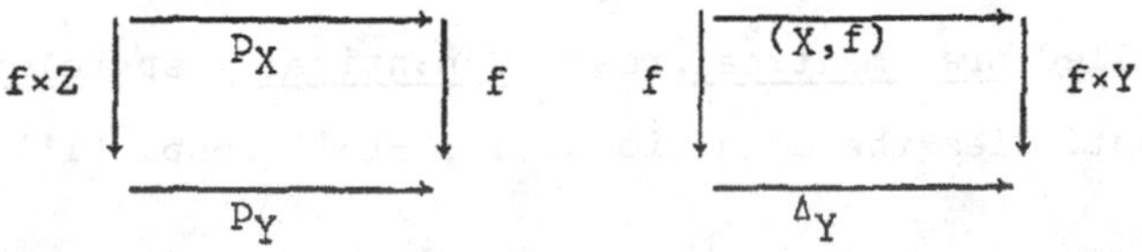

(3.3.)$\underline{C}$ satisfies the axiom of propositional extensionality i.e.

$$\theta_\Omega(\phi_1,\phi_2) = \phi_1 \Leftrightarrow \phi_2$$

(4) For every $f:X\to Y$ in $\underline{C}$ the orderpreserving map $\underline{C}(f,\Omega):\underline{C}(Y,\Omega)\to\underline{C}(X,\Omega)$ has a right adjoint $\forall f:\underline{C}(X,\Omega)\to\underline{C}(Y,\Omega)$ i.e. $\psi\leq\forall f[\phi]$ iff $\psi f<\phi$ for all ϕ and ψ. $\forall f$ is called the universal quantification along f

(5.1.) $\underline{C}$ has an exponentiation i.e. the functor $X\times(-):\underline{C}\longrightarrow\underline{C}$ has a right adjoint $(-)^X:\underline{C}\to C$.-The morphism from Z to Y^X which corresponds to $f:X\times Z\longrightarrow Y$ will be denoted by $\ulcorner f \urcorner$.

(5.2.) $\underline{C}$ satisfies the axiom of extensionality i.e.

$$\theta_\Omega Y(\ulcorner\phi_1\urcorner,\ulcorner\phi_2\urcorner) = \forall p_X[\phi_1 \Leftrightarrow \phi_2] \quad \text{for } \phi_1,\phi_2:Y\times X\longrightarrow\Omega$$

A functor $F:\underline{C}\longrightarrow\underline{C}'$ is called <u>prelogical</u> resp. <u>logical</u> resp. <u>closed logical</u> if it preserves the structure of the categories involved.

Further variants of the above definitions may be obtained by adding first the operation $v:\Omega\times\Omega\longrightarrow\Omega$ and then finite coproducts which distribute over finite products.

Requiring the existence of a subobject classifier rather than a Heyting semi-lattice object, one obtains the concept of a semantical instead of a logical category. Working with subobjects rather than morphisms into Ω, the existential quantification can be replaced by direct image. This motivates the following definition.
A category $\underline{E}$ is called <u>presemantical</u> resp. <u>semantical</u> resp. <u>closed semantical</u> if it satisfies the conditions (1') -(3') resp. (1') - (4') resp. (1') - (5'):

(1') $\underline{E}$ has finite products

(2') $\underline{E}$ has a subobject classifier $I\xrightarrow{1}\Omega$ i.e. there exists a bijection between Sub(X), the subobjects of X, and $\underline{E}(X,\Omega)$ such that $(m,1_{dom(m)}) = pb(m^b,1)$, where $m^b:X\longrightarrow\Omega$ denotes the morphism corresponding to the subobject m. Similarly $\phi^{\#}$ denotes the subobject corresponding to $\phi:X\to\Omega$.

- This implies the existence of arbitrary pullbacks, which can be defined by $(\theta_Y(f_1 \times f_2))^{\#} = pb(f_1,f_2)$, where $\theta_Y = \Delta_Y^{b}$.

(3.1') Every $f:X \longrightarrow Y$ in $\underline{E}$ has an image factorization $f = im(f)f'$.

- As a consequence, $f^{*}:Sub(X) \longrightarrow Sub(Y)$, the direct image under f, is left adjoint to $f^{-1}:Sub(Y) \longrightarrow Sub(X)$, the inverse image under f, i.e. $f^{*}(n) \subseteq m$ iff $n \subseteq f^{-1}(m)$ for all n,m.

(3.2') In $\underline{E}$ image factorization is preserved by pullbacks . - This implies $g_2^{-1} g_1^{*} = f_2^{*} f_1^{-1}$ for a pullback $(g_1,g_2) = pb(f_1,f_2)$

(4') For every $f:X \longrightarrow Y$ in $\underline{E}$ $f^{-1}:Sub(Y) \longrightarrow Sub(X)$ has a right adjoint $f^{o}:Sub(X) \longrightarrow Sub(Y)$ i.e. $m \subseteq f^{o}(n)$ iff $f^{-1}(m) \subseteq n$ for all n and m.

(5') $\underline{E}$ has exponentiation.

The subject classifier becomes a Heyting semi-lattice object by means of $\wedge = (1,1)^{b}:\Omega\times\Omega \longrightarrow \Omega$ and $\Rightarrow = eq(p_1,\wedge)^{b}:\Omega\times\Omega \longrightarrow \Omega$. The existential resp. universal quantification is defined by $\exists f[\phi] = f^{*}(\phi^{\#})^{b}$ resp. $\forall f[\phi] = f^{o}(\phi^{\#})^{b}$. Thus it can be verified that the semantical notions imply the corresponding logical notions.

A functor $G:\underline{E} \longrightarrow \underline{E}'$ is called presemantical resp. semantical resp. closed semantical if it preserves the structur of the categories involved. It should be remarked that a functor preserves the subobject classifier if it preserves the object and the pullbacks required in the definition. Hence the functor preserves arbitrary pullbacks. Mikkelsen [13] showed that (1'),(2') and (5') imply (3'),(4') and the existence of finite colimits. Thus the notions of a topos and a closed semantical category coincide. And a functor which preserves finite products, the subobject classifier and exponentiation (a logical morphism in the sense of Lawvere [11,12])is the same as a closed semantical functor.

The concept of a functional relation in a prelogical category $\underline{C}$ can be made precise as follows.

A morphism $\phi: X \times Y \to \Omega$ is called a *relation* from X to Y and is denoted by $\phi: X \rightharpoonup Y$. The *domain* resp. *codomain* of ϕ is defined by $\mathrm{dom}(\phi) = \exists p_X[\phi]$ resp. $\mathrm{cod}(\phi) = \exists p_Y[\phi]$. The *converse* of ϕ is defined by $\phi^{-1} = \phi(p_X, p_Y): Y \rightharpoonup X$. The *composition* of the relations $\phi: X \rightharpoonup Y$ and $\psi: Y \rightharpoonup Z$ is given by $\psi * \phi = \exists(p_X, p_Z)[\phi(p_X, p_Y) \wedge \psi(p_Y, p_Z)]: X \rightharpoonup Z$. The *identity relation* on X is given by $\theta_X: X \rightharpoonup X$. The restriction $\phi \wedge \lambda p_X$ of ϕ will be denoted by $\phi|\lambda$.

A relation $\phi: X \rightharpoonup Y$ is called *functional* resp. *everywhere defined* if $\phi * \phi^{-1} \leq \theta_Y$ resp. $\mathrm{dom}(\phi) = 1_X$. ϕ is called a *function* if it is functional and everywhere defined. The functional relations of $\underline{C}$ form a category *Fr*($\underline{C}$). The objects of $Fr(\underline{C})$ are pairs $\langle X, \lambda \rangle$ with $\lambda: X \to \Omega$. A morphism from $\langle X, \lambda \rangle$ to $\langle Y, \mu \rangle$ is a functional relation $\phi: X \rightharpoonup Y$ such that $\mathrm{dom}(\phi) = \lambda$ and $\mathrm{cod}(\phi) \leq \mu$ and is denoted by $\langle \mu, \phi, \lambda \rangle$ or by ϕ . The composition in $Fr(\underline{C})$ is the relational composition. The identity on $\langle X, \lambda \rangle$ is given by $\theta_X|\lambda$.

With every morphism $f: X \to Y$ in $\underline{C}$ one can associate the graph $\Gamma(f) = \theta_Y(f p_X, p_Y): \langle X, 1_X \rangle \to \langle Y, 1_Y \rangle$. Together with $\Gamma(X) = \langle X, 1_X \rangle$ this determines a functor $\Gamma: \underline{C} \to Fr(\underline{C})$. $\underline{C}$ is called *functionally complete* if every function in $\underline{C}$ is the graph of an actual morphism in $\underline{C}$ i.e. Γ is full. $\underline{C}$ is called *functionally strict* if $\theta_Y(f,g) = 1_X$ implies $f = g$ or equivalently if Γ is faithful.

It should be remarked that a list of formulas concerning relations can be found in the appendix.

In a presemantical category $\underline{E}$ the same concepts can be expressed using subobjects rather than morphisms into Ω. A subobject $(r_X, r_Y): R \rightarrowtail X \times Y$ is called a *relation* from X to Y. The *domain* resp. *codomain* of (r_X, r_Y) is given by $\mathrm{im}(r_X)$ resp. $\mathrm{im}(r_Y)$.

The <u>converse</u> of (r_X,r_Y) is (r_Y,r_X). The <u>composition</u> of $(r_X,r_Y):R \rightarrowtail X\times Y$ and $(s_X,s_Y):S\to Y\times Z$ is given by $\mathrm{im}(u_X,u_Z)$, where $(u_X,u_Z) = \mathrm{pb}(r_Y,s_Y)$.
The <u>identity relation</u> on X is the diagonal $\Delta_X:X\rightarrowtail X\times X$.
The relation $(r_X,r_Y):R \rightarrowtail X\times Y$ is <u>functional</u> resp. <u>everywhere defined</u> if r_X is monic resp. epic. Thus (r_X,r_Y) is a function if r_X is epic and monic. The graph of a morphism $f:X\to Y$ is defined as $\Gamma(f) = (X,f)$.
In the presemantical category every monomorphism is an equalizer and therefore every epimonomorphism is an isomorphism. Hence $\underline{E}$ is functionally complete, since a function $(r_X,r_Y):R \rightarrowtail X\times Y$ is isomorphic to the graph of $r_Y,r_X^{-1}:X\to Y$. More general , every category with finite limits and images is functionally complete iff every epimonomorphism is an isomorphism. Moreover, $\underline{E}$ is functionally strict because of $\Theta_Y^{\#} = \Delta_Y$.

3. Theorem

Let $\underline{C}$ be a prelogical category. Then $Fr(\underline{C})$, the category of functional relations of $\underline{C}$, and $\Gamma:\underline{C}\to Fr(\underline{C})$, the graph functor, are presemantical. Moreover, $Fr(\underline{C})$ is free over $\underline{C}$ i.e. every prelogical functor $H:\underline{C}\to\underline{E}$. with $\underline{E}$ presemantical can be extended uniquely (up to isomorphism) to a presemantical functor H' such that $H = H'\Gamma$. If $\underline{C}$ is already presemantical, then $Fr(\underline{C})$ is equivalent to $\underline{C}$.
This remains true if prelogical and presemantical is replaced by logical and semantical resp. closed logical and closed semantical. Further variants may be obtained by adding unions and coproducts.

In the following we will present an outline of the proof which contains all the necessary definitions. These can be obtained as follows. Reformulate in a semantical category the required notion using only the language of logical categories. The actual proof requires long computations and the extensive use of the formulas listed in the appendix.

The products are defined by $<X_1,\lambda_1>\times<X_2,\lambda_2> = <X_1\times X_2,\lambda_1\times\lambda_2>$, where $\lambda_1\times\lambda_2 = \lambda_1 p_1 \wedge \lambda_2 p_2$. The projections are given by $\Gamma(p_i)|\lambda_1\times\lambda_2$ for $i = 1,2$. The equalizers are defined by $eq(<\mu,\phi_1,\lambda>\ ,\ <\mu,\phi_2,\lambda>) = <\lambda,\Gamma(id_X)|dom(\phi_1\wedge\phi_2),\ dom(\phi_1\wedge\phi_2)>$. Hence the pullbacks can be described by $pb(<\mu,\phi_1,\lambda_1>,\ <\mu,\phi_2,\lambda>) = (<\lambda_1,\Gamma(p_1)|\phi_2^{-1}*\phi_1,\ \phi_2^{-1}*\phi_1>,\ <\lambda_2,\Gamma(p_2)|\phi_2^{-1}*\phi_1,\ \phi_2^{-1}*\phi_1>)$. This enables us to show that $<\mu,\phi,\lambda>$ is monic iff ϕ^{-1} is functional and that $<\mu,\phi,\lambda>$ is an isomorphism iff ϕ^{-1} is functional and $dom(\phi^{-1}) = \mu$.

Now we have to make $\Gamma(\Omega) = <\Omega,1_\Omega>$ into a subobject classifier. Since we have $(\Gamma(id_X)|\phi^t,\ \Gamma(!_X)|\phi^t) = pb(\phi,\Gamma(1))$ with $\phi^t = \phi^{-1}*\Gamma(1) = \phi(X,1_X)$, we associate with $\phi:<X,\lambda>\longrightarrow<\Omega,1_\Omega>$ the subobject $<1_\Omega,\phi,\lambda>^{\#} = \overline{<\lambda,\Gamma(id_X)|\phi^t,\phi^t>}$. With a subobject $\psi:<Y,\mu>\rightarrowtail<X,\lambda>$ we associate $<\lambda,\psi,\mu>^b = <1_\Omega,\Gamma(cod(\psi))|\lambda,\lambda>$. Then one has to verify $\phi^{\#b} = \phi$ and $\psi^{b\#} = \psi$. The first identity makes use of the propositional extensionality 3.3. . The second identity makes use of the above remark on monomorphisms.

The image factorization of $\phi:<X,\lambda>\longrightarrow<Y,\mu>$ is given by $<\mu,\Gamma(id_Y)|cod(\phi),cod(\phi)>*<cod(\phi),\phi,\lambda>$. Again we have to use the above remark on monomorphisms. Having images we can define the existential quantification of $<1_\Omega,\phi,\lambda>$ along $<\mu,\zeta,\lambda>$ by $\exists\zeta[\phi] = im(\zeta\phi^{\#})^b = \Gamma(\exists p_Y[\zeta\wedge\phi^t p_X])|\mu$. Here it is more convenient to show that the quantification satisfies the condition 3.2. for arbitrary pullbacks than verifying directly that images are preserved by pullbacks.

Using the above definitions one can verify that $\Gamma r(\underline{C})$ is a presemantical category and $\Gamma:\underline{C}\longrightarrow Fr(\underline{C})$ is a presemantical functor. The extension H' of the functor H has to be defined as follows. On objects we have $H'(<X,\lambda>) = H'(dom(\Gamma(\lambda)^{\#})) = dom(H(\lambda)^{\#})$. Since H as a prelogical functor preserves domain, codomain and functionality and $\underline{E}$ as a presemantical category is functionally complete, H sends $\phi:\langle X,\lambda\rangle \longrightarrow \langle Y,\mu\rangle$ into the morphism from $H'(\langle X,\lambda\rangle)$ to $H'(\langle Y,\mu\rangle)$ representing the relation $H(\phi)$. It remains to verify that H' is presemantical, since H is prelogical.

If $\underline{C}$ is already presemantical, then there exists a presemantical functor $\bar{\Gamma}:Fr(\underline{C})\longrightarrow\underline{C}$ such that $\bar{\Gamma}\Gamma = id_{\underline{C}}$ and we have $\bar{\Gamma}(\langle X,\lambda\rangle = dom(\Gamma(\lambda)^{*})$. Γ is full and faithful, since $\underline{C}$ as a presemantical category is functionally complete and strict. Moreover, we have $\langle X,\lambda\rangle \simeq \bar{\Gamma}\Gamma(\langle X,\lambda\rangle)$ since both are a pullback of $\Gamma(\lambda)$ and $\Gamma(1)$. Hence Γ is an equivalence.

If $\underline{C}$ has universal quantification, it can be extended to $Fr(\underline{C})$ by means of $\forall\zeta[\phi] = \Gamma(\forall p_Y[\zeta \Rightarrow \phi^t p_Y])|\mu$ for $\langle 1_\Omega,\phi,\lambda\rangle$ and $\langle\mu,\zeta,\lambda\rangle$. Then $\zeta^o(m) = (\forall\zeta[m^b]^{*})$ defines the required right adjoint to pulling back along ζ. Again one has to verify that Γ and H' preserve universal quantification.

Now assume that $\underline{C}$ has also exponentiation. In the presence of a subobject classifier it suffices to have its powers. This is an observation of Kock (cf. [4], p.5), which uses the graph $\gamma:X^Y \rightarrowtail \Omega^{Y\times X}$. Hence we define $\langle\Omega,1_\Omega\rangle^{\langle Y,\mu\rangle} = \langle\Omega^Y,r_\mu\rangle$, where $r_\mu = \forall!_Y{\times}\Omega^Y[ev_Y](\Rightarrow^Y)(\Omega^Y\times\ulcorner\mu\urcorner)$ expresses the restriction to subobjects contained in μ^{*}. With $\psi:\langle X,\lambda\rangle\times\langle Y,\mu\rangle\longrightarrow\langle\Omega,1_\Omega\rangle$ we associate $\Gamma(\ulcorner\psi^t\urcorner)|\lambda:\langle X,\lambda\rangle\longrightarrow\langle\Omega^Y,r_\mu\rangle$ or equivalently by extensionality $\forall(p_X,p_\Omega Y)[\psi^t(p_X,p_Y) \Leftrightarrow ev_Y(q_\Omega Y,qY)]|\lambda$. Conversely, with $\phi:\langle X,\lambda\rangle\longrightarrow\langle\Omega^Y,r_\mu\rangle$ we associate $(\Gamma(ev_Y)|r_\mu\times\mu)*(\phi\times\langle Y,\mu\rangle):\langle X,\lambda\rangle\times\langle Y,\mu\rangle\longrightarrow\langle\Omega,1_\Omega\rangle$. Finally, it has to be verified that Γ and H' preserve the powers of Ω and hence exponentiation.

4. Applications

As suggested by Lawvere, we will use the theorem to obtain the result of Kock and Mikkelsen in [4] on factorizations of first-order functors, which generalizes the factorization of the ultrapover functor $(-)^* = (-)^K/U: \underline{Sets} \to \underline{Sets}$ from non-standard analysis. There one has the following situation. The functor $(-)^*$ preserves first-order statements i.e. it is semantical. However, it does not preserve in general higher order statements. More precisely, it does not preserve the exponentiation, since $(2^X)^*$ is a subset of 2^{X*} , namely the set of subsets which are internal in the sense of Robinson [15] . If one defines $\underline{Sets}^*(X,Y)$ as the set of internal functions from X to Y, one obtains a new category of sets $\underline{Sets}^*$. The functor (-) can be viewed as a functor into $\underline{Sets}$ and as such it preserves also higher order statements i.e. it is closed semantical. This gives a factorization of $(-)^*$ because of $\underline{Sets}^*(1,X) = \underline{Sets}(1,X^*)=X^*$.

The result can be stated as follows:

Let $\underline{E}$, $\underline{E}'$ be closed semantical categories(=topoi) and let $F:\underline{E} \to \underline{E}'$ be a semantical functor. Then there exists a factorization $F=F^{elt}F^{exp}:\underline{E} \to \underline{E}^* \to \underline{E}'$ such that $\underline{E}$ is closed semantical, F^{exp} is closed semantical and F^{elt} is semantical and preserves elements, i.e. $\underline{E}^*(I,X)=\underline{E}'(F^{elt}(I),F^{elt}(X))$ for every X in E, where I is the terminal object.

This can be proved as follows. Since F is product-preserving functor between the cartesian closed categories $\underline{E}$,$\underline{E}'$, there exists a factorization $F=F_{elt}F_{exp}:\underline{E} \to \underline{E}_* \to \underline{E}'$ such that $\underline{E}_*$ is cartesian closed, F_{exp} is cartesian closed and F_{elt} preserves products and elements. $\underline{E}_*$ is the category of F-internal morphisms of $\underline{E}'$. The objects of $\underline{E}_*$ are the objects of $\underline{E}$. A morphism from X to Y in $\underline{E}$ is a F-internal morphism $g:F(X) \to F(Y)$ in $\underline{E}'$ i.e. $\ulcorner g \urcorner = \phi_Y^X g_o$ for some $g_o: I \to F(Y^X)$, where $\phi_Y^X: F(Y^X) \to F(Y)^{F(X)}$ is the canonical morphism induced by the evaluation.

F_{exp} sends f into the F-standard morphism F(f) and F_{elt} is the obvious embedding.

$\underline{E}_*$ is not sufficient for our purpose, since in general the subobject corresponding to a F-internal morphism into F(Ω) will not be F-internal. Thus Ω is not yet a subobject classifier in $\underline{E}$. However, Ω is still a Heyting semi-lattice object. This suggests the following apprach. Show that $\underline{E}$ is closed logical, F_{exp} is closed logical and F_{elt} is logical. Then $Fr(\underline{E}_*)$ can be used instead of $\underline{E}_*$.

The existential quantification of F-internal morphisms along projections is F-internal, since F as a presemantical functor preserves existential quantification. Hence arbitrary existential quantification preserves F-internality, since it can be reduced to quantification along projections and equality (cf. 5.1.4.). Thus $\underline{E}_*$ inherits the existential quantification of $\underline{E}'$ and F_{exp} and F_{elt} preserve it. The same argument works for universal quantification.

Now $\underline{E}^*$, F^{exp} and F^{elt} are defined as $Fr(\underline{E}_*)$, ΓF_{exp} and the extension of F_{elt}. By the theorem $\underline{E}^*$ is closed semantical and F^{elt} is semantical. Since E' is functionally complete and F_{elt} preserves elements, one can prove that the extension F^{elt} still preserves elements. Since $\underline{E}$ is closed semantical, it is equivalent to $Fr(\underline{E})$ by the theorem. Therefore $F^{exp} = \Gamma F_{exp}$ is up to equivalence the same as $Fr(F_{exp})$ and hence has to be closed semantical.

Another application concerns the construction of a free topos over an arbitrary category. Using the construction of the free topos over a closed logical category, this problem can be reduced to the construction of the free logical category. Since the latter construction does not require any equalizers, the complications introduced by the equalizers can be avoided. More precisely, if one tries to construct the free topos

$F(\underline{C})$ over a given category $\underline{C}$ by means of a deductive system in the sense of Lambek [6], then one has the following situation. The objects of $F(\underline{C})$ are terms generated by means of the necessary operations on objects from the objects in $\underline{C}$. Then derivations of the form $X \longrightarrow Y$, where X and Y are terms, are generated from the morphisms in $\underline{C}$ by means of the necessary operations on morphisms. Finally the morphisms of $F(\underline{C})$ are equivalence classes of derivations. However, in this case one has to introduce a new term $K_{f,g}$ for every pair of derivations $f,g:X \longrightarrow Y$ and one has to introduce a new derivation $E(f,g,h)$ for every derivation $h:Z \longrightarrow X$ such that fh is equivalent to gh. Hence one is forced to define terms, derivations and the equivalence relation simultaneously. Thus the direct construction gets too complicated to be useful. In this situation the above reduction of the problem seems to be useful.

5. Appendix

1.1. $\phi \leq \exists f[\phi]f, \ \exists f[\psi f] \leq \psi$

1.2. $\exists gf[\phi] = \exists g[\exists f[\phi]]$

1.3. $\exists f[\psi f \wedge \phi] = \psi \wedge \exists f[\phi]$

1.4. $\exists f[\phi] = \exists p_Y[\theta_Y(fp_X, p_Y) \wedge \phi p_X]$

1.5. $\theta_Y(f,f) = 1_X$

1.6. $\theta_Y(f,g) = \theta_Y(g,f)$

1.7. $\theta_Y(f,g) \wedge \psi f \leq \psi g$

1.8. $\theta_Y(f,g) \wedge \theta_Y(g,h) \leq \theta_Y(f,h)$

1.9. $\theta_Y(f,g) \leq \theta_Z(hf,hg)$

1.10. $\theta_{Y \times Y'}(f,f',g,g') = \theta_Y(f,g) \wedge \theta_{Y'}(f',g')$

1.11. $\forall f[\phi]f \leq \phi, \ \psi \leq \forall f[\psi f]$

1.12. $\forall gf[\phi] = \forall g[\forall f[\phi]]$

1.13. $\forall f[\psi f \Rightarrow \phi] = \psi \Rightarrow \forall f[\phi]$

1.14. $\forall f[\phi \Rightarrow \psi f] = \exists f[\phi] \Rightarrow \psi$

1.15. $\forall f[\phi] = \forall p_Y[\theta_Y(fp_X, p_Y) \Rightarrow \phi p_X]$

2.1. $\chi*(\psi*\phi)=(\chi*\psi)*\phi$, $\phi*\theta_X=\phi$, $\theta_Y*\phi=\phi$

2.2. if $\phi \leq \phi'$, $\psi \leq \psi'$ then $\psi*\phi \leq \psi'*\phi'$, $\mathrm{dom}(\phi) \leq \mathrm{dom}(\phi')$, $\mathrm{cod}(\phi) \leq \mathrm{cod}(\phi')$

2.3. $(\psi*\phi)^{-1} = \phi^{-1}*\psi^{-1}$, $\mathrm{dom}(\phi^{-1}) = \mathrm{cod}(\phi)$, $\mathrm{cod}(\phi^{-1}) = \mathrm{dom}(\phi)$

2.4. if ϕ,ψ functional then $\psi*\phi$ functional

2.5. if $\mathrm{cod}(\phi) \leq \mathrm{dom}(\psi)$ then $\mathrm{dom}(\psi*\phi) = \mathrm{dom}(\phi)$, $\mathrm{cod}(\psi*\phi) \leq \mathrm{cod}(\psi)$

2.6. $\psi*\Gamma(f) = \psi(f\times Z)$, $\Gamma(g)*\phi = \exists X\times g[\phi]$

2.7. $\Gamma(id_X) = \theta_X$, $\Gamma(gf) = \Gamma(g)*\Gamma(f)$

2.8. $\Gamma(f)|\lambda$ functional, $\mathrm{dom}(\Gamma(f)|\lambda) = \lambda$, $\mathrm{cod}(\Gamma(f)|\lambda) = \exists f[\lambda]$

2.9. if $\phi\leq\psi,\psi$ functional, $\mathrm{dom}(\phi) = \mathrm{dom}(\psi)$ then $\phi=\psi$

2.10. if ϕ functional then $\phi|\mathrm{dom}(\phi\wedge\psi) \leq \psi$

2.11. if $\psi*\phi_1 = \psi*\phi_2$,ψ functional, $\mathrm{cod}(\psi) \leq \mathrm{dom}(\phi_1)$, $\mathrm{dom}(\phi_2)$ then $\mathrm{cod}(\psi) \leq \mathrm{dom}(\phi_1\wedge\phi_2)$

2.12. $\Gamma(f_1)|\lambda = \Gamma(f_2)|\lambda$ iff $\lambda \leq \theta_Y(f_1,f_2)$

2.13. $(\psi*\zeta)^t = \exists p_X[\zeta \wedge \psi^t p_Y] = \mathrm{cod}(\zeta^{-1}|\psi^t)$

2.14. $\exists\zeta[\phi]^t = \exists p_Y[\zeta\wedge\phi^t p_X] = \mathrm{dom}(\zeta|\phi^t)$

2.15. $\forall\zeta[\phi]^t = \forall p_Y[\xi\Rightarrow \phi^t p_X]$

2.16. $\Gamma(\alpha)^t = \alpha$

2.17. $r_\mu{}^{\ulcorner}\lambda^{\urcorner} = \forall!_Y[\lambda\Rightarrow \mu]$

6. Bibliography

[1] Barr, M.: Exact categories, Springer Lecture Notes 236, (1971), p. 1-119

[2] Freyd, P.: Aspects of topoi, Bull. Austral. Math. Soc. 7 (1972), p. 1-76

[3] Joyal, A.: The Gödel-Kripke-Mitchell-Barr-embedding theorem, Talk at the conference on category theory in Oberwolfach,July 72

[4] Kock, A.; Mikkelsen, C.J.: Non-standard extensions in the theory of toposes, Aarhus Universitet Preprint Series 25 (1971/72)

[5] Kock, A.; Wraith, G.C.: Elementary toposes, Aarhus Universitet Lecture Notes Series 30 (1971)

[6] Lambek, J.: Deductive systems and categories I,II Math. Systems Theory 2 (1968), p. 287-318; Springer Lecture Notes 86 (1969), p. 76-122

[7] Lambek, J.: Deductive systems and categories, Springer Lecture Notes 274 (1972), p. 57-82

[8] Lawvere, F.W.: Functorial semantics of elementary theories, J. of Symb. Logic 31 (1966), p. 294 (abstract)

[9] Lawvere, F.W.: Theories as categories and the completeness theorem, J. of Symb. Logic 32 (1967), p. 562 (abstract)

[10] Lawvere, F.W.: Equality in hyperdoctrines and comprehension scheme as an adjoint functor, Proc. of the AMS, Symposia in Pure Math. 17, (1970), p. 1-14

[11] Lawvere, F.W.: Quantifier and sheaves, Actes du Congrès Intern. des Math. , Nice (1970) , tome 1, p. 329-334

[12] Lawvere, F.W.: Introduction to Springer Lecture Notes 274 (1972), p. 1-12

[13] Mikkelsen, C.J.: Finite colimits in toposes, Talk at the conference on category theory in Oberwolfach, July 1972

[14] Reyes, G.E.: Logique via faisceaux, Talk at the Congrès de Logique in Orléans, Sept. 1972

[15] Robinson, A.: Non-standard analysis, North-Holland (1966)

[16] Tierney, M.: Sheaf theory and the continuum hypothesis, Springer Lecture Notes 274 (1972), p. 13-42

[17] Volger, H.: Logical categories, dissertation, Dalhousie University, Halifax, N.S. (1971)

[18] Volger, H.: Completeness theorem for logical categories

PART II

(Presented at a conference in Bangor, organized by R. Brown, in September 1973; manuscripts received by the editors in November 1973)

INTERNAL CATEGORIES AND CLASSIFICATION THEOREMS

P.T. Johnstone

0. INTRODUCTION

The object of these notes is to develop methods for constructing internal categories which will enable us to prove classification theorems for algebraic theories, using an arbitrary topos with natural number object as a base. The limitation of the classifying toposes described in [2] is that they are necessarily defined over Sets; so one cannot hope, say, to classify rings in a topos $\mathcal{E}$ by mapping it into Rings, unless $\mathcal{E}$ is itself defined over Sets. The theorem of Diaconescu [3] gives us a method of proving classification theorems which are not set-based; we give here a couple of examples of how this method may be applied.

In §1, we summarize those properties of the natural number object and of finite cardinals to which we will need to appeal later. The construction of "internal full subcategories" described in §2 was first given by J. Benabou, and its application to the category $\mathbf{Fin}(\mathcal{E})$ in §3 was developed by G.C. Wraith. In §4 we show how the same ideas may be used to construct classifying toposes for algebraic theories other than the trivial one.

1. SOME ELEMENTARY ARITHMETIC

1.1 Definition Recall that a natural number object in a topos $\mathcal{E}$ is an object N with maps $1 \xrightarrow{o} N \xrightarrow{\sigma} N$ having the property that, for any diagram $1 \xrightarrow{x} X \xrightarrow{t} X$ in $\mathcal{E}$, there exists a unique $f: N \longrightarrow X$ such that $fo = x$ and $f\sigma = tf$.

In other words, morphisms whose domain is N (or, using the exponential adjunction, an object of the form $N \times X$) may be defined "by induction". In particular, we use induction to define the binary operations of addition, multiplication and exponentiation on N, and to prove that they satisfy the usual identities.

<u>1.2 Definition</u> By a <u>natural number</u> in a topos with N, we mean an element (global section) of N. With each natural number $1 \xrightarrow{p} N$, we associate an object $[p]$, the <u>cardinal</u> of p, by forming the pullback

$$\begin{array}{ccc} [p] & \longrightarrow & 1 \\ \downarrow & & \downarrow p \\ N\times N \xrightarrow{+} N & \xrightarrow{\sigma} & N \end{array}$$

It is readily seen that in $\mathcal{S}$ the cardinal of p is the set $\{(a,b) \in N\times N \mid a+b+1 = p\}$, which has precisely p elements.

<u>1.3 Lemma</u> If $\mathcal{F} \xrightarrow{f} \mathcal{E}$ is a geometric morphism and $\mathcal{E}$ has a natural number object, then f^* preserves it (and hence any inductively defined morphisms). Moreover, if p is a natural number in $\mathcal{E}$, then $f^*[p] \cong [f^*(p)]$.

<u>Proof</u> The first part is immediate from the existence of a right adjoint for f^*; the second follows from left exactness. □

In particular, the natural number object in $\mathcal{E}/N$ is $(N\times N \xrightarrow{\pi_2} N)$, and it has a distinguished element, called the <u>generic natural number</u> n, which is just the diagonal map $N \longrightarrow N\times N$. This has the property that, for any natural number $1 \xrightarrow{p} N$ in $\mathcal{E}$, the image of n under the pullback functor p^* is p itself - a property which we shall find very useful. The cardinal $[n]$ is readily seen to be the object $(N\times N \xrightarrow{\sigma+} N)$ of $\mathcal{E}/N$.

<u>1.4 Lemma</u> The following square is a pullback:

$$\begin{array}{ccc} (N\times N\times N) \amalg (N\times N\times N) & \xrightarrow{\begin{pmatrix} id\times + \\ +\times id \end{pmatrix}} & N\times N \\ \Big\downarrow \begin{pmatrix} \sigma+\times id \\ id\times\sigma+ \end{pmatrix} & & \Big\downarrow \sigma+ \\ N\times N & \xrightarrow{+} & N \end{array}$$

or equivalently, $+^*(\sigma+) \cong \pi_1^*(\sigma+) \amalg \pi_2^*(\sigma+)$, where the π_i are the product projections $N\times N \longrightarrow N$.

<u>Proof</u> In $\mathcal{S}$, this lemma says that a quadruple of natural numbers

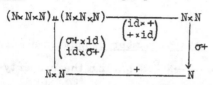

(a,b,c,d) satisfying a+b = c+d+1 may be represented uniquely by a triple either of the form (c,a−c−1,b) or of the form (a,c−a,d), since exactly one of (a−c−1) and (c−a) is nonnegative. An internal version of this argument can be given in an arbitrary topos. □

<u>1.5 Corollary</u> For any pair of natural numbers (p,q), we have $[p+q] \cong [p] \amalg [q]$.

<u>Proof</u> The functor $(p,q)^* : \mathcal{E}/N\times N \longrightarrow \mathcal{E}$ preserves coproducts. So applying it to the isomorphism of 1.4 we have $(p+q)^*(\sigma+) \cong p^*(\sigma+) \amalg q^*(\sigma+)$, as required. □

We may similarly prove that $[pq] \cong [p] \times [q]$ and $[p^q] \cong [p]^{[q]}$.

2. INTERNAL FULL SUBCATEGORIES

One method of forming internal categories which we have available in $\mathcal{S}$ is as follows: given a set-indexed family of sets $(A_i)_{i\in I}$, we form the full subcategory of $\mathcal{S}$ on the A_i as objects, and it is a small category, i.e. an object of Cat($\mathcal{S}$).

To perform the same construction internally in an arbitrary topos $\mathcal{E}$, we replace the family $(A_i)_{i\in I}$ by an $\mathcal{E}$-morphism $A \xrightarrow{f} I$, and proceed as follows:

<u>2.1 Definition</u> The internal category $\underline{\mathrm{Full}}_{\mathcal{E}}(A \xrightarrow{f} I)$ is defined in the following way:

$(\mathrm{Full}_{\mathcal{E}}(f))_0 = I$.

$(\mathrm{Full}_{\mathcal{E}}(f))_1 \xrightarrow{(d_0,d_1)} I\times I$ is obtained by constructing the objects $\pi_1^*(f)$, $\pi_2^*(f)$ of $\mathcal{E}/I\times I$ and forming the exponential $\pi_2^*f^{(\pi_1^*f)}$.

Since pullback functors preserve exponentials, defining the multiplication on $\underline{\mathrm{Full}}_{\mathcal{E}}(f)$ amounts to giving a morphism

$\pi_2^* f^{(\pi_1^* f)} \times \pi_3^* f^{(\pi_2^* f)} \longrightarrow \pi_3^* f^{(\pi_1^* f)}$ in $\mathcal{E}/I \times I \times I$; and for this we take the internal composition map.

The inclusion of identities $(\mathrm{Full}_{\mathcal{E}}(f))_0 \longrightarrow (\mathrm{Full}_{\mathcal{E}}(f))_1$ is similarly defined.

<u>2.2 Lemma</u> $\underline{\mathrm{Full}}_{\mathcal{E}}(f)$ is equipped with a canonical (covariant) <u>inclusion functor</u> $\underline{U}:\underline{\mathrm{Full}}_{\mathcal{E}}(f) \longrightarrow \mathcal{E}$, which may be described as the object $(A \xrightarrow{f} I)$ of $\mathcal{E}/I$, equipped with a structure map for the appropriate monad which is basically the evaluation map $\pi_2^* f^{(\pi_1^* f)} \times \pi_1^* f \longrightarrow \pi_2^* f$ in $\mathcal{E}/I \times I$.

Similarly, given a set-indexed family of groups $(G_i)_{i \in I}$, we can form the full subcategory of $\underline{Gp}$ $(= Gp(\mathcal{S}))$ on the G_i as objects. In order to do this internally, we need the following definition:

<u>2.3 Definition</u> Let G, H be group objects in a topos, with multiplication maps m_G, m_H. The <u>object of homomorphisms</u> $Gp(G,H)$ is the subobject of H^G defined by the equalizer of

$$H^G \xrightarrow{H^{m_G}} H^{G \times G}, \qquad H^G \xrightarrow{sq.} (H \times H)^{(G \times G)} \xrightarrow{m_H^{(G \times G)}} H^{G \times G} .$$

Clearly, the evaluation and composition maps defined on exponentials can be restricted to objects of homomorphisms.

<u>2.4 Definition</u> Given a group object $(G \longrightarrow I)$ in the topos $\mathcal{E}/I$, we define $\underline{\mathrm{Full}}_{Gp(\mathcal{E})}(G \longrightarrow I)$ in the same way as 2.1, but using objects of homomorphisms instead of exponentials throughout. The associated inclusion functor $\underline{U}$ is defined as in 2.2, but this time it is readily seen to be a group object in the functor category $\mathcal{E}^{\underline{\mathrm{Full}}}$, i.e. we can think of it as a group-valued functor.

<u>2.5 Remark</u> We can repeat the construction of 2.4 with "group" replaced by any finitely-presented, finitary algebraic theory. (The finite presentation is necessary in order

to define the appropriate objects of homomorphisms as finite intersections of equalizers.)

3. THE CATEGORY Fin($\mathcal{E}$)

We now turn our attention to the construction of classifying toposes for algebraic theories. We consider first the trivial theory; its models are simply objects, and so finitely-presented models in are finite sets. Henceforth $\mathcal{E}$ will always be a topos with N.N.O.

3.1 Definition $\underline{Fin}(\mathcal{E}) = \underline{Full}_{\mathcal{E}}(N\times N \xrightarrow{\sigma+} N)$, the internal category of finite objects of $\mathcal{E}$.

3.2 Lemma If $\mathcal{F} \xrightarrow{f} \mathcal{E}$ is a geometric morphism, then $f^*(\underline{Fin}(\mathcal{E}))$ is isomorphic to $\underline{Fin}(\mathcal{F})$.

Proof We know f^* preserves N and all constructions involved, except possibly the exponential constructed in 2.1.
But this exponential is an object of $\mathcal{E}/N\times N$, and we can show that it is preserved by the following argument:
(i) The exponential $Fin(\mathcal{E})_1 \longrightarrow N\times N$ is a solution of the recursion problem in $\mathcal{E}/N$ defined by
$(o\times id)^*(Fin(\mathcal{E})_1 \longrightarrow N\times N) \cong N \xrightarrow{id} N$ in $\mathcal{E}/N$, and
$(\sigma\times id)^*(Fin(\mathcal{E})_1 \longrightarrow N\times N) \cong Fin(\mathcal{E})_1 \times [\pi_2^*(n)]$ in $\mathcal{E}/N\times N$.
(ii) f^* preserves the data for this recursion problem, by 1.3 and left exactness.
(iii) It can be shown that the solution to a recursion problem of this type, if it exists, is unique up to isomorphism. (We will give the proof of this result below, as Proposition 3.6.) Hence f^* preserves the solution of the recursion problem. □

3.3 Theorem $\mathrm{Sex}(\underline{Fin}(\mathcal{E})^{op}, \mathcal{E}) \simeq \mathcal{E}$.

Proof We set up an equivalence of categories as follows:

To a discrete fibration $\underline{F}\longrightarrow\underline{Fin}(\mathcal{E})$ we associate the object $(\sigma o)^*(F_0\longrightarrow N)$.

To an object X of $\mathcal{E}$ we associate the fibration defined on objects by $(X\times N)^{[n]}\longrightarrow N$, where the exponential is computed in $\mathcal{E}/N$; this is readily seen to be a flat presheaf.

And it is not hard to check that these two operations are mutually inverse. □

3.4 Corollary The topos $\mathcal{E}^{\underline{Fin}(\mathcal{E})}$ is an object classifier for toposes defined over $\mathcal{E}$, i.e. for any $\mathcal{F}\xrightarrow{f}\mathcal{E}$ we have equivalences of categories

$$\begin{aligned} Top/\mathcal{E}(\mathcal{F},\mathcal{E}^{\underline{Fin}(\mathcal{E})}) &\simeq Sex(f^*(\underline{Fin}(\mathcal{E})^{op}),\mathcal{F}) && \text{by [3]} \\ &\simeq Sex(\underline{Fin}(\mathcal{F})^{op},\mathcal{F}) && \text{by 3.2} \\ &\simeq \mathcal{F} && \text{by 3.3.} \end{aligned}$$

Moreover, the universal object of $\mathcal{E}^{\underline{Fin}(\mathcal{E})}$ classified by the identity geometric morphism is the functor $\underline{U}$ of 2.2. □

It remains to prove the uniqueness theorem required for the proof of 3.2. We proceed by way of the following lemma:

3.5 Lemma Suppose given an object $(X\longrightarrow N)$ of $\mathcal{E}/N$ and

(i) an element $x_0:1\longrightarrow o^*(X)$ in $\mathcal{E}$,

(ii) a morphism $t:X\longrightarrow\sigma^*(X)$ in $\mathcal{E}/N$.

Then there exists a unique section $x:N\longrightarrow X$ such that $o^*(x) = x_0:1\longrightarrow o^*(X)$ and $\sigma^*(x) = tx:N\longrightarrow X\longrightarrow\sigma^*(X)$.

Proof Consider the following diagram:

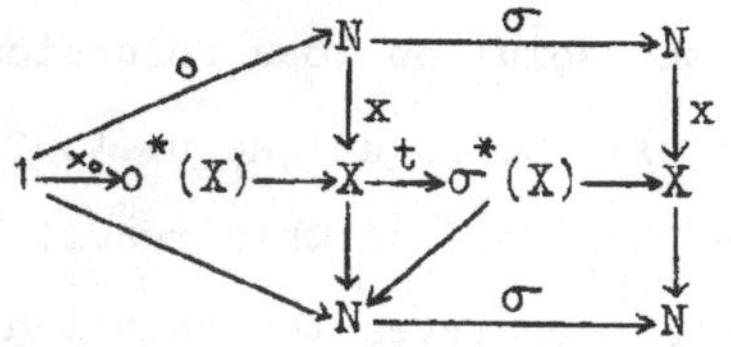

x exists uniquely by the definition of N, and $N\xrightarrow{x}X\longrightarrow N$ is the identity by the uniqueness clause of 1.1. □

<u>3.6 Proposition</u> Let X_o be an object of $\mathcal{E}$, and $T:\mathcal{E}/N \longrightarrow \mathcal{E}/N$ a <u>strong functor</u>, i.e. a functor together with maps $T_{X,Y}:Y^X \longrightarrow TY^{TX}$ for X,Y in $\mathcal{E}/N$, satisfying the obvious composition and compatibility conditions. Then an object X of $\mathcal{E}/N$ satisfying $o^*(X) \cong X_o$ and $\sigma^*(X) \cong TX$, if it exists, is unique up to canonical isomorphism.

<u>Proof</u> Suppose X,X' are two objects satisfying the given recursion data. Consider the <u>object of isomorphisms</u> Iso(X,X') in $\mathcal{E}/N$; we have maps $1 \xrightarrow{\ulcorner id \urcorner} Iso(X_o,X_o) \cong o^*(Iso(X,X'))$ in $\mathcal{E}$, and $Iso(X,X') \xrightarrow{T_{X,X'}} Iso(TX,TX') \cong \sigma^*(Iso(X,X'))$ in $\mathcal{E}/N$. So we can use 3.5 to construct a section $N \longrightarrow Iso(X,X')$; but such a section corresponds by definition to an isomorphism $X \xrightarrow{\cong} X'$ in $\mathcal{E}/N$. □

Finally, we may observe that a functor of the form $(-)\times X:\mathcal{E} \longrightarrow \mathcal{E}$ (for some fixed $X \in \mathcal{E}$) is always strong, and so we can apply 3.6 in the situation of 3.2.

4. THE CATEGORY <u>FPGp</u>($\mathcal{E}$)

In this paragraph we repeat the arguments of §3 for a nontrivial theory, namely that of groups. It should be noted, however, that groups have been chosen merely as an example, and the basic arguments will work equally well for other theories (with minor modifications as indicated in 4.9 below).

<u>4.1 Proposition</u> In any topos $\mathcal{E}$ with N.N.O., there exists a <u>free monoid functor</u> $M:\mathcal{E} \longrightarrow Mon(\mathcal{E})$, which is left adjoint to the forgetful functor.

<u>Proof</u> Let $X \in \mathcal{E}$. Construct the exponential $N^*(X)^{[n]}$ in $\mathcal{E}/N$, where N^* denotes pullback along $N \longrightarrow 1$; we will show that (the domain of) this object is M(X).

The unit of MX is given by the pullback $1 \cong X^{[o]} \longrightarrow MX$;

$$\begin{array}{ccc} X^{[o]} & \longrightarrow & MX \\ \downarrow & & \downarrow \\ 1 & \xrightarrow{o} & N \end{array}$$

the multiplication by $\pi_1^*(MX)\times\pi_2^*(MX) \longrightarrow MX$ (using 1.4 and

$$\begin{array}{ccc} \pi_1^*(MX)\times\pi_2^*(MX) & \longrightarrow & MX \\ \downarrow & & \downarrow \\ N\times N & \xrightarrow{+} & N \end{array}$$

the isomorphism $A^{(B\amalg C)} \cong A^B\times A^C$). The fact that these definitions make MX into a monoid follows from the fact that $(N,+,o)$ is a monoid.

The front adjunction $X \longrightarrow MX$ is given by pullback along $1 \xrightarrow{\sigma o} N$; and if Y is a monoid, we construct the end adjunction $MY \longrightarrow Y$ by inductively defining an element of $N^*(Y)^{MY}$ in $\mathcal{E}/N$, using the method of 3.5. Similar inductive arguments show that the end adjunction is a monoid homomorphism, and that the "triangular identities" are satisfied. □

<u>4.2 Corollary</u> In any topos $\mathcal{E}$ with N.N.O., there exists a <u>free group functor</u> $F:\mathcal{E} \longrightarrow Gp(\mathcal{E})$.

<u>Proof</u> Let $X \in \mathcal{E}$. In the free monoid $M(X\amalg X)$, let R be the submonoid generated by the subobject $X\amalg X \xrightarrow{\begin{pmatrix} i_2 & i_1 \\ i_1 & i_2 \end{pmatrix}} (X\amalg X)^2 \longrightarrow M(X\amalg X)$,

$$\begin{array}{ccc} (X\amalg X)^2 & \longrightarrow & M(X\amalg X) \\ \downarrow & & \downarrow \\ 1 & \xrightarrow{\sigma^2 o} & N \end{array}$$

where $i_1, i_2 : X \longrightarrow X\amalg X$ are the coproduct inclusions.

Then it is easy to show that $M(X\amalg X)/R$ is a group, and that it is the required free group. □

<u>4.3 Remark</u> In fact we can "internalize" the adjunctions of 4.1 and 4.2; i.e. given $X \in \mathcal{E}$, $G \in Gp(\mathcal{E})$, we have $Gp(FX,G) \cong G^X$.

<u>4.4 Lemma</u> If $\mathcal{F} \xrightarrow{f} \mathcal{E}$ is a geometric morphism, then the functors F and f^* commute up to natural isomorphism.

<u>Proof</u> Their right adjoints (i.e. the forgetful functor and f_*) clearly commute, so this follows from uniqueness of adjoints. □

Now consider the free group $F[n]$ in $\mathcal{E}/N$, where n is the generic natural number. By 1.2 and 4.4, the pullback of $F[n] \longrightarrow N$ along a natural number $1 \xrightarrow{p} N$ is $F[p]$; so we can think of $F[n] \longrightarrow N$ as the indexed union of all finitely-generated free groups in $\mathcal{E}$. Thus $\underline{\text{Full}}_{Gp(\mathcal{E})}(F[n] \longrightarrow N)$ is the category of finitely-generated free groups of $\mathcal{E}$.

Now we are interested in finitely-presented groups, which are cokernels of homomorphisms from one finitely-generated free group to another; so the indexing object I which we want is simply the object of maps of $\underline{\text{Full}}_{Gp(\mathcal{E})}(F[n] \longrightarrow N)$. We have a map $I \xrightarrow{(d_0, d_1)} N \times N$, so we can form the free groups $d_0^*(F[n])$ and $d_1^*(F[n])$ in $\mathcal{E}/I$. And the definition of I gives us a homomorphism $d_0^*(F[n]) \longrightarrow d_1^*(F[n])$ in $Gp(\mathcal{E}/I)$ (this is just the statement that $\underline{U}$ is a covariant group-valued functor), so we can form its cokernel $G \longrightarrow I$.

<u>4.5 Definition</u> $\underline{\text{FPGp}}(\mathcal{E}) = \underline{\text{Full}}_{Gp(\mathcal{E})}(G \longrightarrow I)$.

<u>4.6 Lemma</u> If $\mathcal{F} \xrightarrow{f} \mathcal{E}$ is a geometric morphism, then

$f^*(\underline{\text{FPGp}}(\mathcal{E})) \cong \underline{\text{FPGp}}(\mathcal{F})$.

<u>Proof</u> As in 3.2, we need only concern ourselves with the objects of homomorphisms which occur in the definition.

The first one is $I \longrightarrow N \times N = Gp(\pi_1^*(F[n]), \pi_2^*(F[n]))$

$$\cong \pi_2^*(F[n])^{\pi_1^*[n]} \text{ using 4.3.}$$

And we can describe this exponential recursively as in 3.2. The second is $H = Gp(\pi_1^*(G), \pi_2^*(G))$ in $\mathcal{E}/I \times I$; the argument here is more complicated, since $\pi_1^*(G)$ is not a free group, but in fact we can still give a recursive description of H as follows:

Recalling that I is an object over $N \times N$, we have $(o \times id)^*(G \longrightarrow I \longrightarrow N \times N) \cong (F[n] \longrightarrow N \xrightarrow{id} N)$; so $(o \times id^3)^*(H \longrightarrow I \times I \longrightarrow N^4) \cong (\pi_2^* G^{\pi_1^*[n]} \longrightarrow N \times I \longrightarrow N^3)$, and a further recursive argument will show that this exponential is preserved by f^*.

And $(\sigma\times id)^*(G \longrightarrow I \longrightarrow N\ N) \cong (Q \longrightarrow I\times F[\pi_2^* n] \longrightarrow N\times N)$, where Q is the cokernel of a certain homomorphism $F(1) \longrightarrow G\times F[\pi_2^* n]$ in $Gp(\mathcal{E}/I\times F[\pi_2^* n])$. Hence $(\sigma\times id^3)^*(H \longrightarrow N^4)$ is the kernel of the corresponding map of pointed objects

$Gp(\pi_1^*(G\ F[\pi_2^* n]),\pi_2^*(G)) \longrightarrow Gp(\pi_1^*(F(1)),\pi_2^*(G))$.

And this expression involves only finite limits and colimits; so it is preserved by f^*. □

<u>4.7 Theorem</u> $Sex(\underline{FPGp}(\mathcal{E})^{op},\mathcal{E}) \simeq Gp(\mathcal{E})$.

<u>Proof</u> Once again, we set up the equivalence by sending a flat presheaf $\underline{F} \longrightarrow \underline{FPGp}(\mathcal{E})$ to the object $i^*(F_0 \longrightarrow I)$, where i is the element of I corresponding to F(1); it is readily checked that this object has a natural group structure in $\mathcal{E}$. And we send $X \in Gp(\mathcal{E})$ to the presheaf defined on objects by $Gp(G,I^*(X)) \longrightarrow I$. □

<u>4.8 Corollary</u> The topos $\mathcal{E}^{\underline{FPGp}(\mathcal{E})}$ is a <u>group classifier</u> for toposes defined over $\mathcal{E}$, i.e. for any $\mathcal{F} \xrightarrow{f} \mathcal{E}$ we have $Top/\mathcal{E}(\mathcal{F},\mathcal{E}^{\underline{FPGp}(\mathcal{E})}) \simeq Gp(\mathcal{F})$.

And the universal group object of $\mathcal{E}^{\underline{FPGp}(\mathcal{E})}$ classified by the identity geometric morphism is the inclusion functor $\underline{U}$. □

<u>4.9 Remark</u> The arguments of 4.5 - 4.8 may be repeated with "group" replaced by any finitely-presented, finitary algebraic theory for which we have a free functor. The only modification needed is that, in the case of a theory whose hom-objects are not pointed (so that we must take coequalizers rather than cokernels to define finitely-presented models), the indexing object I must be replaced by the pullback $I\times_{(N\times N)}I$.

It is of interest to ask whether the method outlined above can be adapted for even more general theories. For example, it should be possible to define a topology on the ring classifier for Top/$\mathcal{E}$, such that the corresponding sheaf category is a local-ring classifier. (See [1] for a description of this topology in the case $\mathcal{E}$ = Sets.)

REFERENCES

[1] M. Hakim: Topos annelés et schémas relatifs, Springer Ergebnisse vol. 64.

[2] C.J. Mulvey: Toposes, logic and ring theory (to appear).

[3] G.C. Wraith: Lectures on elementary topoi (this volume), chapter 9.

Lectures on Elementary Topoi

G. C. Wraith

Given at the University College of North Wales

Bangor

September 1973

Contents

Introduction

These notes are based on the text of ten lectures given at the University College of North Wales, Bangor in September 1973. As far as I know, apart from Professor S. MacLane's, these were the first lectures on elementary topoi to be given in Britain, so I was at pains to avoid getting entangled in detailed proofs, in order to concentrate on the main aspects of the subject. In ten lectures it is impossible to be comprehensive so these notes must of necessity reflect a personal bias. In fact, these notes are rather more detailed than the lectures, but even so, a great many statements and examples are left unproved. In many places the reader is urged to seek the proof elsewhere, in Freyd's Aspects of Topoi, or Elementary Toposes by Kock and Wraith. The aim is not to provide the reader with an exhaustive and complete text, but to give him some sort of idea as to what has happened in the subject so far, and where it is likely to go. Indeed, this may well be up to the reader himself. In my opinion, the subject has exploded so fast, since Lawvere and Tierney's first work at Halifax in 1969, that it is hard for anybody not in at the beginning to swallow all the new material so suddenly available. The subject is now ripe for application, I believe; certainly it is such a pretty subject that it would be most disappointing if it were not good for anything - all my instincts tell me that it will be useful, and not just for applications in logic.

There are certain new developments, due to J. Benabou, which I should have liked to have included. Until recently, when one wished to carry out a construction in an elementary topos that was well enough understood in S , the category of sets and functions, one had to wrestle with pullback diagrams and the like. Benabou's formal language permits one to dispense with these problems, and to proceed directly to the construction from its formal description. I believe that these methods must displace the older, clumsier ones.

The interest which the audience expressed during the lectures I take to be a tribute to F. W. Lawvere's deep insights. It is often easier to express the flavour of an idea with the spoken word than with the printed, and I fear that these notes do not really do justice to some of the most important underlying ideas.

I have added a short preface to the notes on the development of the concept of topos. This was written over a year ago, so the references need revision. There are many people I should like to thank for their help, encouragement, conversations or communications on the the subject of elementary topoi. I would also like to thank Professor R. Brown and I. Morris for organizing the Bangor Conference, and all my fellow lecturers, C. J. Mulvey, B. Tennison, P. Johnston, M. Reid, and A. Thomas.

I used to hold that too strong a leaning to proper classical endings was an affectation, but weight of usage goes against me; so toposes now becomes topoi.

The Development of the Concept of Topos

Section 1

The subject of toposes really has two beginnings. The curtain rises in the early 60's, the scene algebraic geometry. The modern approach to algebraic geometry is founded on the idea of a sheaf. A presheaf on a topological space is a contravariant set valued functor on the category of open sets and inclusions, and a sheaf is a presheaf satisfying some extra conditions, of the form, 'given an open set, for every open covering of it, it is the case that...'. Grothendieck's idea was to replace the category of open sets and inclusions of a topological space by an arbitrary category. Thus, a presheaf on a category is a contravariant set valued functor on it. To define a sheaf we have to say what we mean by a covering of an object. A Grothendieck topology on a category is defined by saying which families of maps into an object are to constitute a covering of the object; the family of coverings has to satisfy certain axioms which we will not go into here. Be warned that the terminology 'Grothendieck topology' is rather misleading - it has little to do with topology in the usual sense of the word. A category together with a Grothendieck topology on it is called a site. To every site we can assign a certain full subcategory of the category of presheaves, called the category of sheaves, by analogy with the definition of sheaves on topological spaces. This is the raison d'etre of the concept of Grothendieck topology. The Grothendieck topologies on a category form a lattice - we may talk of the finest Grothendieck topology on a category such that a given class of presheaves are sheaves. In particular, for any category, we define the canonical Grothendieck topology to be the finest for which the representable presheaves are sheaves.

J. Giraud discovered a remarkable theorem which bears his name. From any site we may construct a new one by considering the category of sheaves for the site with its canonical Grothendieck topology. Giraud's theorem asserts that the category of sheaves of the latter site is equivalent to the category of sheaves of the former.

In consequence, the special name of topos was given to those categories which were equivalent to the category of sheaves for the canonical topology on them. Giraud's theorem may then be stated; a category is a topos if and only if it is the category of sheaves on a site. (Actually, there are a few foundational points that need clearing up here - usually, recourse is had to 'Grothendieck universes'.) Internal conditions were found for a category to be a topos, stating with certain limits and colimits must exist, with certain properties.

Let me pause to summarise: a topos is a category satisfying certain conditions, whose details I will not bother to describe here. These conditions were concocted to describe categories of sheaves on a site, so that one could carry through certain constructions (chiefly, cohomology) that one can perform for the category of sheaves of sets on a topological space. It is worth saying that it was soon realized that toposes are more important than sites. Different sites may give rise to the same topos. For example, the category of open inclusions and the category of local homeomorphisms into a fixed space, with their canonical topologies, give rise to two distinct sites which have the same topos.

1. M. Artin, Grothendieck Topologies. Harvard University Press. 1962.

2. M. Artin, A. Grothendieck, J. Verdier, Theoreie des Topos et Cohomology Etale des Schemas. Springer L. N. 269 and 270 (revised version of SGA4 1963/64).

3. J. Giraud. Analysis situs. Sem. Bourbaki. 1962/63.

4. J. Giraud, Methode de la descente. Mem. Soc. Math. France. 1964.

5. J. Ciraud, Cohomologie non abelienne. Springer 1971.

6. M. Hakim. Topos Anneles et schemas relatifs. Springer 1971.

7. D. Mumford. Picard group of moduli problems. Proceedings of the conference on arithmetical algebraic geometry of Purdue 1963.

<u>Section 2</u> The scene now changes to a borderland between logic and category theory, being explored by F. W. Lawvere. He had observed many formal similarities between rules of logic and the calculus of adjoint functors. He realized that it is possible to axiomatise category theory without using sets, so that it may be possible to avoid the problems of set theory. Anything defined by adjoint functors will be an <u>elementary</u> notion in the formal language of categories. The problem, therefore, is to pinpoint elementary properties of the category of sets and functions which are good enough for reconstructing as much set theory as one needs.

8. J. C. Cole. Categories of sets and models of set theory. Thesis (Sussex). Aarhus preprint No. 52 (1971).

9. F. W. Lawvere. An elementary theory of the category of sets. Proc. Nat. Acad. Sci. 52 (1964). pp. 1506-1511.

10. F. W. Lawvere. Adjointness in Foundations. Dialectica 23 (1969) pp. 281-296.

11. F. W. Lawvere. Equality in hyperdoctrines and comprehension as an adjoint functor. Symposia in pure maths. Vol. XVII A.M.S. (1970).

<u>Section 3</u> In 1969, at the University of Dalhousie, F. W. Lawvere and M. Tierney began to investigate the consequences of the following three axioms for categories:-

T(i) finite completeness and finite cocompleteness

T(ii) Cartesian closedness

T(iii) the existence of a subobject classifier

These three are all elementary axioms, and they are satisfied by the category of sets and functions. For example, any two element set acts as a subset classifier. It was soon found that any topos satisfies the above three axioms.

For this reason, any category satisfying these three axioms was called an elementary topos. To distinguish them from elementary topoi topoi in the old sense are now called Grothendieck topoi. The category of finite sets and functions is an elementary topos but it is not Grothendieck.

The definition of an elementary topos is much simpler than that of a Grothendieck topos. Recently, A. Kock and C. Juul Mikkelsen have shown that it can be simplified even more. In any category with finite limits we may define Rel(A, B), the set of relations from A to B, to be the set of subobjects of $A \times B$. By using pullback, we can make this into a functor Rel(-,B) for any fixed B. The simplified axioms are

T"(i) finite completeness.

T"(ii) for any B, Rel(-, B) is representable.

It is a remarkable fact that these axioms imply those above. Elementary toposes are to abelian categories what sets are to abelian groups. P. Freyd considers the development of elementary topoi to be the most important event in the history of categorical algebra.

For Grothendieck topoi the emphasis had been on cohomology, and on generalizing ideas of topology. One may, of course, still consider these notions in the context of elementary topoi. However, the elementary nature of the axioms brings out a new and fundamental feature, that had not been exploited before - the concept of internalization.

It has long been realized that any category with finite limits admits the interpretation of universal sentences (this is the fundamental idea behind universal algebra - one considers only sentences using '=' and '∀'). An elementary topos admits the interpretation of any sentence in the higher order predicate calculus. That is to say, a topos may be considered as a universe of discourse. Constructions normally carried out 'within' the category of sets and functions may be carried out 'within' an elementary topos. Let me give a very basic example: suppose X^Y denotes the exponential, so that maps $Z \longrightarrow X^Y$ are in bijective correspondence with maps $Z \times Y \longrightarrow X$, and suppose that Ω denotes the subobject classifier. To any pair of functions $S \rightrightarrows T$ we may assign the subobject of S on which they agree, their equalizer. Corresponding to this set-theoretic construction there will be a map in the topos

$$X^Y \times X^Y \longrightarrow \Omega^Y$$

which is the internalization of the construction which assigns to a pair of maps $Y \rightrightarrows X$ their equalizer.

We can define the notion of a category object in an elementary topos - we have an object-object, a maps object and a pair of maps called 'domain' and 'codomain' together with certain other maps defining composition, etc. (actually, since categories are defined by universal sentences, we only need left limits in our category to define category objects). We may also define 'internal presheaves' on a category object. These form an elementary topos. This underlines another point; the property of being an elementary topos is stable under a wide variety of categorical constructions. It is easy to construct new topoi out of old.

Of particular value is the interplay between the topological aspect, and the logical. For example, the subobjects of the terminal object in a topos may be interpreted topologically as open sets, and logically as truth values.

The fundamentals of the theory have begun to crystallize. A large number of questions to be resolved remains.

12. P. Freyd. Aspects of topoi. Bull. Australian. Math. Soc. (1972) Vol. 7. pp. 1-76.

13. G. M. Kelly and R. Street. Abstracts of the Sydney Category Theory Seminar 1972. University of New South Wales.

14. A. Kock and G. Wraith. Elementary Toposes. Aarhus Lecture Notes No. 30.

15. F. W. Lawvere. Quantifiers and Sheaves. Actes du Congres International des Mathematicians. Nice. 1970. pp. 329-334.

16. F. W. Lawvere and M. Tierney. Summary by J. Gray. Springer L.N. 195.

§1. Elementary topoi.

Let us consider some of the properties of S, the category of sets and functions.

(i) Finite limits.

The category S has finite limits. That is to say, it has a terminal object; any singleton set will do - we pick one and call it 1. The elements of a set X are given by the maps

$$1 \longrightarrow X.$$

It also has pullbacks. For any two functions

$$A \xrightarrow{f} C, \quad B \xrightarrow{g} C$$

with common codomain, we may form the set

$$P = \{(a,b) \in A \times B \mid f(a) = g(b)\}.$$

If p_1, p_2 denote the obvious projections

$$\begin{array}{ccc} P & \xrightarrow{p_1} & A \\ {\scriptstyle p_2}\downarrow & & \downarrow{\scriptstyle f} \\ B & \xrightarrow{g} & C \end{array}$$

is a pullback diagram.

(ii) Power sets.

For any set X, let P(X) denote the set of subsets of X. Then P(X) has the following property:- For any set Y, the set of functions

$$Y \longrightarrow P(X)$$

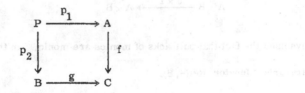

is in bijective correspondence with the set of relations from Y to X. To be precise, a map $Y \xrightarrow{f} P(X)$ and a relation $R \subseteq X \times Y$ are said to correspond if

$$xRy \iff x \in f(y) \qquad \forall x \in X, \quad \forall y \in Y .$$

In any category $\underline{C}$ with finite limits we define a relation from an object Y to an object X to be a subobject of $X \times Y$. We denote by $\mathrm{Rel}(X, Y)$ the class of subobjects of $X \times Y$. If $A' \xrightarrow{a} A$ is a map and

$$R \rightarrowtail A \times B$$

denotes an element of $\mathrm{Rel}(A, B)$, we obtain an element $R' \rightarrowtail A' \times B$ of $\mathrm{Rel}(A', B)$ by forming the pullback diagram

$$\begin{array}{ccc} R' & \longrightarrow & R \\ \downarrow & & \downarrow \\ A' \times B & \xrightarrow{a \times 1} & A \times B \end{array}$$

We have used the fact that pullbacks of monics are monic. In this way we get a contravariant functor $\mathrm{Rel}(-, B)$.

<u>Definition 1</u>. A category $\underline{E}$ is an <u>elementary topos</u> if

i) $\underline{E}$ has finite limits,

(ii) for every object A of $\underline{E}$ there is an object $P(A)$ of $\underline{E}$ and a monic map

$$\epsilon_A \rightarrowtail A \times P(A)$$

with the property that for any object B of $\underline{E}$ and monic map

$$R \rightarrowtail A \times B$$

there is a unique map

$$B \xrightarrow{r} P(A)$$

such that

$$\begin{array}{ccc} R & \longrightarrow & \in_A \\ \downarrow & & \downarrow \\ A \times B & \xrightarrow{1 \times r} & A \times P(A) \end{array}$$

is a pullback diagram.

We may paraphrase condition ii) by saying that for every object X of $\underline{E}$ the functor $\mathrm{Rel}(-,X)$ is representable, i.e. we have a natural isomorphism

$$\mathrm{Rel}(-,X) \simeq \mathrm{Hom}_{\underline{E}}(-,P(X)) .$$

The natural isomorphism $A \times B \simeq B \times A$ sets up a natural isomorphism

$$\mathrm{Hom}_{\underline{E}}(A,P(B)) \simeq \mathrm{Hom}_{\underline{E}}(B,P(A))$$

which tells us that P is a contravariant functor from $\underline{E}$ to itself, which is adjoint to itself on the right.

We denote by 1 a terminal object of $\underline{E}$, and by analogy with the case for S we call a map

$$1 \longrightarrow X$$

an element of X. We call an element of P(1) a truth-value of $\underline{E}$. Each truth-value corresponds to a subobject of 1.

Examples

i) S, the category of sets and functions.

ii) $\mathcal{S}_{fin}$ the category of finite sets and functions.

There are only two truth values in S and in $\mathcal{S}_{fin}$.

iii) $\mathbb{1}$, the category having only one map. This has only one truth value.

iv) S × S, the category of pairs of sets and pairs of functions. This has four truth values, given by the subobjects (ϕ,ϕ), $(1,\phi)$, $(\phi,1)$, $(1,1)$ of the terminal object $(1,1)$. Note that

$$(\phi\,,\,\phi) \neq (\phi,1)$$

and that $(\phi,1)$ has no elements. We see that an object in an elementary topos is not determined by its elements.

We may think of S × S as a pair of "non-interacting universes". As a generalization, the reader can easily verify that if $\underline{E}_1$ and $\underline{E}_2$ are elementary topoi, then $\underline{E}_1 \times \underline{E}_2$ is an elementary topos.

v) Let G be a group. A G-set is a set together with an action of G on it by permutations. A G-function between G-sets is a function which preserves G-action. The category of G-sets and G-functions is an elementary topos. It has two truth-values. The functor **P** assigns to a G-set its set of subsets (not sub-G-sets) which is given a G-action via the notion of inverse image, i.e. if X is a G-set, $A \subseteq X$, $g \in G$ define $g.(A) = \{x \in X \mid g.x \in A\}$.

vi) Consider a simplified model of time with just two states of existence - "then" and "now". We have a category (usually denoted by 2) described by the diagram "then" $\longrightarrow$ "now"

A functor X from 2 to S we might call a "set in time"; it gives a diagram

$$X(\text{then}) \longrightarrow X(\text{now})$$

in S. A "function in time" is to be a natural map. Sets and functions in time form an elementary topos, which has three truth-values, given by the subobjects

$\phi \longrightarrow \phi$	(always false)
$\phi \longrightarrow 1$	(false then, true now)
$1 \longrightarrow 1$	(always true)

of the terminal object $1 \longrightarrow 1$.

Of course, "time" may be construed as any partially ordered set, or, indeed, as any small category $\underline{C}$. As a common generalization of v) and vi) we may show that

$$S^{\underline{C}^0}$$

the category of functors $\underline{C}^0 \longrightarrow S$ and natural maps, is an elementary topos. This is known as the category of presheaves on $\underline{C}$. For any presheaf $F : \underline{C}^0 \longrightarrow S$, the presheaf $P(F) : \underline{C}^0 \longrightarrow S$ is given by taking $(P(F))(X)$, for X an object of $\underline{C}$, to be the set of subfunctors of $F \times \mathrm{Hom}_{\underline{C}}(-,X)$.

vii) A continuous map $Y \xrightarrow{p} X$ between topological spaces is a local homeomorphism if it is an open map such that for every $y \in Y$ there is an open neighbourhood U of y mapped homeomorphically by p onto p(U).

Let Top(X) denote the category whose objects are local homeomorphisms $Y \xrightarrow{p} X$ and whose maps are commutative triangles of continuous maps

$$\begin{array}{ccc} Y & \longrightarrow & Y' \\ & {\scriptstyle p}\searrow \quad \swarrow{\scriptstyle p'} & \\ & X & \end{array}$$

Then Top(X) is an elementary topos (we call it a spatial topos) whose truth-values correspond to the open sets of X.

The monic map $A \xrightarrow{<1_A, 1_A>} A \times A$ (the diagonal map) gives the identity relation from A to A, and corresponds to a map

$$A \xrightarrow{\{\cdot\}} P(A) \quad ,$$

for any object A in an elementary topos $\underline{E}$. If $\underline{E} = S$, then $\{\cdot\}$ is the function $a \longmapsto \{a\}$.

Proposition 1.1 The map $A \xrightarrow{\{\cdot\}} P(A)$ is monic.

Proof.

For any map $X \xrightarrow{u} A$, the diagram

$$\begin{array}{ccc} X & \xrightarrow{\quad u \quad} & A \\ {\scriptstyle <u, 1_X>}\downarrow & & \downarrow{\scriptstyle <1_A, 1_A>} \\ A \times X & \xrightarrow{\quad 1_A \times u \quad} & A \times A \end{array}$$

is a pullback. Hence, if $u, u' : X \longrightarrow A$ are such that $\{\cdot\}u = \{\cdot\}u'$, then $<u, 1_X> = <u', 1_X>$, and so $u = u'$.

The identity map $A \xrightarrow{1_A} A \simeq 1 \times A$, considered as a relation from 1 to A, gives rise to a map

$$1 \xrightarrow{\ulcorner A \urcorner} P(A) .$$

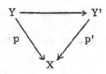

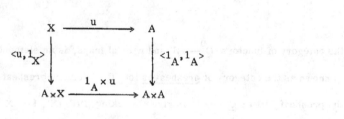

For any objects A, B of an elementary topos, the identity map

$$P(A \times B) \xrightarrow{1_{P(A \times B)}} P(A \times B)$$

corresponds to a subobject of $A \times B \times P(A \times B)$, and hence to a map

$$P(A \times B) \times A \longrightarrow P(B) .$$

Let

$$\begin{array}{ccc} P(A \times B) \times A & \longrightarrow & P(B) \\ \uparrow & & \uparrow \{\cdot\} \\ Q & \longrightarrow & B \end{array}$$

be a pullback diagram and let

$$q : P(A \times B) \longrightarrow P(A)$$

correspond to the subobject Q of $P(A \times B) \times A$.

The interpretation of q in S is :-

given $R \subseteq A \times B$, then

$$q(R) = \{a \in A \mid \exists b \in B, \{b\} = \{b' \in B \mid (a, b') \in R\}\}.$$

Define the object B^A by defining

$$\begin{array}{ccc} B^A & \longrightarrow & 1 \\ \downarrow & & \downarrow \ulcorner A \urcorner \\ P(A \times B) & \xrightarrow{q} & P(A) \end{array}$$

to be a pullback diagram.

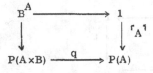

<u>Proposition 1.2</u> For any object D, there is a natural isomorphism

$$\mathrm{Hom}_{\underline{E}}(D \times A, B) \simeq \mathrm{Hom}_{\underline{E}}(D, B^A).$$

We leave the proof as an exercise in diagram chasing for the reader. I believe I am correct in crediting this result to C. J. Mikkelsen. We may interpret B^A as the object of maps from A to B, and $P(A \times B)$ as the object of relations between A and B. The construction of B^A from $P(A \times B)$ follows precisely the procedure for sets.

Proposition 2 is summarized by saying that an elementary topos is Cartesian-closed; that is to say, for every object B the functor $B \times (-)$ has a right adjoint $(-)^B$.

It is conventional to denote the object P(1) by Ω, and to denote by

$$1 \xrightarrow{\ t\ } \Omega$$

the map corresponding to the maximal relation, namely

$$1 \rightarrowtail{\sim} 1 \times 1 \ .$$

The defining property of the functor P implies that for any monic map $A \rightarrowtail X$ there is a unique map $X \longrightarrow \Omega$, which we call the classifying map of $A \rightarrowtail X$, such that

$$\begin{array}{ccc} A & \longrightarrow & 1 \\ \downarrow & & \downarrow t \\ X & \longrightarrow & \Omega \end{array}$$

is a pullback diagram. For this reason we call Ω a <u>subobject classifier.</u>

<u>Proposition</u> **1.3** **A category $\underline{E}$ is an elementary topos if and only if it satisfies the following conditions:**

i) $\underline{E}$ has finite limits,

ii) $\underline{E}$ is Cartesian-closed,

iii) $\underline{E}$ has a subobject classifier.

Proof: we have already seen that an elementary topos satisfies the above three conditions. Conversely, if a category $\underline{E}$ satisfies these conditions, for any object B define P(B) to be Ω^B. Then

$$\mathrm{Hom}_{\underline{E}}(A, \Omega^B) \simeq \mathrm{Hom}_{\underline{E}}(A \times B, \Omega) \simeq \mathrm{Rel}(A, B),$$

so $\underline{E}$ is an elementary topos.

§2. Exactness properties of elementary topoi

The original formulation of the axioms for elementary topoi contained the condition that finite colimits should exist. C.J. Mikkelsen showed that this condition is in fact a consequence of the axioms we have given in 1. We sketch here very briefly part of an elegant paper by Robert Pare, which shows that the functor

$$\underline{E}^{o} \xrightarrow{P} \underline{E}$$

makes $\underline{E}^{o}$ tripleable over $\underline{E}$. Since tripleable functors preserve, reflect and create limits, it follows that $\underline{E}^{o}$ has all the limits which exist in $\underline{E}$.

Def. 2. A pair of maps $B \overset{f}{\underset{g}{\rightrightarrows}} A$ is reflexive if there exists a map $A \xrightarrow{d} B$ such that $fd = gd = 1_A$.

A version of the tripleability theorem (CTT) of Jon Beck asserts that if

$$\underline{F} \xrightarrow{U} \underline{E}$$

is a functor having a left adjoint, then U is tripleable if

i) $\underline{F}$ has coequalizers of reflexive pairs;

ii) U preserves these coequalizers;

iii) U reflects isomorphisms.

Let the end adjunction

$$\Omega^{A} \times A \xrightarrow{ev} \Omega$$

classify the monic

$$\epsilon_A \rightarrowtail \Omega^{A} \times A.$$

If $A \rightarrowtail^{i} B$ is monic, we get a monic

$$\in_A \rightarrowtail \Omega^A \times A \xrightarrow{1 \times i} \Omega^A \times B$$

whose classifying map $\Omega^A \times B \longrightarrow \Omega$ is exponentially adjoint to a map we call

$$\Omega^A \xrightarrow{\exists_i} \Omega^B \ .$$

<u>Proposition 2.1</u> Let

$$\begin{array}{ccc} A' & \xrightarrow{f} & A \\ {\scriptstyle i'}\downarrow & & \downarrow{\scriptstyle i} \\ B' & \longrightarrow & B \end{array}$$

be a pullback diagram in an elementary topos, with i (and therefore i') monic. Then the diagram

$$\begin{array}{ccc} \Omega^A & \xrightarrow{\Omega^f} & \Omega^{A'} \\ {\scriptstyle \exists_i}\downarrow & & \downarrow{\scriptstyle \exists_{i'}} \\ \Omega^B & \xrightarrow{\Omega^g} & \Omega^{B'} \end{array}$$

commutes.

The proof amounts to checking that the two maps

$$\Omega^A \times B' \longrightarrow \Omega$$

exponentially adjoint to the maps obtained by going round the diagram in either way, classify the same subobject.

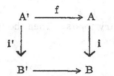

<u>Proposition 2.2</u> Let $A \overset{i}{\rightarrowtail} B$ be monic. Then

$$\Omega^A \xrightarrow{\exists_i} \Omega^B \xrightarrow{\Omega^i} \Omega^A = \Omega^A \xrightarrow{1} \Omega^A .$$

Proof: Apply proposition 2.1 to the pullback diagram

$$\begin{array}{ccc} A & \xrightarrow{1} & A \\ {\scriptstyle 1}\downarrow & & \downarrow{\scriptstyle i} \\ A & \xrightarrow{i} & B \end{array}$$

<u>Theorem 2.3</u> Let $\underline{E}$ be an elementary topos. Then the functor

$$\underline{E}^o \xrightarrow{P} \underline{E}$$

satisfies the criteria of CTT.

Proof: We have already seen that P has a left adjoint (namely, itself).

i) Since $\underline{E}$ has equalizers, $\underline{E}^o$ has coequalizers.

ii) Let

$$A \underset{g}{\overset{f}{\rightrightarrows}} B \xrightarrow{h} C$$

be a coequalizer diagram in $\underline{E}^o$, where (f, g) is a reflexive pair. In $\underline{E}$ this means that

$$C \xrightarrow{h} B \underset{g}{\overset{f}{\rightrightarrows}} A$$

is an equalizer diagram and that there is a map $A \xrightarrow{d} B$ such that $df = dg = 1_B$.
It follows that f, g, h are monic and that

$$\begin{array}{ccc} C & \xrightarrow{h} & B \\ {\scriptstyle h}\downarrow & & \downarrow{\scriptstyle f} \\ B & \xrightarrow{g} & A \end{array}$$

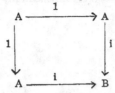

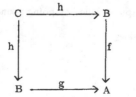

is a pullback diagram. By propositions 2.1 and 2.2 it follows that the diagrams

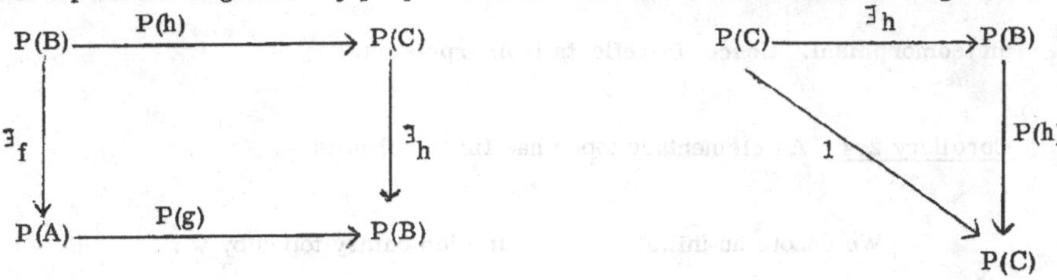

commute, so that

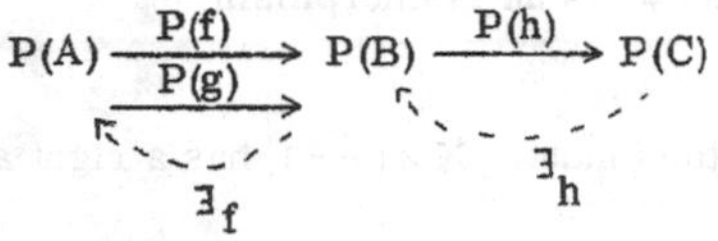

is a contractible coequalizer diagram.

iii) For any map $B \xrightarrow{f} A$, the composite

$$A \xrightarrow{\{\cdot\}} P(A) \xrightarrow{P(f)} P(B)$$

corresponds to the monic

$$B \rightarrowtail^{\langle f, 1_B \rangle} A \times B \ .$$

Hence $P(f) = P(f')$ implies $< f, 1_B > \; = \; < f', 1_B >$ which implies $f = f'$. Hence P is a faithful functor, and so reflects monics and epics.

If $A \overset{i}{\rightarrowtail} X$ is monic, and has classifying map

$$X \xrightarrow{\phi} \Omega$$

then

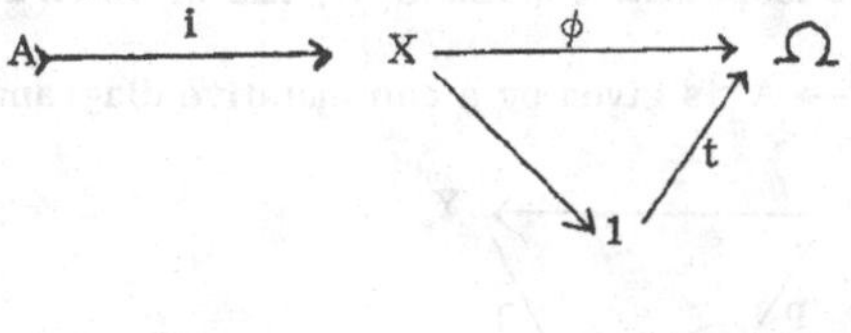

is an equalizer diagram. Hence, in an elementary topos every monic is an equalizer. It follows that any map in a topos which is both monic and epic is an isomorphism. Hence P reflects isomorphisms.

Corollary 2.4 An elementary topos has finite colimits.

We denote an initial object of an elementary topos by ϕ .

Proposition 2.5 Any map into ϕ is an isomorphism.

Proof: For any object X, the functor $X \times (\text{—})$ has a right adjoint and so preserves colimits. So

$$X \times \phi \simeq \phi \; .$$

Any map $X \xrightarrow{f} \phi$ has an inverse

$$\phi \simeq X \times \phi \xrightarrow{p_1} X .$$

One of the primary uses of sets in mathematics is to formulate the notion of an indexed collection of things. If an elementary topos is to be a useful generalization of S, we must know how to express the concept of an indexing over an object in it. To see how to do this, we remind the reader of some elementary category theory.

For any category $\underline{C}$ and object A of $\underline{C}$, define $\underline{C}/A$ to be the category whose objects are maps with codomain A, and in which a map $p \longrightarrow q$ from $X \xrightarrow{p} A$ to $Y \xrightarrow{q} A$ is given by a commutative diagram

$$\begin{array}{ccc} X & \longrightarrow & Y \\ & {\scriptstyle p}\searrow \quad \swarrow{\scriptstyle q} & \\ & A & \end{array}$$

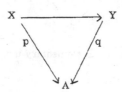

If $\underline{C} = S$, we may interpret S/A as the category of A-indexed sets and functions as follows:-

From an object $X \xrightarrow{p} A$ in S/A we get the A-indexed family

$$\{p^{-1}(a)\}_{a \in A}$$

and from a map

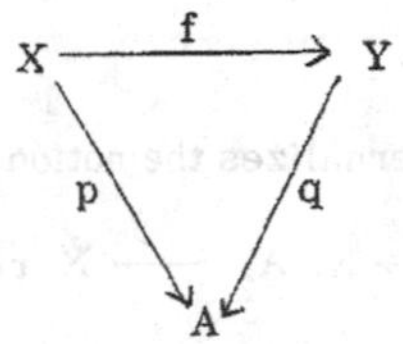

in S/A we get an A-indexed family of maps

$$\{f_a : p^{-1}(a) \longrightarrow q^{-1}(a)\}_{a \in A}$$

where f_a is the restriction of f to $p^{-1}(a)$.

Conversely, given an A-indexed family $\{X_a\}_{a \in A}$ of sets, we get an object $X \xrightarrow{p} A$ of S/A by taking

$$X = \bigcup_{a \in A} (X_a \times \{a\}) = \coprod_{a \in A} X_a$$

and $p(x, a) = a$. If $\{f_a : X_a \longrightarrow Y_a\}_{a \in A}$ is an A-indexed family of maps, we get a map

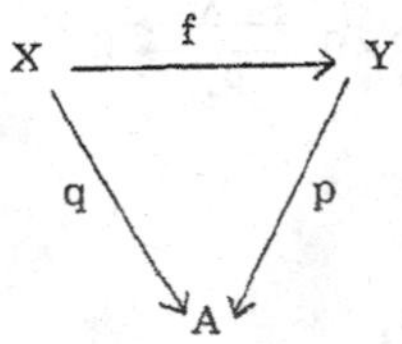

in S/A by taking $f = \coprod_{a \in A} f_a$, i.e. f is given by

$$f(x, a) = (f_a(x), a).$$

We have the slogan, therefore, that maps into an object A correspond to A-indexed objects.

For any elementary topos $\underline{E}$, let us define

$$\Omega \times \Omega \xrightarrow{\wedge} \Omega$$

to be the classifier of $1 \xrightarrow{<t,t>} \Omega \times \Omega$.

Proposition 2.6 The map $\wedge$ internalizes the notion of intersection of subobjects. That is to say, if $A_1 \rightarrowtail X$, $A_2 \rightarrowtail X$ represent subobjects of X, with classifying maps $X \xrightarrow{\phi_1} \Omega$, $X \xrightarrow{\phi_2} \Omega$ respectively, then

$$X \xrightarrow{<\phi_1, \phi_2>} \Omega \times \Omega \xrightarrow{\wedge} \Omega$$

classifies $A_1 \cap A_2 \rightarrowtail X$, given by the pullback diagram

$$\begin{array}{ccc} A_1 \cap A_2 & \rightarrowtail & A_1 \\ \downarrow & & \downarrow \\ A_2 & \rightarrowtail & X \end{array}$$

We leave the easy verification of this to the reader.

Let

$$\subseteq \rightarrowtail \Omega \times \Omega \underset{p_1}{\overset{\wedge}{\rightrightarrows}} \Omega$$

be an equalizer diagram.

Proposition 2.7 Let $X \xrightarrow{\phi_i} \Omega$ $(i = 1,2)$ classify subobjects $A_i \subseteq X$ $(i = 1,2)$. Then $A_1 \subseteq A_2$ if and only if $X \xrightarrow{<\phi_1, \phi_2>} \Omega \times \Omega$ factors through the subobject $\subseteq$ of $\Omega \times \Omega$.

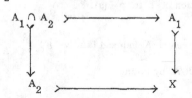

Proposition 2.7 is an easy consequence of proposition 2.6.

Theorem 2.8 If A is an object in an elementary topos $\underline{E}$, then $\underline{E}/A$ is an elementary topos.

Proof: The object $A \xrightarrow{1_A} A$ is terminal in $\underline{E}/A$.

Pullbacks in $\underline{E}/A$ may be constructed "in $\underline{E}$", so that $\underline{E}/A$ has finite limits. Given an object $X \xrightarrow{p} A$ in $\underline{E}/A$, let

$$A \xrightarrow{\tilde{p}} P(X)$$

correspond to the monic $X \xrightarrow{\langle 1_X, p\rangle} X \times A$.

Let

$$\begin{array}{ccc} R & \longrightarrow & \subseteq_X \\ \downarrow & & \downarrow \\ P(X) \times A & \xrightarrow{1 \times \tilde{p}} & P(X) \times P(X) = (\Omega \times \Omega)^X \end{array}$$

be a pullback diagram, and define

$$P(X \xrightarrow{p} A)$$

to be

$$R \rightarrowtail P(X) \times A \xrightarrow{p_2} A .$$

It remains to check that $P(X \xrightarrow{p} A)$ does what it should. We leave this to the reader. This construction is due to Kelly and Street.

In any category $\underline{E}$ with finite limits, pullback along a map

$$A \xrightarrow{f} B$$

induces a functor

$$f^* : \underline{E}/B \longrightarrow \underline{E}/A$$

which has a left adjoint

$$\Sigma_f : \underline{E}/A \longrightarrow \underline{E}/B$$

given by $\Sigma_f(X \xrightarrow{p} A) = (X \xrightarrow{p} A \xrightarrow{f} B)$

It is instructive to interpret what f^* and Σ_f mean for indexed families of sets. The reader will soon convince himself that f^* signifies "relabelling along f", i.e.

$$f^*(\{Y_b\}_{b \in B}) = \{Y_{f(a)}\}_{a \in A}$$

and that Σ_f signifies "coproduct over the fibres of f", i.e.

$$\Sigma_f(\{X_a\}_{a \in A}) = \{\coprod_{f(a)=b} X_a\}_{b \in B}.$$

<u>Theorem 2.9</u> Let $A \xrightarrow{f} B$ be a map in an elementary topos $\underline{E}$. Then the functor

$$f^* : \underline{E}/B \longrightarrow \underline{E}/A$$

has a right adjoint

$$\Pi_f : \underline{E}/A \longrightarrow \underline{E}/B$$

Proof: By working in the elementary topos $\underline{E}/_B$ we may suppose without loss of generality that $B = 1$. Let

$$1 \xrightarrow{\ulcorner 1_A \urcorner} A^A$$

be exponentially adjoint to 1_A. For any object $X \xrightarrow{p} A$ of $\underline{E}/_A$ let

$$\begin{array}{ccc} \Pi_f(p) & \longrightarrow & 1 \\ \downarrow & & \downarrow{\scriptstyle \ulcorner 1_A \urcorner} \\ X^A & \xrightarrow{p^A} & A^A \end{array}$$

be a pullback diagram. It is now routine to check that this gives a functor Π_f right adjoint to f^*.

Of course, for sets Π_f signifies "product over the fibres of f",

i.e.
$$\Pi_f(\{X_a\}_{a \in A}) = \{\prod_{f(a)=b} X_a\}_{b \in B} .$$

<u>Corollary 2.10</u> In an elementary topos, pullbacks preserve epics and colimit diagrams.

This follows from the fact that the functors Σ_f preserve and reflect colimit diagrams, and the fact that the functors f^* must preserve them, as they have right adjoints.

The <u>kernel pair</u> $K \overset{k_1}{\underset{k_2}{\rightrightarrows}} A$ of a map $A \xrightarrow{f} B$ is a pair of maps such that $fk_1 = fk_2$ and such that if $X \overset{x_1}{\underset{x_2}{\rightrightarrows}} A$ is any pair of maps such that $fx_1 = fx_2$, then there is a unique map $X \xrightarrow{h} K$ such that $x_i = k_i h$ $(i = 1, 2)$. Any category with pullbacks has kernel pairs. The kernel pair $K \overset{k_1}{\underset{k_2}{\rightrightarrows}} A$ of

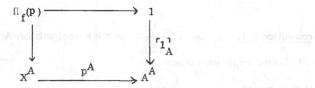

$A \xrightarrow{f} B$ is given by the pullback diagram

$$\begin{array}{ccc} K & \xrightarrow{k_1} & A \\ {\scriptstyle k_2}\downarrow & & \downarrow{\scriptstyle f} \\ A & \xrightarrow{f} & B \end{array}$$

<u>Proposition 2.11</u> Let $K \underset{k_2}{\overset{k_1}{\rightrightarrows}} A$ be the kernel pair of $A \xrightarrow{f} B$. Then the following imply each other:-

i) f is monic,

ii) $k_1 = k_2$,

iii) at least one of k_1 or k_2 is an isomorphism.

Because kernel pairs are defined by pullback diagrams, pullbacks of kernel pairs are kernel pairs.

<u>Proposition 2.12</u> In an elementary topos, pullback along epics reflects monics, epics and isomorphisms. That is to say, if

$$\begin{array}{ccc} X & \xrightarrow{p} & A \\ {\scriptstyle g}\downarrow & & \downarrow{\scriptstyle f} \\ Y & \xrightarrow{q} & B \end{array}$$

is a pullback diagram in an elementary topos, with f epic, then p epic, monic, iso implies that q is epic, monic, iso .

Proof: If p and f are epic, then clearly so is q. Let

$$K \underset{k_2}{\overset{k_1}{\rightrightarrows}} Y$$

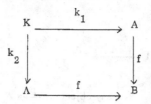

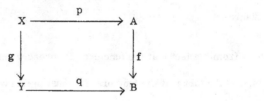

be the kernel pair of q. Let $L \overset{\ell_1}{\underset{\ell_2}{\rightrightarrows}} X$ be the pullback of this kernel pair along g. Then this is the kernel pair of p, and if p is monic, $\ell_1 = \ell_2$. We get a pullback diagram

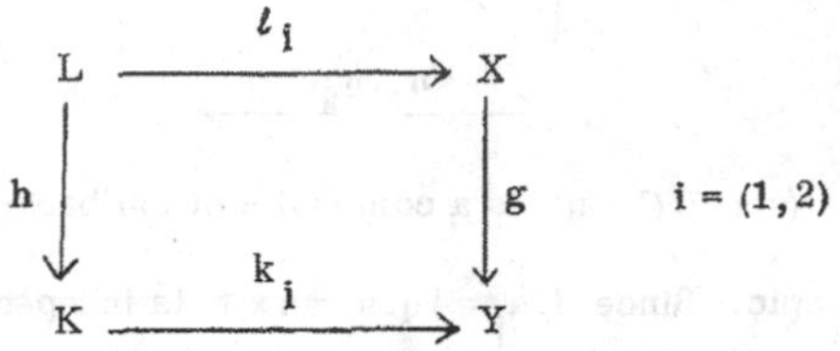

where g and h are epic, since pullbacks of epics are epic. Hence $k_1 h = k_2 h$, so $k_1 = k_2$ and so q is monic if p is. We have already seen, in the proof of theorem 2.3, that monic epics are isos in an elementary topos, so p iso implies that q is an iso.

<u>Proposition 2.13</u> Any map in an elementary topos can be factored as an epic followed by a monic. Such a factorization is unique up to a commuting isomorphism.

Proof: Let $A \xrightarrow{f} B$ be the map to be factorized. Let $K \overset{k_1}{\underset{k_2}{\rightrightarrows}} A$ be its kernel pair, and let $A \xrightarrow{q} Q$ be the coequalizer of the kernel pair. Then f factors as iq in the diagram below. We shall prove that i is monic.

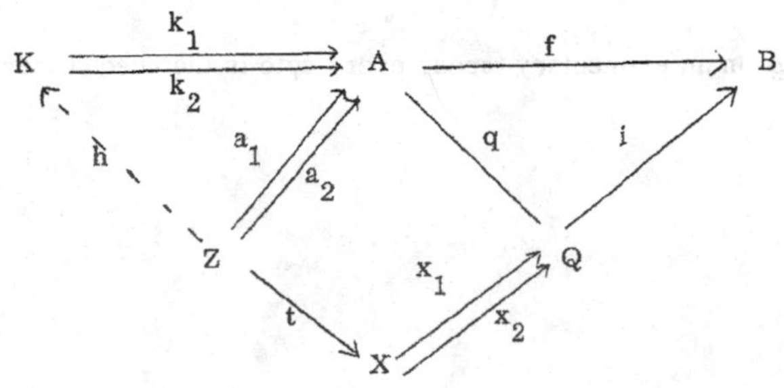

Suppose $ix_1 = ix_2$. For the pullback diagram

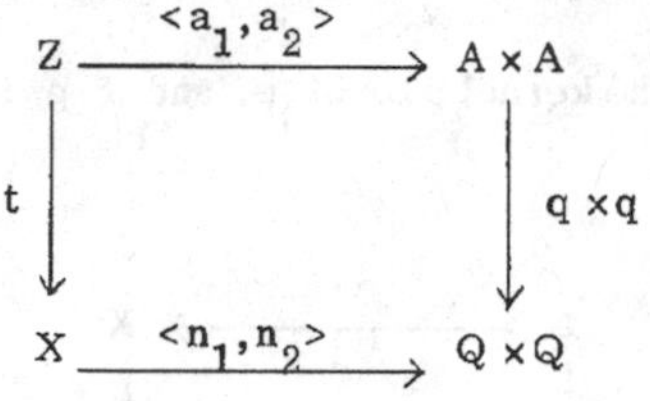

Since $q \times q = (q \times 1).(1 \times q)$ is a composite of pullbacks of epics, it is epic, and so t is epic. Since $f.a_j = iq.a_j = ix_jt$ is independent of j $(j = 1,2)$, there exists $Z \xrightarrow{h} K$ such that $a_j = k_j.h$. Hence $x_j.t = q.a_j = q.k_j.h$ is independent of j. But t is epic, so $x_1 = x_2$. So i is monic.

Suppose $A \xrightarrow{u} I \xrightarrow{v} B$ is any other factorization with u epic and v monic.

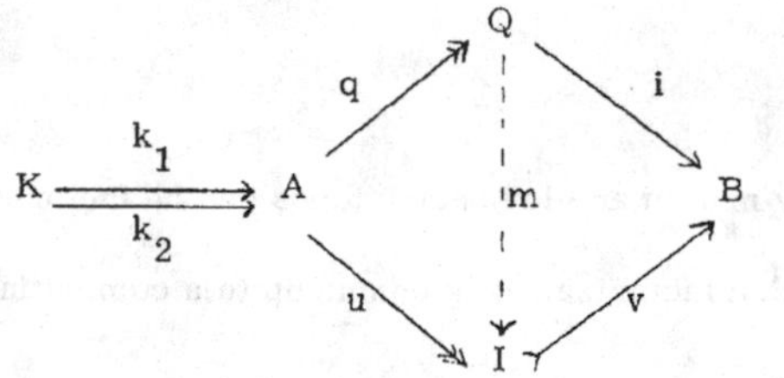

Since $uk_1 = uk_2$, there is a unique $Q \xrightarrow{m} I$ such that $u = mq$, so m is epic. Since $iq = f = vu = vmq$ we have $vm = i$, so m is monic. Hence m is an isomorphism.

<u>Corollary 2.14</u> In an elementary topos, every epic is the coequalizer of its kernel pair.

§3. Geometric morphisms

We saw in example vii) of §1 that a topological space X gives rise to an elementary topos Top(X), whose objects are local homeomorphisms into X. In the category of topological spaces and continuous maps, pullbacks of local homeomorphisms are local homeomorphisms, so a continuous map

$$X \xrightarrow{f} Y$$

gives rise to a functor

$$f^* : \mathrm{Top}(Y) \longrightarrow \mathrm{Top}(X) .$$

We shall use the term left exact for a finite limit preserving functor. The functor f^* is left exact and has a right adjoint f_*. This motivates

Definition 3.1 A geometric morphism

$$\underline{E} \xrightarrow{f} \underline{F}$$

between two elementary topoi is a functor

$$f_* : \underline{E} \longrightarrow \underline{F}$$

which has a left exact left adjoint

$$f^* : \underline{F} \longrightarrow \underline{E} .$$

We call f_* the direct image part of f and f^* the inverse image part of f.

We give some examples to show that this notion is not as unnatural as at first appears.

Examples

i) We have already seen that a continuous map $X \xrightarrow{f} Y$ between topological spaces gives a geometric morphism

$$\text{Top}(X) \xrightarrow{f} \text{Top}(Y) .$$

ii) A map $A \xrightarrow{f} B$ in an elementary topos $\underline{E}$ gives a geometric morphism

$$\underline{E}/_A \xrightarrow{f} \underline{E}/_B$$

where $f_* = \Pi_f$.

iii) Let $G \xrightarrow{\phi} H$ be a homomorphism of groups. If M is an H-set, let $\phi^*(M)$ denote the G-set whose underlying set is the same as that of M, but with G-action defined by ϕ, i.e. $g.m = \phi(g).m$ for $g \in G$, $m \in M$. This gives a left exact functor

$$\phi^* : \text{H-sets} \longrightarrow \text{G-sets} .$$

If N is a G-set, let $\phi_*(N)$ be the set of functions $q : H \longrightarrow N$ such that for $g \in G$, $h \in H$

$$q(\phi(g).h) = g.q(h).$$

We make $\phi_*(N)$ into an H-set by defining $h.q.$ for $h \in H$, $q \in \phi_*(N)$ to be the function given by $h' \longrightarrow q(h'.h)$. Then ϕ_* is a functor right adjoint to ϕ^* and we have a geometric morphism

$$\text{G-sets} \xrightarrow{\phi} \text{H-sets} .$$

If $\underline{E} \overset{f}{\underset{g}{\rightrightarrows}} \underline{F}$ are a pair of geometric morphisms, a map $f \longrightarrow g$ is to mean a natural map $f^* \longrightarrow g^*$ (and so, by adjointness, a natural map $g_* \longrightarrow f_*$). Thus, for any two elementary topoi $\underline{E}$, $\underline{F}$ we get a category (in general, illegitimate, i.e. the hom-classes need not be sets)

$$\mathrm{Top}\,(\underline{E},\ \underline{F})$$

of geometric morphisms from $\underline{E}$ to $\underline{F}$ and maps between them.

If $\underline{E}$ is an elementary topos, an $\underline{E}$-topos is a pair $(\underline{F}, f)$ where $\underline{F} \xrightarrow{f} \underline{E}$ is a geometric morphism. We will usually abuse language by referring to the $\underline{E}$-topos $\underline{F}$, leaving f understood. We call f the <u>structural morphism</u> of $\underline{F}$.

If $\underline{F}_1$, $\underline{F}_2$ are $\underline{E}$-topoi, a morphism of $\underline{E}$-topoi is a geometric morphism $\underline{F}_1 \longrightarrow \underline{F}_2$ making the diagram

$$\begin{array}{ccc} \underline{F}_1 & \longrightarrow & \underline{F}_2 \\ & \searrow \quad \swarrow & \\ & \underline{E} & \end{array}$$

commute up to natural isomorphism. If $\underline{F}_1 \overset{g}{\underset{g'}{\rightrightarrows}} \underline{F}_2$ are morphisms of $\underline{E}$-topoi, a map $\alpha : g \longrightarrow g'$ is a natural map

$$\alpha : g^* \longrightarrow g'^*$$

such that $\alpha \bullet f_2^*$ is a natural isomorphism, where $\underline{F}_2 \xrightarrow{f_2} \underline{E}$ is the structural morphism of $\underline{F}_2$. We obtain the (illegitimate) category

$$\mathrm{Top}_{\underline{E}}(\underline{F}_1,\ \underline{F}_2)$$

of morphisms of $\underline{E}$-topoi $\underline{F}_1 \longrightarrow \underline{F}_2$.

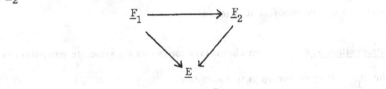

For any object X in an elementary topos $\underline{E}$, the unique map $X \longrightarrow 1$ in $\underline{E}$, gives a geometric morphism

$$\underline{E}/_X \longrightarrow \underline{E}/_1 \simeq \underline{E}$$

by which $\underline{E}/_X$ is made into an $\underline{E}$-topos. Clearly, any geometric morphism

$$\underline{E}/_X \longrightarrow \underline{E}/_Y$$

induced by a map $X \longrightarrow Y$ in $\underline{E}$, is a morphism of $\underline{E}$-topoi.

If $*$ denotes a topological space with one point, for any topological space X, the unique map $X \longrightarrow *$ induces a geometric morphism

$$\mathrm{Top}(X) \longrightarrow \mathrm{Top}(*) = S$$

so that a spatial topos is an S-topos.

Proposition 3.2 If an elementary topos has a geometric morphism to S or S_{fin}, it is unique up to isomorphism.

Proof: Consider a geometric morphism

$$\underline{E} \xrightarrow{f} S .$$

Since f^* is left exact, $f^*(1) \simeq 1$. Since it has a right adjoint,

$$f^*(s) \simeq f^*(\coprod_s 1) \simeq \coprod_s f^*(1) \simeq \coprod_s 1.$$

It follows that $f_* \simeq \mathrm{Hom}_{\underline{E}}(1, -)$. A similar argument holds for S_{fin}.

Corollary 3.3 An elementary topos is an S_{fin}-topos if and only if it has finite hom-sets.

Proof: If $\underline{E}$ has finite hom-sets, the functor

$$\mathrm{Hom}_{\underline{E}}(1,-) : \underline{E} \longrightarrow S_{fin}$$

has a left exact left adjoint $s \longrightarrow \coprod_{s} 1$.

Conversely, if

$$\underline{E} \xrightarrow{f} S_{fin}$$

is a geometric morphism, proposition 3.2 shows that $f_* \simeq \mathrm{Hom}_{\underline{E}}(1,-)$, so every object of $\underline{E}$ has a finite number of elements. But the maps $X \rightarrow Y$ in $\underline{E}$ are given by the elements of Y^X.

§4. Sober spaces

In this chapter we investigate how much information is lost in passing from a topological space X to the elementary topos Top(X). The material of this chapter is to be found in SGA 4.

Definition 4.1 A topological space is irreducible if the intersection of two non-empty open sets is non-empty.

Example: For any point x in a topological space X, $\overline{\{x\}}$ is a closed irreducible subspace of X, because any nonempty open set of $\overline{\{x\}}$ must contain x.

Definition 4.2 A point x of a topological space X is generic if $X = \overline{\{x\}}$.

Thus, any space with a generic point is irreducible.

Definition 4.3 A topological space is sober if every irreducible closed subspace has a unique generic point.

Examples: i) A Hausdorff space is sober. The irreducible closed subspaces are the singleton subsets.

ii) For any commutative ring R, spec(R) is sober. The prime ideal p is the unique generic point of the closed irreducible subspace spec(R/p) consisting of all the prime ideals containing p.

For any topological space X, let $\hat{X}$ be the set of irreducible closed subspaces of X. For any open set U of X, let $\hat{U}$ be the subset of $\hat{X}$ of all the irreducible closed subspaces of X which have non-empty intersection with U.

<u>Proposition 4.4</u> The subsets $\hat{U}$ of $\hat{X}$ form a topology.

We define a map $\eta : X \longrightarrow \hat{X}$ by $x \longrightarrow \overline{\{x\}}$.

<u>Proposition 4.5</u> The function η is continuous and induces a bijection $U \longleftrightarrow \hat{U}$ between the open set lattices of X and $\hat{X}$.

The well-known result that Top(X) is equivalent to the category of sheaves on X implies that η induces an equivalence of categories

$$\mathrm{Top}(X) \xrightarrow{\sim} \mathrm{Top}(\hat{X}) ,$$

since a sheaf on a topological space may be defined purely in terms of the open set lattice.

<u>Proposition 4.6</u> For any topological space X, the space $\hat{X}$ is sober. Any continuous map from X to a sober space factors uniquely through $\eta : X \longrightarrow \hat{X}$. In consequence $X \longmapsto \hat{X}$ defines a functor left adjoint to the inclusion of sober spaces in the category of all topological spaces.

The remark above shows that the functor Top factors through the soberification functor $X \longmapsto \hat{X}$.

If X is a sober space, we define a partial order on X as follows:

$$x_1 \leq x_2 \iff x_1 \in \overline{\{x_2\}}.$$

If $f, g : Y \longrightarrow X$ are continuous maps into a sober space X, define

$$f \leq g \iff \forall y \in Y,\ f(y) \leq g(y).$$

Proposition 4.7 Let $f, g : Y \longrightarrow X$ be continuous maps into a sober space. Any two natural maps $f^* \longrightarrow g^*$ agree. There exists a natural map $f^* \longrightarrow g^*$ if and only if $f \leq g$.

Definition 4.8 A point of an S-topos $\underline{E}$ is a geometric morphism $S \longrightarrow \underline{E}$.

The class of points of an S-topos may not form a set.

Definition 4.9 An open of an S-topos $\underline{E}$ is a subobject of 1 in $\underline{E}$.

We may put a topology on the class of points of an S-topos $\underline{E}$, Points ($\underline{E}$), as follows: if $U \rightarrowtail 1$ defines an open of $\underline{E}$ and $S \xrightarrow{p} \underline{E}$ is a point, since p^* is left exact $p^*(U)$ is either ϕ or 1. We write $p \in U$ if $p^*(U) = 1$ and $p \notin U$ if $p^*(U) = \phi$. We take for the open subclasses of Points ($\underline{E}$)

$$\{p \in \text{Points}\ (\underline{E}) \mid p \in U\}$$

for U an open of $\underline{E}$.

A geometric morphism $\underline{E} \longrightarrow \underline{F}$ induces a continuous map

$$\text{Points}\ (\underline{E}) \longrightarrow \text{Points}\ (\underline{F})\ .$$

Proposition 4.10 Points $(\text{Top}(X)) \simeq \hat{X}$.

Proof: In a spatial topos every object is a colimit of opens. Hence, if $S \xrightarrow{p} \mathrm{Top}(X)$ is a point, p is determined by the restriction of p^* to the opens. Let U be the union of all the opens V not containing p, i.e. such that $p^*(V) = \phi$. Since p^* is left exact, X - U is an irreducible closed subset of X, i.e. a point of $\hat{X}$. Conversely, a point $* \longrightarrow \hat{X}$ determines a geometric morphism $S \simeq \mathrm{Top}(*) \longrightarrow \mathrm{Top}(\hat{X}) \simeq \mathrm{Top}(X)$.

Corollary 4.11 The category of spatial topoi and geometric morphisms is equivalent to the category of sober spaces and continuous maps.

One may now play the game of extending the definition of various topological properties to elementary topoi, or at least to S-topoi.

Examples:

i) An S-topos is connected if it satisfies one of the following equivalent conditions:-

a) $\mathrm{Hom}_{\underline{E}}(1,-)$ preserves coproducts;

b) The functor $S \mapsto \coprod_S 1$ is full;

c) The object 1 is coproduct-irreducible, i.e. if $1 = \coprod_{i \in S} U_i$, then $U_i = \phi$ for all $i \in S$ except one value.

ii) An S-topos is locally connected if the functor $S \to \coprod_S 1$ has a left adjoint, π_0. We interpret $\pi_0(X)$ as the set of connected components of X.

iii) Let $\underline{E}$ be an S-topos. Since $\mathrm{Hom}_{\underline{E}}(1,-)$ is left exact it takes abelian group objects in $\underline{E}$ to abelian groups, and so defines a left exact functor

$$\Gamma_{\underline{E}} : \mathrm{Ab}\,\underline{E} \longrightarrow \mathrm{Ab}\ .$$

The category $Ab\,\underline{E}$ is abelian and has enough injectives, so we may define the right derived functors of $\Gamma_{\underline{E}}$, $R^n\,\Gamma_{\underline{E}}$. The n-th Grothendieck cohomology functor of $\underline{E}$ is $H^n(\underline{E}, -) = R^n\,\Gamma_{\underline{E}}$.

Let $\underline{E} \xrightarrow{f} \underline{F}$ be a geometric morphism. We have a commutative diagram of functors

$$\begin{array}{ccc} Ab\,\underline{E} & \xrightarrow{f_*} & Ab\,\underline{F} \\ & {\scriptstyle \Gamma_{\underline{E}}}\searrow \quad \swarrow{\scriptstyle \Gamma_{\underline{F}}} & \\ & Ab & \end{array}$$

Since f_* has a left exact left adjoint, it preserves injectives. Hence there is a spectral sequence

$$R^p\,\Gamma_{\underline{F}} \cdot R^q f_* \Longrightarrow R^n\,\Gamma_{\underline{E}}$$

giving the Leray spectral sequence

$$H^p(\underline{F}, R^q f_*(-)) \Longrightarrow H^n(\underline{E}, -).$$

The front adjunction $\eta : id_{\underline{F}} \longrightarrow f_* f^*$ gives a natural map

$$\Gamma_{\underline{F}}\,\eta : \Gamma_{\underline{F}} \longrightarrow \Gamma_{\underline{F}} f_* f^* \simeq \Gamma_{\underline{E}} f^*.$$

Since f^* is exact, $R^n(\Gamma_{\underline{E}} f^*) \simeq (R^n\,\Gamma_{\underline{E}}) f^*$, so we get the map in cohomology induced by f

$$H^n(f, -) : H^n(\underline{F}, -) \longrightarrow H^n(\underline{E}, f^*(-)).$$

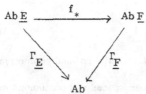

§5. Left exact comonads

Recall that a comonad on a category $\underline{E}$ is a functor

$$C : \underline{E} \longrightarrow \underline{E}$$

together with natural maps

$$C \xrightarrow{\epsilon} id_{\underline{E}} \qquad \text{(the co-unit)}$$
$$C \xrightarrow{\delta} C^2 \qquad \text{(the co-multiplication)}$$

satisfying the usual axioms for two-sided co-unit and co-associativity. We call the comonad left exact if the functor C is left exact.

A C-coalgebra is a pair (X, ξ) where X is an object of $\underline{E}$ and $X \xrightarrow{\xi} C(X)$ is a map of $\underline{E}$ (the co-structure) satisfying the standard identities. We have the appropriate notion of a map of C-coalgebras, and we denote the category of C-coalgebras by $\underline{E}_C$. The forgetful functor

$$\underline{E}_C \longrightarrow \underline{E} \; : \; (X, \xi) \longrightarrow X$$

has a right adjoint - "cofree" - which assigns to an object Y of $\underline{E}$ the C-coalgebra (CY, δ_Y).

Theorem 5.1 If $\underline{E}$ is an elementary topos and C is a left exact comonad on $\underline{E}$, then $\underline{E}_C$ is an elementary topos.

For the details of the proof, and for a more precise treatment of left exact comonads we refer the reader to page 39 of "Elementary Toposes", Kock and Wraith.

Because the forgetful functor $\underline{E}_C \longrightarrow \underline{E}$ is left exact and has a right adjoint, we get a geometric morphism

$$\underline{E} \longrightarrow \underline{E}_C$$

which we call the canonical geometric morphism associated to C.

Example Let G be a monoid object in an elementary topos $\underline{E}$. Then $G \times (-)$ has a monad structure, and so the exponentially adjoint functor $(-)^G$ has the structure of a left exact comonad. A G-action on an object X,

$$G \times X \longrightarrow X$$

corresponds by adjointness to a $(-)^G$-coalgebra costructure

$$X \longrightarrow X^G .$$

It follows that the category of G-objects in $\underline{E}$ form an elementary topos.

If $\underline{E} \xrightarrow{f} \underline{F}$ is a geometric morphism, the adjoint pair f_*, f^* gives a left exact comonad $C = f^*f_*$ on $\underline{E}$. Observe that f^* satisfies all the conditions of the dual of Beck's crude tripleability theorem (see §2) except the condition of reflecting isomorphisms.

Theorem 5.2 Let $\underline{E} \xrightarrow{f} \underline{F}$ be a geometric morphism such that f^* reflects isomorphisms. Then f^* is cotripleable, i.e. $\underline{F}$ is equivalent to the category of f^*f_*-coalgebras on $\underline{E}$, with f^* for forgetful functor.

Theorem 5.2 characterizes geometric morphisms f for which f^* reflects isomorphisms. We shall call them cotripleable geometric morphisms.

Proposition 2.12 gives the following examples.

i) If $A \longrightarrow B$ is an epic map in an elementary topos $\underline{E}$, the induced geometric morphism

$$\underline{E}/_A \longrightarrow \underline{E}/_B$$

is cotripleable.

ii) If $X \longrightarrow Y$ is a surjective continuous map between topological spaces, then

$$\mathrm{Top}(X) \longrightarrow \mathrm{Top}(Y)$$

is cotripleable.

In example (iv) of §1 we remarked that the Cartesian product $\underline{E}_1 \times \underline{E}_2$ of two elementary topoi was an elementary topos. Unfortunately for the notation, in the category of topoi and geometric morphisms $\underline{E}_1 \times \underline{E}_2$ is the coproduct of $\underline{E}_1$ and $\underline{E}_2$ with canonical injections

$$\underline{E}_1 \xrightarrow{i_1} \underline{E}_1 \times \underline{E}_2 \xleftarrow{i_2} \underline{E}_2$$

given as follows:

i_1^*, i_2^* are the projection functors, and $i_{1*}(X) = (X, 1)$, $i_{2*}(Y) = (1, Y)$.

A pair of geometric morphisms

$$\underline{E}_1 \xrightarrow{f} \underline{F} \xleftarrow{g} \underline{E}_2$$

gives a unique geometric morphism $\underline{E}_1 \times \underline{E}_2 \xrightarrow{h} \underline{F}$ such that the diagram

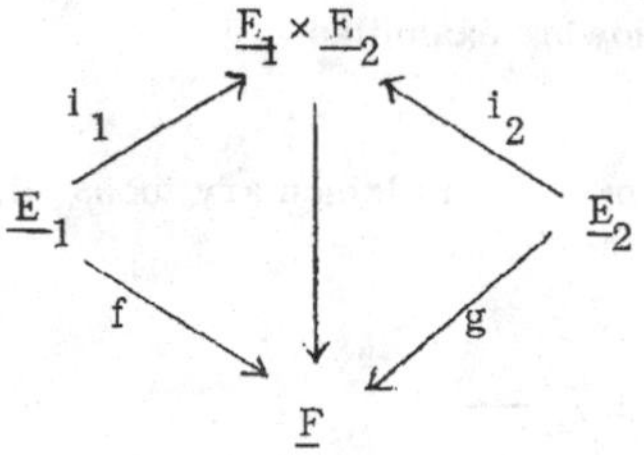

commutes, given by $h_*(X,Y) = f_*(X) \times g_*(Y)$ and $h^*(Z) = (f^*(Z), g^*(Z))$.

Suppose that $\underline{E}_1 \longrightarrow \underline{E}_2$ is a left exact functor between elementary topoi. Define a left exact functor

$$C\ .\ \underline{E}_1 \times \underline{E}_2 \xrightarrow{\partial} \underline{E}_1 \times \underline{E}_2$$

by

$$C(X,Y) = (X,\ \partial(X) \times Y).$$

It has a comonad structure given by

$$\varepsilon_{(X,Y)} = (1_X, p_2) \ :\ (X,\ \partial(X) \times Y) \longrightarrow (X,Y)$$

$$\delta_{(X,Y)} = (1_X, <1_{\partial(X)}, 1_{\partial(X)}> \times 1_Y) : (X,\ \partial(X) \times Y) \longrightarrow (X,\ \partial(X) \times \partial(X) \times Y).$$

The elementary topos $(\underline{E}_1 \times \underline{E}_2)_C$ is called the topos obtained by glueing along ∂. It is equivalent to the comma category $(id_{\underline{E}_2}, \partial)$
The glueing process can be generalized to arbitrary finite 2-diagrams of left exact functors. 2-colimits of 2-diagrams of geometric morphisms can be obtained by glueing along the direct image parts.

Let X be a topological space, $U \subseteq X$ an open subspace, and $X-U$ its closed complement. Denote by

$$i : U \longrightarrow X$$
$$j : X-U \longrightarrow X$$

the inclusion maps. Let ∂ denote the composite

$$\mathrm{Top}(U) \xrightarrow{i_*} \mathrm{Top}(X) \xrightarrow{j^*} \mathrm{Top}(X - U).$$

We may call this the "fringe" functor, because for any object F of Top(U), $\partial(F)$ is trivial everywhere on X - U except on the boundary of U. The other composite, $i^* j_*$, is a functor of little interest since it takes all objects of Top(X - U) to the terminal object.

<u>Proposition 5.3</u> Top(X) is equivalent to the elementary topos obtained by glueing along

$$\mathrm{Top}(U) \xrightarrow{\partial} \mathrm{Top}(X - U).$$

It is quite possible to glue two spatial topoi together to get a non-spatial one.

§6. Topologies

A Heyting algebra is a category which is

(i) a partially ordered set,

(ii) has finite limits and finite colimits,

(iii) is Cartesian closed.

As usual, we write $a \leq b$ for a map $a \longrightarrow b$ in the Heyting algebra. It is conventional also to write $a \wedge b$ in place of $a \times b$, and $a \vee b$ in place of $a \amalg b$. We write t (= "true") for the terminal object and f (= "false") for the initial object. It is conventional to write $a \Rightarrow b$ in place of b^a, so that the Cartesian closedness is expressed by the adjunction

$$\frac{a \wedge b \leq c}{a \leq (b \Rightarrow c)} .$$

We write $\neg a$ for $a \Rightarrow f$. This gives a unary operation $\neg$ which is called "negation". A Heyting algebra can be presented purely in terms of the operations $\wedge, \vee, \Rightarrow$, t, f, subject to appropriate axioms. Among the theorems we may deduce are, for example,

$$\neg\neg\neg a = \neg a,$$

$$(\neg\neg a) \wedge (\neg\neg b) = \neg\neg(a \wedge b),$$

$$\neg\neg t = t .$$

In general, a Heyting algebra does not satisfy the identity

$$\neg\neg a = a .$$

If it does, it is a Boolean algebra. Intuitionistic logic corresponds to Heyting algebras in the same way that classical logic corresponds to Boolean algebras.

Proposition 6.1 Let Ω be the subobject classifier in an elementary topos $\underline{E}$. Then Ω is a Heyting algebra object, with t, f, $\wedge$, $\vee$, $\Rightarrow$, $\neg$ interpreted as follows:

$1 \xrightarrow{t} \Omega$ classifies the maximal subobject $1 \rightarrowtail 1$,

$1 \xrightarrow{f} \Omega$ classifies the minimal subobject $\phi \rightarrowtail 1$,

$\Omega \times \Omega \xrightarrow{\wedge} \Omega$ classifies $1 \xrightarrow{\langle t,t\rangle} \Omega \times \Omega$,

$\Omega \times \Omega \xrightarrow{\vee} \Omega$ classifies the image of $\Omega \amalg \Omega \xrightarrow{\begin{pmatrix} 1 & t \\ t & 1 \end{pmatrix}} \Omega \times \Omega$,

$\Omega \times \Omega \xrightarrow{\Rightarrow} \Omega$ classifies $\leq \rightarrowtail \Omega \times \Omega$, the equalizer of $\wedge$ and p_1

$\Omega \xrightarrow{\neg} \Omega$ classifies $1 \xrightarrow{f} \Omega$.

For the details of the proof we refer the reader to Aspects of Topoi, P. Freyd or Elementary Toposes, A. Kock and G. Wraith.

We call an elementary topos Boolean if, in it, we have the identity

$$\Omega \xrightarrow{\neg} \Omega \xrightarrow{\neg} \Omega = \Omega \xrightarrow{1_\Omega} \Omega .$$

Proposition 6.2 The following statements for an elementary topos $\underline{E}$ are equivalent.

i) $\underline{E}$ is Boolean.

ii) For every object X of $\underline{E}$, the subobject lattice of X is Boolean.

iii) Subobjects of objects in $\underline{E}$ have complements.

iv) The map $1 \amalg 1 \xrightarrow{\binom{t}{f}} \Omega$ in $\underline{E}$ is an isomorphism.

We leave the proof to the reader.

In general, a spatial topos is not Boolean, for if U is an open subset of a topological space X, then $\neg U$ is the exterior of U. Hence $\neg\neg U$ is the interior of the closure of U.

<u>Definition 6.3</u> A <u>topology</u> on an elementary topos $\underline{E}$ is an endomorphism $\Omega \xrightarrow{j} \Omega$ of the subobject classifier of Ω such that

i) $j^2 = j$,

ii) $j.t = t$

iii) $j. \wedge = \wedge.(j \times j)$.

In terms of diagrams, these conditions express the commutativity of

$$\begin{array}{ccc} \Omega & \xrightarrow{j} & \Omega \\ & {\scriptstyle j}\searrow \quad \swarrow{\scriptstyle j} & \\ & \Omega & \end{array} \qquad \begin{array}{ccc} & 1 & \\ & {\scriptstyle t}\swarrow \quad \searrow{\scriptstyle t} & \\ \Omega & \xrightarrow{j} & \Omega \end{array} \qquad \begin{array}{ccc} \Omega \times \Omega & \xrightarrow{\wedge} & \Omega \\ {\scriptstyle j \times j}\downarrow & & \downarrow{\scriptstyle j} \\ \Omega \times \Omega & \xrightarrow{\wedge} & \Omega \end{array}$$

If we think of Ω as a category object, then j is simply a left exact monad on Ω. It determines a closure operator on the subobject lattice of each object of $\underline{E}$; i.e. if X is an object of $\underline{E}$, and A is a subobject of X classified by $X \xrightarrow{\phi} \Omega$, we denote by $\bar{A}$ the subobject classified by $X \xrightarrow{\phi} \Omega \xrightarrow{j} \Omega$.

Condition ii) gives $A \subseteq \bar{A}$,

Condition i) gives $\bar{\bar{A}} = \bar{A}$

Condition iii) gives $\overline{A_1 \cap A_2} = \bar{A}_1 \cap \bar{A}_2$.

We call subobject A of X, <u>j-dense</u>, if $\bar{A} = X$.

<u>Definition 6.4</u> If j is a topology on an elementary topos $\underline{E}$, an object X of $\underline{E}$ is a <u>j-sheaf</u> if for every <u>j-dense</u> monic $A' \xrightarrow{i} A$, the function

$$\mathrm{Hom}_{\underline{E}}(i, 1_X) : \mathrm{Hom}_{\underline{E}}(A, X) \longrightarrow \mathrm{Hom}_{\underline{E}}(A', X)$$

is bijective.

In other words, an object X is a j-sheaf if every map into it from a j-dense subobject of an object A, lifts uniquely to the whole of A. We denote by $sh_j(\underline{E})$ the full subcategory of $\underline{E}$ of j-sheaves.

Theorem 6.5 Let j be a topology on an elementary topos $\underline{E}$. Then $sh_j(\underline{E})$ is an elementary topos, and the inclusion functor

$$sh_j(\underline{E}) \longrightarrow \underline{E}$$

has a left exact adjoint (the sheafification functor). Thus j determines a geometric morphism

$$sh_j(\underline{E}) \longrightarrow \underline{E} .$$

For the proof we again refer the reader to Aspects of Topoi, P. Freyd or Elementary Toposes, A. Kock and G. Wraith. Freyd's elegant use of injectives renders the category of fraction techniques in Elementary Toposes unnecessary. Their sole purpose was to show the left exactness of the sheafification functor. In the context of Grothendieck topoi, the construction of the sheafification functor, as given, say, in SGA 4, involved the use of infinite limits and colimits. It must be stressed that in the context of elementary topoi, the sheafification functor only involves elementary operations, i.e. finite limits and exponentiation. The novel feature which permits this, is, of course, the possibility of exponentiation. Somehow, all the colimits needed for the Grothendieck approach sum up to give exponentials. P. Johnstone has given a different construction of sheafification from that of Lawvere and Tierney, which mirrors more closely that given in SGA 4, but in elementary terms.

Examples

i) The maximal topology $\Omega \xrightarrow{1_\Omega} \Omega$.

In this case $sh_j(\underline{E}) = \underline{E}$.

ii) The minimal topology $\Omega \longrightarrow 1 \xrightarrow{t} \Omega$.

In this case $sh_j(\underline{E}) = \{1\}$.

iii) The double negation topology $\Omega \xrightarrow{\neg} \Omega \xrightarrow{\neg} \Omega$.

In this case $sh_j(\underline{E})$ is Boolean.

iv) If U is a subobject of 1 in $\underline{E}$, the unary operation $U \Rightarrow (-) : \Omega \longrightarrow \Omega$, i.e. the composite

$$\Omega \xrightarrow{\ulcorner U \urcorner \times 1_\Omega} \Omega \times \Omega \xrightarrow{\Rightarrow} \Omega$$

where $1 \xrightarrow{\ulcorner U \urcorner} \Omega$ classifies $U \rightarrowtail 1$, is a topology. There is an equivalence of categories

$$sh_j(\underline{E}) \simeq \underline{E}/U$$

in this case, making the diagram

$$\begin{array}{ccc} sh_j(\underline{E}) & \xrightarrow{\sim} & \underline{E}/U \\ & \searrow \quad \swarrow & \\ & \underline{E} & \end{array}$$

commute. We call a topology of this form open.

v) If U is a subobject of 1 in $\underline{E}$, the unary operation given by $U \vee (-)$, i.e. the map

$$\Omega \xrightarrow{\ulcorner U \urcorner \times 1_\Omega} \Omega \times \Omega \xrightarrow{\vee} \Omega$$

is a topology. We call it the closed complement to the topology of example iv).

If $\underline{E} = Top(X)$ and j is the closed complement to the topology whose sheaves give $Top(X)/U \simeq Top(U)$, for U an open subspace, then $sh_j(Top(X)) \simeq Top(X-U)$. In general, if j is an open topology on $\underline{E}$ and j' is its closed complement, with geometric morphisms

$$sh_j(\underline{E}) \xrightarrow{u} \underline{E} \xleftarrow{u'} sh_{j'}(\underline{E})$$

where u_*, u'_* are the inclusion functors, then $\underline{E}$ is equivalent to the topos obtained by glueing along the left exact functor $u'^* u_*$.

vi) Let X be a topological space and let $\underline{E}$ be the category of presheaves on X. For any open set U of X, $\Omega(U)$ is the set of cribles of U, i.e. families of open subsets of U closed under taking open subsets. A crible is called <u>principal</u> if it consists of all the open subsets of some given open subset. Define a function $j_U : \Omega(U) \longrightarrow \Omega(U)$ by sending each crible on U to the principal crible defined by the union of all its members. We obtain a map $j : \Omega \longrightarrow \Omega$ which is a topology on $\underline{E}$. The j-sheaves are precisely the sheaves on X.

We define a partial order on topologies on an elementary topos $\underline{E}$ by writing $j \leq j'$ if $sh_j(\underline{E}) \leq sh_{j'}(\underline{E})$

For any topology $\Omega \xrightarrow{j} \Omega$, let

$$1 \xrightarrow{\ulcorner j \urcorner} \Omega^{\Omega}$$

denote the exponential adjoint. If j_{max}, j_{min} denote the maximal and minimal topologies, we write $Int(j)$ and $Ext(j)$ for the equalizers of $(\ulcorner j \urcorner, \ulcorner j_{max} \urcorner)$ and $(\ulcorner j \urcorner, \ulcorner j_{min} \urcorner)$ respectively. The open topologies associated with $Int(j)$ and $Ext(j)$ we call the <u>interior</u> of j and the <u>exterior</u> of j.

By internalizing the three conditions of definition 6.3 we may define a subobject $top(\underline{E})$ of Ω^{Ω}, whose elements correspond to topologies on $\underline{E}$. In fact, we get that $Hom_{\underline{E}}(X, top(\underline{E}))$ is in bijective correspondence with the topologies on $\underline{E}/X$.

The notion of open topology and interior give rise to maps

$$\Omega \underset{\longleftarrow}{\longrightarrow} top(\underline{E})$$

which are adjoint functors in an internal sense. The map $\Omega \longrightarrow top(\underline{E})$ arises from the exponential adjoint to $\Omega \times \Omega \xrightarrow{\Rightarrow} \Omega$. We leave the reader to formulate similar notions for closed topologies and the closure of a topology.

§ 7. Factorization of geometric morphisms.

Let $\mathbb{T} = (T, \eta, \mu)$ be a left exact monad on an elementary topos $\underline{E}$. Since T is left exact, $T(t)$ defines a subobject of $T(\Omega)$, whose classifying map we call

$$T(\Omega) \xrightarrow{\lambda} \Omega .$$

Let us write $\Omega \xrightarrow{j} \Omega$ for the composite

$$\Omega \xrightarrow{\eta_\Omega} T(\Omega) \xrightarrow{\lambda} \Omega .$$

Proposition 7.1 The map j is a topology on $\underline{E}$.

We call it the topology induced by $\mathbb{T}$. For the details of the proof, see pp. 68-70 of Elementary Toposes.

Recall that a subcategory is wide if any object isomorphic to one in the subcategory belongs to the subcategory, and reflective if the inclusion functor has a left adjoint. A monad $\mathbb{T} = (T, \eta, \mu)$ is idempotent if the multiplication $T^2 \xrightarrow{\mu} T$ is a natural isomorphism. If $\mathbb{T}$ is idempotent, an object which has a $\mathbb{T}$-algebra structure has a unique $\mathbb{T}$-algebra structure. These objects are precisely those isomorphic to objects in the image of T. They form a full wide reflective subcategory. Conversely, any full wide reflective subcategory gives rise to an idempotent monad, given by the adjoint pair consisting of the inclusion functor and its left adjoint.

Proposition 7.2 Let $\mathbb{T}$ be an idempotent left exact monad on an elementary topos $\underline{E}$, and let j be the topology on $\underline{E}$ induced by $\mathbb{T}$. Then the full subcategory of $\mathbb{T}$-algebras is equal to $sh_j(\underline{E})$.

The proof is given on pp. 70-72 of Elementary Toposes.

Corollary 7.3 A subcategory of an elementary topos $\underline{E}$ is of the form $sh_j(\underline{E})$ if and only if it is a full wide reflective subcategory with a left exact reflection functor.

Corollary 7.4 Let $\underline{F} \xrightarrow{f} \underline{E}$ be a geometric morphism with f_* full and faithful. Then there is an equivalence

$$\underline{F} \simeq sh_j(\underline{E})$$

making the diagram

$$\begin{array}{ccc} \underline{F} & \xrightarrow{\sim} & sh_j(\underline{E}) \\ & {\scriptstyle f}\searrow & \swarrow \\ & \underline{E} & \end{array}$$

commute, where j is the topology on $\underline{E}$ induced by the left exact monad f_*f^*.

Examples

i) If $A \xrightarrow{f} B$ is a map in $\underline{E}$, then $\underline{E}/_A \longrightarrow \underline{E}/_B$ has f_* full and faithful if and only if f is monic.

ii) If Y is a topological space, and $X \subseteq Y$ a subspace, the induced geometric morphism $Top(X) \xrightarrow{f} Top(Y)$ has f_* full and faithful.

Theorem 7.5 Let j be a topology on an elementary topos $\underline{E}$, with canonical geometric morphism

$$sh_j(\underline{E}) \xrightarrow{i} \underline{E}\,.$$

Then a geometric morphism $\underline{F} \xrightarrow{f} \underline{E}$ factors through i if and only if f^* takes j-dense monics to isomorphisms.

Proof. Let $K \overset{g}{\rightarrowtail} L$ be a j-dense monic in $\underline{E}$, and let X be an object of $\underline{F}$. We have a commutative diagram

$$\begin{array}{ccc} \mathrm{Hom}_{\underline{E}}(L, f_*(X)) & \longrightarrow & \mathrm{Hom}_{\underline{E}}(K, f_*(X)) \\ \downarrow & & \downarrow \\ \mathrm{Hom}_{\underline{F}}(f^*(L), X) & \longrightarrow & \mathrm{Hom}_{\underline{F}}(f^*(K), X) \end{array}$$

where the top map is induced by g and the bottom by $f^*(g)$. The top map is an isomorphism for all j-dense monics g if and only if $f_*(X)$ is a j-sheaf. The bottom map is an isomorphism for all objects X of $\underline{F}$ if and only if $f^*(g)$ is an isomorphism. It follows that if f factors through i, f^* takes j-dense monics to isomorphisms. Conversely, if f^* does this, then f_* factors through i_*, say

$$f_* = i_* u_*$$

where $u_* : \underline{F} \longrightarrow sh_j(\underline{E})$.

Let $u^* = f^* i_*$. Then u^* is left exact and left adjoint to u_* in virtue of the natural bijections.

$$\mathrm{Hom}_{sh_j(\underline{E})}(Y, u_*(X)) \simeq \mathrm{Hom}_{\underline{E}}(i_*(Y), f_*(X)) \simeq$$

$$\simeq \mathrm{Hom}_{\underline{F}}(f^* i_*(Y), X) \simeq \mathrm{Hom}_{\underline{F}}(u^*(Y), X).$$

Corollary 7.6 Let X be a Hausdorff space with no isolated points. Then

$$\mathrm{Points}\ (sh_{\neg\neg}(\mathrm{Top}(X))) = \phi\ .$$

Proof. For any $x \in X$, $X - \{x\}$ is open and dense in X. Hence the inclusion map [illegible] $\rightarrowtail X$ is a $\neg\neg$-dense monic. Now, if

$$\mathcal{S} \xrightarrow{x} \mathrm{Top}(X)$$

is the geometric morphism corresponding to the insertion of x, we have $x^*(X) = 1$, $x^*(X - \{x\}) = \phi$, so it cannot factor through

$$sh_{\neg\neg}(Top(X)) \longrightarrow Top(X).$$

Proposition 7.7 Let j be the topology on $\underline{E}$ induced by the geometric morphism $\underline{F} \xrightarrow{f} \underline{E}$. Then a monic $K \xrightarrow{g} L$ in $\underline{E}$ is j-dense if and only if $f^*(g)$ is an isomorphism.

Proof: As a corollary of theorem 7.5, since

$$sh_j(\underline{E}) \longrightarrow \underline{E}$$

factors through itself, the sheafification of a dense monic is an isomorphism, so g j-dense implies $f^*(g)$ is an isomorphism. Conversely, suppose that $f^*(g)$ is an isomorphism and that g has a classifying map $L \xrightarrow{\phi} \Omega$. We must show that $L \xrightarrow{\phi} \Omega \xrightarrow{j} \Omega$ factors through $1 \xrightarrow{t} \Omega$. Consider the commutative diagram

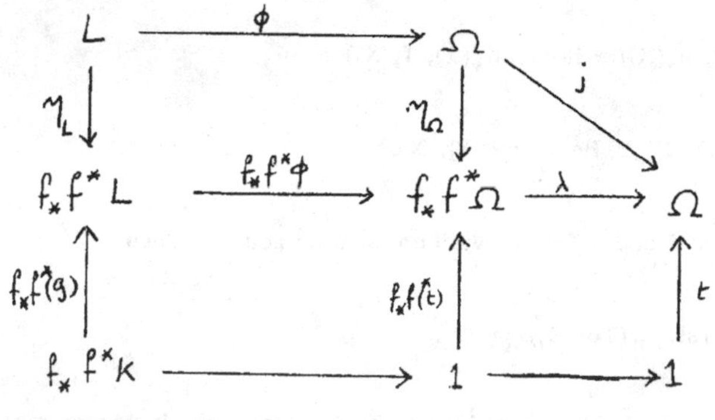

Since $f_*f^*(g)$ is an isomorphism, it is clear that $j.\phi$ factors through $1 \xrightarrow{t} \Omega$.

Theorem 7.8 Every geometric morphism

$$\underline{F} \xrightarrow{f} \underline{E}$$

can be factorized $\underline{F} \xrightarrow{a} \underline{H} \xrightarrow{b} \underline{E}$ where a^* reflects isomorphisms and b_* is full and faithful.

Proof. Let $\underline{H} \xrightarrow{b} \underline{E}$ be $sh_j(\underline{E}) \longrightarrow \underline{E}$ where j is the topology on $\underline{E}$ induced by f. By theorem 7.5 and proposition 7.7 f factorizes as ba. Suppose g is a monic in $\underline{H}$ such that $a^*(g)$ is an isomorphism. By proposition 7.7 $b_*(g)$ is j-dense, and so $g = b^*b_*(g)$ is an isomorphism. Let $X \xrightarrow{m} Y$ be any map in $\underline{H}$. Let

$$X \twoheadrightarrow I \xrightarrow{g_1} Y$$

be an epi-mono factorization of m, and let $K \underset{k_1}{\overset{k_0}{\rightrightarrows}} X$ be the kernel pair of m. As a subobject of $X \times X$ $K \xrightarrow{\langle k_0, k_1 \rangle} X \times X$ contains the diagonal $X \xrightarrow{\langle 1_X, 1_X \rangle} X \times X$. Let $X \xrightarrow{g_2} K$ be the inclusion. Then m is an isomorphism if and only if g_1 and g_2 are isomorphisms. Now a^* preserves epi-mono factorizations and kernel pairs, so if $f^*(m)$ is an isomorphism, so are $f^*(g_1)$ and $f^*(g_2)$. Hence g_1, g_2 are isomorphisms, so m is an isomorphism. We conclude that a^* reflects isomorphisms.

Note that the topologies on $\underline{E}$ induced by f and by b are the same, and that the topology induced by a on H is trivial. Dually, the left exact comonads on $\underline{F}$ induced by f and by a agree, and the left exact comonad induced by b on $\underline{H}$ is trivial. In the factorization we may regard $\underline{H}$ either as a category of coalgebras on $\underline{F}$ for the left exact comonad $f^* f_*$, or as $sh_j(\underline{E})$ where j is the topology on $\underline{E}$ induced by f.

<u>Proposition 7.9</u> Let f be a geometric morphism for which f^* reflects isomorphisms and f_* is full and faithful. Then f_*, f^* are adjoint equivalences.

Proof. Let μ , ε be the front and end adjunctions. Since f_* is full and faithful, ε is an isomorphism, so $f^*(\mu)$ is an isomorphism. As f^* reflects isomorphisms, μ is also an isomorphism.

As an immediate corollary of proposition 7.7 we have: -

<u>Proposition 7.10</u> Let f_1, f_2 be geometric morphisms with the same codomain. If f_{2*} is full and faithful, a necessary and sufficient condition that f_1 factor through f_2 is that f_1^* should invert every map inverted by f_2^*.

<u>Proposition 7.11</u> Let

$$\begin{array}{ccccc}
\underline{E}_1 & \xrightarrow{a_1} & \underline{F}_1 & \xrightarrow{b_1} & \underline{G}_1 \\
{\scriptstyle u}\downarrow & & \vdots{\scriptstyle v} & & \downarrow{\scriptstyle w} \\
\underline{E}_2 & \xrightarrow{a_2} & \underline{F}_2 & \xrightarrow{b_2} & \underline{G}_2
\end{array}$$

be a diagram of geometric morphisms, commuting up to natural isomorphisms, such that a_1^* , a_2^* reflect isomorphisms, and b_{1*}, b_{2*} are full and faithful. Then there is a geometric morphism v making the whole diagram commute up to natural isomorphism.

Proof: Apply proposition 7.10 to wb_1 and b_2. If α is a map in $\underline{G}_2$ such that $b_2^*(\alpha)$ is iso, then $u^*a_2^*\, b_2^*(\alpha) \simeq a_1^*\, b_1^*\, w^*(\alpha)$ is iso. Hence $b_1^*\, w^*(\alpha)$ is iso.

Corollary 7.12 The factorization of geometric morphisms into cotripleable morphisms followed by sheaf-inclusions is unique up to isomorphism.

Proof: Take u and w to be identity morphisms in proposition 7.11.

Examples

i) If $A \longrightarrow B$ is a map in $\underline{E}$ with epi-mono factorization $A \twoheadrightarrow I \rightarrowtail B$, then

$$\underline{E}/_A \longrightarrow \underline{E}/_I \longrightarrow \underline{E}/_B$$

is the factorization of $\underline{E}/_A \longrightarrow \underline{E}/_B$.

ii) If $X \xrightarrow{f} Y$ is a continuous map between topological spaces, and I denotes Im(f) with the subspace topology, then

$$\text{Top}(X) \longrightarrow \text{Top}(Y)$$

factorizes

$$\text{Top}(X) \longrightarrow \text{Top}(I) \longrightarrow \text{Top}(Y).$$

A historic example of factorization is given by that for the geometric morphism

$$S/_X \xrightarrow{\ f\ } \text{Presheaves}(X)$$

for a topological space X, where f^* assigns to a presheaf on X the X-indexed family of stalks, and f_* associates to a discrete space over X the presheaf of its sections. The factorization is

$$\begin{array}{ccc} S/_X & \longrightarrow & \text{Presheaves}(X)\,. \\ & \searrow \quad \text{Top}(X) \quad \nearrow & \end{array}$$

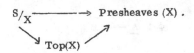

§ 8. Internal categories

In any category with finite limits we can define the notions of internal category and internal profunctor. An internal category $\underline{A}$ in $\underline{E}$ is given by objects A_0, A_1 (object of objects, object of maps), maps $A_1 \rightrightarrows A_0$ (domain, codomain), a map $A_0 \longrightarrow A_1$ (identity assignment) which splits domain and codomain, and a map (composition) $A_2 \xrightarrow{\mu} A_1$, where

$$\begin{array}{ccc} A_2 & \xrightarrow{p_2} & A_1 \\ {\scriptstyle p_1}\big\downarrow & & \big\downarrow{\scriptstyle \mathrm{dom}} \\ A_1 & \xrightarrow[\mathrm{cod}]{} & A_0 \end{array}$$

is a pullback diagram defining A_2 as the object of pairs of composable maps, such that

$$\begin{array}{ccccc} A_1 & \xleftarrow{p_1} & A_2 & \xrightarrow{p_2} & A_1 \\ {\scriptstyle \mathrm{dom}}\big\downarrow & & \big\downarrow{\scriptstyle \mu} & & \big\downarrow{\scriptstyle \mathrm{cod}} \\ A_1 & \xleftarrow[\mathrm{dom}]{} & A_1 & \xrightarrow[\mathrm{cod}]{} & A_0 \end{array}$$

commutes, and such that certain other diagrams commute, expressing associativity of composition and the laws satisfied by identity maps. We shall not dirty our hands here with the details. A smoother definition in terms of "spans" is given in Elementary Toposes § 5, page 85.

If $\underline{A}$ and $\underline{B}$ are internal categories in $\underline{E}$, an internal functor

$$\phi : \underline{A} \longrightarrow \underline{B}$$

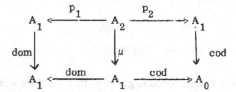

is given by maps $\phi_0 : A_0 \longrightarrow B_0$, $\phi_1 : A_1 \longrightarrow B_1$ such that appropriate diagrams commute. Again, we leave the details for the reader to make explicit himself.

Examples An internal category in

i) $\underline{S}$, is a small category ;

ii) $\underline{S}_{fin}$, is a finite category ;

iii) Top(X) , is a sheaf of categories ;

iv) G-sets , is a small category with G acting by automorphisms on it.

We are faced with a problem; how do we internalize the notion of a presheaf on a category? If $\underline{A}$ is an internal category in $\underline{E}$ what should we mean by a functor

$$\underline{A}^0 \longrightarrow \underline{E} \ ?$$

To answer this question, we first recall some category theoretic preliminaries. For any category $\underline{E}$ with finite limits, let Cat($\underline{E}$) denote the category of internal categories and internal functors in $\underline{E}$. In particular, Cat($\underline{S}$) we write as Cat.

For any $\underline{A} \in$ Cat, and functor $\underline{B} : \underline{A}^0 \longrightarrow$ Cat we construct a category

$$\mathcal{F}_{\underline{A}}(\underline{B})$$

as follows: -

The objects of $\mathcal{F}_{\underline{A}}(\underline{B})$ are pairs (A, X) where A is an object of $\underline{A}$ and X is an object of $\underline{B}$(A). A map

$$(A, X) \longrightarrow (A', X')$$

in $\mathcal{F}_{\underline{A}}(\underline{B})$ is a pair (a, x) where $A \xrightarrow{a} A'$ is a map in $\underline{A}$ and

$$X \xrightarrow{x} (\underline{B}(a))(X')$$

is a map in $\underline{B}(A)$. Maps in $\mathcal{F}_{\underline{A}}(\underline{B})$ are to be composed by the rule

$$(a', x').(a, x) = (a'a, (\underline{B}(a))(x').x).$$

This formula should remind the reader of that for semi-direct products of groups. Indeed, if $\underline{A}$ is a group and $\underline{B}$ takes values in groups, then $\underline{B}$ is simply a group with a homomorphism $\underline{A} \longrightarrow \mathrm{Aut}(\underline{B})$, and $\mathcal{F}_{\underline{A}}(\underline{B})$ is the semi-direct product.

We have a functor

$$p : \mathcal{F}_{\underline{A}}(\underline{B}) \longrightarrow \underline{A} \; : \; (a, x) \longrightarrow a$$

which we call the split fibration associated to $\underline{B} : \underline{A}^0 \longrightarrow \mathrm{Cat}$. We call $\mathcal{F}_{\underline{A}}(\underline{B})$ the total category of the split fibration.

A natural map $\underline{B} \longrightarrow \underline{B}'$ gives rise in an obvious way to a commutative diagram

$$\begin{array}{ccc} \mathcal{F}_{\underline{A}}(\underline{B}) & \longrightarrow & \mathcal{F}_{\underline{A}}(\underline{B}') \\ & {\scriptstyle p}\searrow \quad \swarrow{\scriptstyle p'} & \\ & \underline{A} & \end{array}$$

so that we have a functor

$$\mathrm{Cat}^{\underline{A}^0} \longrightarrow \mathrm{Cat}/_{\underline{A}}$$

generally known as "the Grothendieck construction".

A functor

$$\underline{A}' \xrightarrow{F} \underline{A}$$

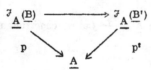

gives rise to a pullback diagram in Cat

$$\begin{array}{ccc} \mathcal{F}_{\underline{A}'}(B\ F) & \longrightarrow & \mathcal{F}_{\underline{A}}(B) \\ \downarrow & & \downarrow \\ \underline{A}' & \xrightarrow{\quad F \quad} & \underline{A} \end{array} ,$$

from which it follows that split fibrations are preserved by pullback.

A presheaf on $\underline{A}$, $\underline{A}^0 \xrightarrow{k} S$, gives rise to a functor $\underline{A}^0 \xrightarrow{k} S \longrightarrow \mathrm{Cat}$, where $S \longrightarrow \mathrm{Cat}$ is the functor which associates to a set the corresponding discrete category. By abuse of language, we call this functor k. The corresponding split fibration

$$\mathcal{F}_{\underline{A}}(k) \longrightarrow \underline{A}$$

we call a <u>discrete fibration</u>. Clearly, a split fibration

$$\underline{B} \xrightarrow{\phi} \underline{A}$$

is discrete if and only if the fibres of ϕ are discrete categories, i.e. if for every $A \in \underline{A}$, $\phi^{-1}(1_A)$ is a discrete category.

<u>Proposition 8.1</u> A functor $\underline{B} \xrightarrow{\phi} \underline{A}$ is a discrete fibration if and only if

$$\begin{array}{ccc} B_1 & \xrightarrow{\mathrm{cod}} & B_0 \\ {\scriptstyle \phi_1}\downarrow & & \downarrow{\scriptstyle \phi_0} \\ A_1 & \xrightarrow{\mathrm{cod}} & A_0 \end{array}$$

is a pullback diagram.

This proposition is very convenient because it enables us to define discrete fibrations in any category with finite limits.

Proposition 8.2 The category of presheaves on $\underline{A}$ is equivalent to the full subcategory of $Cat/\underline{A}$ of discrete fibrations.

To prove proposition 8.2 we need to show how to associate a presheaf on $\underline{A}$ to any discrete fibration $\underline{B} \xrightarrow{\phi} \underline{A}$. We define $\underline{A}^0 \xrightarrow{k} S$ as follows: $k(A) = \{B \in \underline{B}/\phi(B) = A\}$; for any map $A' \xrightarrow{a} A$ in $\underline{A}$ and $B \in k(A)$ there is a unique element $b \in B_1$ such that $cod(b) = B$ and $\phi_1(b) = a$, in virtue of proposition 8.1. We define $k(a)(B)$ to be $dom(b)$.

We have now answered the question we posed above. If $\underline{A}$ is an internal category in $\underline{E}$, a functor $\underline{A}^0 \xrightarrow{k} \underline{E}$ is to be interpreted as a discrete fibration $\underline{B} \xrightarrow{\phi} \underline{A}$, i.e. an internal functor for which the diagram of proposition 8.1 is a pullback. We denote by

$$\underline{E}^{\underline{A}^0}$$

the full subcategory of $Cat(\underline{E})\ \underline{A}$ of discrete fibrations.

Theorem 8.3 If $\underline{A}$ is an internal category in an elementary topos $\underline{E}$, then $\underline{E}^{\underline{A}^0}$ is an elementary topos.

Proof: In Elementary toposes, it is shown how the category structure of $\underline{A}$ makes the composite

$$\underline{E}/A_0 \xrightarrow{(cod)^*} \underline{E}/A_1 \xrightarrow{\Sigma_{dom}} \underline{E}/A_0$$

into a monad on $\underline{E}/_{A_0}$. It has a right adjoint

$$E/_{A_0} \xrightarrow{(\mathrm{dom})^*} \underline{E}/_{A_1} \xrightarrow{\Pi\,\mathrm{cod}} \underline{E}/_{A_0}$$

which is therefore a left exact comonad on $\underline{E}/_{A_0}$. Let us denote it by C. A discrete fibration $\underline{B} \xrightarrow{\phi} \underline{A}$ is determined by the object

$$B_0 \xrightarrow{\phi_0} A_0$$

in $\underline{E}/_{A_0}$ together with the map $B_1 \xrightarrow{\mathrm{dom}} B_0$, such that various diagrams commute, where B_1 is defined by the pullback diagram

$$\begin{array}{ccc} B_1 & \xrightarrow{\mathrm{cod}} & B_0 \\ {\scriptstyle \phi_1}\downarrow & & \downarrow{\scriptstyle \phi_0} \\ A_1 & \xrightarrow{\mathrm{cod}} & A_0 \end{array}$$

i.e. $\phi_1 = \mathrm{cod}^*(\phi_0)$. But these conditions state precisely that (ϕ_0, dom) be an algebra for the monad mentioned above, or equivalently that ϕ_0 be given a C-coalgebra structure. Thus $\underline{E}^{\underline{A}^0}$ is equivalent to $(\underline{E}/_{A_0})_C$, and so by theorem 5.1 is an elementary topos.

Examples

i) For any object X of $\underline{E}$ we have the discrete category $\underline{X}$ given by $X_0 = X_1 = X$, with 1_X for both domain and codomain maps. Clearly we have

$$\underline{E}^{\underline{X}^0} \simeq \underline{E}/_X .$$

ii) For any monad object G of $\underline{E}$ we have the internal category $\underline{G}$ given by $G_0 = 1$, $G_1 = G$. We may identify $\underline{E}^{\underline{G}^0}$ with the category of left G-objects and $\underline{E}^{\underline{G}}$ with the category of right G-objects.

We saw above that split fibrations were preserved under pullback. A minor modification to the argument shows that discrete fibrations are preserved under pullback Hence, if

$$\underline{A} \xrightarrow{F} \underline{B}$$

is an internal functor, we get a functor

$$\underline{E}^{\underline{B}^0} \xrightarrow{F^*} \underline{E}^{\underline{A}^0}$$

by pullback along F.

<u>Theorem 8.4</u> The functor $\underline{E}^{\underline{B}^0} \xrightarrow{F^*} \underline{E}^{\underline{A}^0}$ has a left adjoint $F_!$ and a right adjoint F_*.

The proof follows from what is set out in the appendix of Elementary Toposes. This appendix constructs the bicategory Prof($\underline{E}$) of internal categories and internal profunctors and shows that it is biclosed, i.e. that profunctor composition has a right adjoint. The category of profunctors from $\underline{A}$ to $\underline{B}$ is simple $\underline{E}^{(\underline{A}^0 \times \underline{B})}$. Thus, an internal functor $\underline{A} \xrightarrow{F} \underline{B}$ gives a geometric morphism

$$\underline{E}^{\underline{A}^0} \xrightarrow{F} \underline{E}^{\underline{B}^0} \quad .$$

In particular, for any internal category $\underline{A}$ we have the internal functor $\underline{A} \xrightarrow{c} \underline{1}$ to the discrete category on 1, which is terminal. This gives a geometric morphism

$$\underline{E}^{\underline{A}^0} \xrightarrow{c} \underline{E}$$

whereby we consider $\underline{E}^{\underline{A}^0}$ as an $\underline{E}$ - topos. The assignment

$$(\underline{A} \xrightarrow{F} \underline{B}) \longmapsto (\underline{E}^{\underline{A}^0} \xrightarrow{F} \underline{E}^{\underline{B}^0})$$

gives a functor

$$\mathrm{Cat}(\underline{E}) \longrightarrow \mathrm{Top}_{\underline{E}} \; .$$

For any object X of $\underline{E}$, $c^*(X)$ is the discrete fibration $\underline{A} \times X \xrightarrow{p_1} \underline{A}$ representing the constant presheaf on $\underline{A}$ taking the value X. We may thus interpret the left and right adjoints $c_!$, c_* of c^* as $\varinjlim_{\underline{A}^0}$ and $\varprojlim_{\underline{A}^0}$ respectively.

Proposition 8.5 If $\underline{B} \xrightarrow{\phi} \underline{A}$ is a discrete fibration representing an internal presheaf $\underline{A}^0 \xrightarrow{K} \underline{E}$ then $\varinjlim_{\underline{A}^0}(K)$ is the coequalizer of

$$B_1 \underset{\text{cod}}{\overset{\text{dom}}{\rightrightarrows}} B_0 \; .$$

we leave the proof to the reader.

The Grothendieck construction gave for any $\underline{A} \in$ Cat, a functor

$$\mathrm{Cat}(S^{\underline{A}^0}) \longrightarrow \mathrm{Cat}(S)/\underline{A} \, ,$$

since a functor $\underline{A}^0 \longrightarrow$ Cat is simply an internal category in $S^{\underline{A}^0}$. It is not hard to see that for any category $\underline{E}$ with finite limits, the Grothendieck construction generalizes to

$$\mathrm{Cat}(\underline{E}^{\underline{A}^0}) \longrightarrow \mathrm{Cat}(\underline{E})/_{\underline{A}}$$

for any internal category $\underline{A}$ in $\underline{E}$. Suppose

$$\underline{B} \in \mathrm{Cat}(\underline{E}^{\underline{A}^0})$$

i.e. that we have a diagram

$$\begin{array}{ccc} \underline{B}_1 & \overset{\text{dom}}{\underset{\text{cod}}{\rightrightarrows}} & \underline{B}_0 \\ & {\scriptstyle \phi_1} \searrow \quad \swarrow {\scriptstyle \phi_0} & \\ & \underline{A} & \end{array}$$

of discrete fibrations over $\underline{A}$. Then $\mathcal{F}_{\underline{A}}(\underline{B}) \longrightarrow \underline{A}$ is given by

$$\mathcal{F}_{\underline{A}}(\underline{B})_0 = (B_0)_0$$

$$\mathcal{F}_{\underline{A}}(\underline{B})_1 = (B_1)_1$$

and we may write down the maps defining the category structure of $\mathcal{F}_{\underline{A}}(\underline{B})$ in terms of the data for $\underline{B}$. For those who like simplicial objects, identifying a category with a simplicial object via the nerve functor, gives that $\underline{B}$ is a bisimplicial object augmented toward $\underline{A}$. Taking the diagonal simplicial object of $\underline{B}$ gives $\mathcal{F}_{\underline{A}}(\underline{B})$.

Proposition 8.6 Let $\underline{A}$ be an internal category in an elementary topos $\underline{E}$, and let $\underline{B}$ be an internal category in $\underline{E}^{\underline{A}^0}$. Let $\mathcal{F}_{\underline{A}}(\underline{B}) \xrightarrow{p} \underline{A}$ be the associated split fibration in $\mathrm{Cat}(\underline{E})$. Then there is an equivalence of categories

$$(\underline{E}^{\underline{A}^0})^{\underline{B}^0} \simeq \underline{E}^{\mathcal{F}_{\underline{A}}(\underline{B})^0}$$

such that the diagram

$$\begin{array}{ccc} (\underline{E}^{\underline{A}^0})^{\underline{B}^0} & \xrightarrow{\sim} & \underline{E}^{\mathcal{F}_{\underline{A}}(\underline{B})^0} \\ & \searrow \quad \swarrow {\scriptstyle p} & \\ & \underline{E}^{\underline{A}^0} & \end{array}$$

commutes.

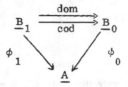

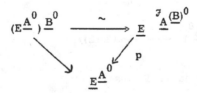

We omit the proof. The only difficulties are ones of formalism. The method of bisimplicial objects probably gives the neatest proof. Alternatively, prove it for $\underline{E} = S$, where it is straightforward, and then note that all the constructions involve nothing worse than pullback diagrams.

For any internal category $\underline{A}$ in $\underline{E}$ we have a special functor

$$\underline{A}^0 \times \underline{A} \xrightarrow{\text{Hom}_{\underline{A}}} \underline{E}$$

given by a discrete fibration

$$\underline{\text{Hom}}_{\underline{A}} \xrightarrow{h} \underline{A} \times \underline{A}^0$$

where $(\underline{\text{Hom}}_{\underline{A}})_0 = A_1$ and $(\underline{\text{Hom}}_{\underline{A}})_1 = A_3$, the object of triples of composable maps. The map

$$\text{dom} : (\underline{\text{Hom}}_{\underline{A}})_1 \longrightarrow (\underline{\text{Hom}}_{\underline{A}})_0$$

is given by the map $A_3 \longrightarrow A_1$ which composes all the maps together, and the map

$$\text{cod} : (\underline{\text{Hom}}_{\underline{A}})_1 \longrightarrow (\underline{\text{Hom}}_{\underline{A}})_0$$

is given by the projection $A_3 \longrightarrow A_1$ to the middle factor.

The map $h_0 : (\underline{\text{Hom}}_{\underline{A}})_0 \longrightarrow (\underline{A} \times \underline{A}^0)_0$ is

$$A_1 \xrightarrow{\langle \text{dom}, \text{cod} \rangle} A_0 \times A_0$$

and $h_1 : (\underline{\text{Hom}}_{\underline{A}})_1 \longrightarrow (A \times \underline{A}^0)_1$ is given by the map $A_3 \longrightarrow A_1 \times A_1$ projecting onto the first and third factors.

$\underline{\mathrm{Hom}}_{\underline{A}}$ is the "twisted morphism" category, and $\underline{\mathrm{Hom}}_{\underline{A}} \xrightarrow{p} \underline{A} \times \underline{A}^0$ is the identity profunctor from $\underline{A}$ to itself.

Consider the commutative diagram

$$\begin{array}{ccc} \underline{\mathrm{Hom}}_{\underline{A}} & \xrightarrow{\quad h \quad} & \underline{A} \times \underline{A}^0 \\ & {\scriptstyle U_{\underline{A}}} \searrow \quad \swarrow {\scriptstyle p_1} & \\ & \underline{A} & \end{array}$$

Now $U_{\underline{A}}$ and p_1 are split fibrations; in fact p_1 represents the category object $c^*(\underline{A}^0)$ in $\underline{E}^{\underline{A}^0}$. We assert that in $\underline{E}^{\underline{A}^0}$,

$$\underline{\mathrm{Hom}}_{\underline{A}} \xrightarrow{h} \underline{A} \times \underline{A}^0$$

defines a discrete fibration

$$U_{\underline{A}} \xrightarrow{h} c^*(\underline{A}^0)$$

and hence a functor

$$c^*(\underline{A}) \longrightarrow \underline{E}^{\underline{A}^0} .$$

This functor "is" the Yoneda embedding.

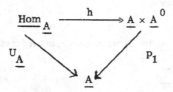

§ 9. The Diaconescu Theorem

A category $\underline{A}$ is called filtered if

(i) it is nonempty,

(ii) for every pair of objects A_1, A_2 of $\underline{A}$ there is a diagram

$$A_1 \longrightarrow A_3 \longleftarrow A_2 ,$$

(iii) for every pair of maps $A \overset{a_1}{\underset{a_2}{\rightrightarrows}} A'$ of $\underline{A}$ there is a map $A' \xrightarrow{a} A''$ such that

$$aa_1 = aa_2 .$$

We call $\underline{A}$ cofiltered if $\underline{A}^0$ is filtered.

The condition that a category be filtered is an elementary statement in the first order language of category theory, and so is interpretable in any elementary topos. In fact, each of the conditions above can be expressed by saying that a certain map is epic: -

(i) $A_0 \longrightarrow 1$

(ii) $P \longrightarrow A_1 \times A_1 \xrightarrow{\text{dom} \times \text{dom}} A_0 \times A_0$ where $P \rightrightarrows A_1$ is the kernel pair of $A_1 \xrightarrow{\text{cod}} A_0$,

(iii) we leave as an exercise for the reader.

Notice that if $\underline{F} \xrightarrow{f} \underline{E}$ is a geometric morphism, and $\underline{A} \in \text{Cat}(\underline{E})$, then $f^*(\underline{A}) \in \text{Cat}(\underline{F})$, and if $\underline{A}$ is filtered, so is $f^*(\underline{A})$.

We call a presheaf $\underline{A}^0 \longrightarrow \underline{E}$ flat if the total category of the associated discrete fibration is filtered.

Example The internal category

$$\underline{\mathrm{Hom}}_{\underline{A}} \xrightarrow{U_{\underline{A}}} \underline{A}$$

in $\underline{E}^{\underline{A}^0}$ is cofiltered. To see this, note that for $\underline{E} = S$, each fibre of $U_{\underline{A}}$ has an initial object and so is cofiltered.

It follows that the Yoneda embedding

$$U_{\underline{A}} \xrightarrow{h} c^*(\underline{A}^0)$$

is flat.

If $\underline{F} \xrightarrow{p} \underline{E}$ is an $\underline{E}$-topos, we denote by

$$\mathrm{Mod}(\underline{A}, \underline{F})$$

the full subcategory of $\underline{F}^{p^*(\underline{A})}$ of flat $p^*(\underline{A})^0$ - presheaves. We call the objects of $\mathrm{Mod}(\underline{A}, \underline{F})$ $\underline{A}$ - models in $\underline{F}$. We call the Yoneda embedding

$$c^*(\underline{A}) \longrightarrow \underline{E}^{\underline{A}^0}$$

the universal $\underline{A}$ - model. It lives in $\underline{E}^{\underline{A}^0}$.

A morphism of $\underline{E}$-topoi

$$\underline{F}_1 \xrightarrow{g} \underline{F}_2$$

induces, via g^*, a functor

$$\mathrm{Mod}(\underline{A}, \underline{F}_2) \longrightarrow \mathrm{Mod}(\underline{A}, \underline{F}_1) .$$

Theorem 9.1 (Diaconescu)

Let $\underline{E}$ be an elementary topos, and $\underline{A} \in \mathrm{Cat}(\underline{E})$. For any $\underline{E}$-topos $\underline{F}$ and $\underline{A}$-model X in $\underline{F}$, there is a unique morphism of $\underline{E}$-topoi

$$\underline{F} \xrightarrow{\phi} \underline{E}^{\underline{A}^0}$$

such that $X = \phi^*(U_{\underline{A}})$, where $U_{\underline{A}}$ denotes the universal $\underline{A}$-model. In other words, there is an equivalence of categories

$$\mathrm{Top}_{\underline{E}}(\underline{F}, \underline{E}^{\underline{A}^0}) \simeq \mathrm{Mod}(\underline{A}, \underline{F}) .$$

We call the morphism of $\underline{E}$-topoi ϕ the classifying morphism of X. The theorem states that $\underline{E}^{\underline{A}^0}$ classifies $\underline{A}$-models for $\underline{E}$-topoi.

For the proof we refer the reader to Diaconsecu's thesis. It is based on the fact that $\mathrm{Hom}_{\underline{A}} \xrightarrow{h} \underline{A} \times \underline{A}^0$ is the unit profunctor and the proposition that a presheaf is flat if and only if profunctor composition with it is a left exact process.

Examples

(i) Let $\underline{X}$ be a discrete internal category in $\underline{E}$, on an object X. Then for any $\underline{E}$-topos $\underline{F} \xrightarrow{p} \underline{E}$ we find

$$\mathrm{Mod}(\underline{X}, \underline{F}) = \mathrm{Hom}_{\underline{F}}(1, p^*(X)).$$

This gives the well known result

$$\mathrm{Top}_{\underline{E}}(\underline{F}, \underline{E}/X) \simeq \mathrm{Hom}_{\underline{F}}(1, p^*(X)) .$$

The universal $\underline{X}$-model is the global section of the object $X \times X \xrightarrow{p_1} X$ in $\underline{E}/X$ given by the diagonal map $X \xrightarrow{<1_X, 1_X>} X \times X$.

(ii) If $\underline{E} = S$ and T denotes a finitary algebraic theory, let f. p. T-mod denote the category of finitely presented models of T. Note that this is a small category. Let

$$\underline{\underline{\text{T - mod}}}$$

denote the category of functors and natural maps

$$\text{f. p. T-mod} \longrightarrow S$$

and let $U_T \in \underline{\underline{\text{T-mod}}}$ denote the forgetful functor. Clearly, U_T is a T-model in $\underline{\underline{\text{T-mod}}}$. It is the universal T-model, and $\underline{\underline{\text{T-mod}}}$ classifies T-models for S-topoi.

The case for T = (commutative rings) is dealt with by M. Hakim in her book "Topos Anneles et schemas Relatifs".

(iii) A particularly interesting case of (ii) arises by considering the initial theory, i. e. the trivial theory whose models are simply objects with no further structure. A finitely presented model in S is simply a finite set. We get that

$$S^{S_{fin}}$$

is an object classifier for S-topoi.

A natural number object (NNO) in an elementary topos $\underline{E}$ is an object N together with maps $1 \xrightarrow{0} N$, $N \xrightarrow{s} N$ such that given any diagram

$$1 \xrightarrow{x} X \xrightarrow{t} X$$

in $\underline{E}$, there exists a unique map

$$N \xrightarrow{h} X$$

making the diagram

$$\begin{array}{ccccc} & & N & \xrightarrow{s} & N \\ & \overset{0}{\nearrow} & \Big\downarrow h & & \Big\downarrow h \\ 1 & & & & \\ & \underset{x}{\searrow} & & & \\ & & X & \xrightarrow{t} & X \end{array}$$

commute.

J. Benabou has shown how to construct in a topos $\underline{E}$ with an NNO an internal category $\underline{E}_{fin}$, which plays for $\underline{E}$ the same role that S_{fin} plays for S. The author has shown that $\underline{E}^{\underline{E}_{fin}}$ is an object classifier for $\underline{E}$-topoi. More recently, P. Johnstone has shown that if $\underline{E}$ is a elementary topos with an NNO and if T is a finitary finitely presented algebraic theory (i.e. described by a finite number of generating operations, satisfying a finite number of axioms) then one may construct in $\underline{E}$ the internal category of finitely presented T-models in $\underline{E}$. That the theory be finitary is necessary, since inverse image parts of geometric morphisms only preserve finite limits. That the theory should be finitely presented is not surprising - we would expect only those infinities which are "internal to $\underline{E}$" to be allowed.

<u>Corollary 9.2</u> (Diaconescu)

Let $\underline{F} \xrightarrow{f} \underline{E}$ be a geometric morphism and $\underline{A} \in \mathrm{Cat}(\underline{E})$. Then

$$\begin{array}{ccc} \underline{F}^{f^*(\underline{A})^0} & \xrightarrow{f'} & \underline{E}^{\underline{A}^0} \\ c' \Big\downarrow & & \Big\downarrow c \\ \underline{F} & \xrightarrow{f} & \underline{E} \end{array}$$

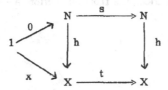

is a pullback diagram in the category of $\underline{E}$-topoi. The geometric morphism f' is defined as follows: $f'*$ "is" $f*$. If $\underline{B} \xrightarrow{\phi} f*(\underline{A})$ is a discrete fibration in $\underline{F}$, then $f'_*(\phi)$ is obtained by pulling back $f_*(\phi)$ along the front adjunction $\underline{A} \longrightarrow f_* f^*(\underline{A})$.

Proof:

Let $\underline{H} \xrightarrow{a} \underline{F}$, $\underline{H} \xrightarrow{b} \underline{E}^{\underline{A}^0}$ be geometric morphisms such that $fa = cb$. Then b defines an $\underline{A}$ - model in $\underline{H}$,

$$a*f*(\underline{A}) \longrightarrow \underline{H}$$

so there exists a unique geometric morphism $\underline{H} \xrightarrow{q} \underline{F}^{f*(\underline{A})^0}$ classifying it, such that $c'q = a$. To prove that $f'q = b$ it is enough to remark that

$$f'*(U_{\underline{A}}) = U_{f*(\underline{A})} .$$

In his thesis Diaconescu also shows that given a geometric morphism $\underline{F} \xrightarrow{f} \underline{E}$ and a topology j on $\underline{E}$, then there exists a topology j' on $\underline{F}$, definable in terms of f and j, giving a pullback diagram

$$
\begin{array}{ccc}
sh_{j'}(\underline{F}) & \longrightarrow & sh_j(\underline{E}) \\
\downarrow & & \downarrow \\
\underline{F} & \xrightarrow{f} & \underline{E}
\end{array}
$$

in the category of $\underline{E}$ - topoi.

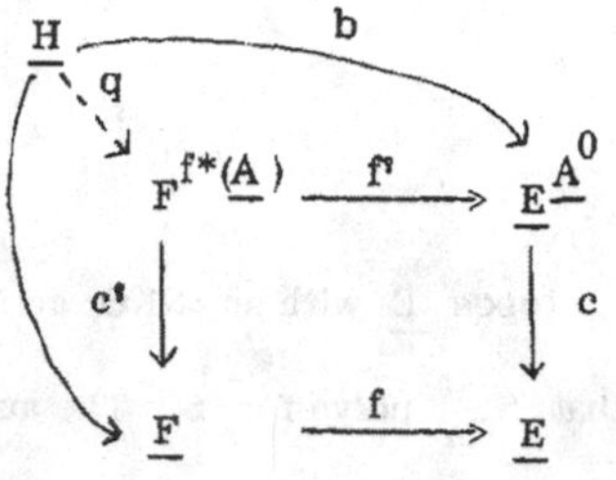

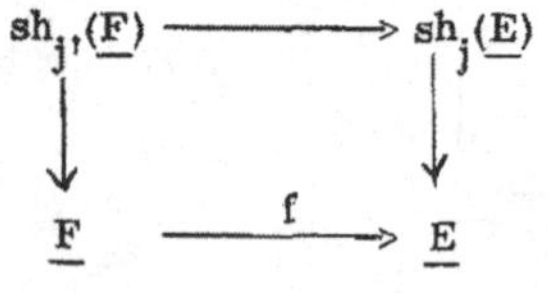

Definition 9.3 A geometric morphism

$$\underline{F} \xrightarrow{\ f\ } \underline{E}$$

is bounded if there exists $\underline{A} \in \mathrm{Cat}(\underline{E})$ and a factorization of f

$$\begin{array}{ccc} \underline{F} & \longrightarrow & \underline{E} \\ & {\scriptstyle i}\searrow \quad \swarrow & \\ & \underline{E}^{\underline{A}^0} & \end{array}$$

where i_* is full and faithful. We say that f makes $\underline{F}$ into a bounded $\underline{E}$-topos.

Thus, an $\underline{E}$-topos is bounded if it is equivalent to one of the form $sh_j(\underline{E}^{\underline{A}^0})$ for some $\underline{A} \in \mathrm{Cat}(\underline{E})$ and some topology j on $\underline{E}^{\underline{A}^0}$. Thus, a Grothendieck topos is a bounded S-topos.

Example

Let G be an infinite profinite group. Then the category of finite G-sets with continuous action is an S_{fin}-topos but is not a bounded S_{fin}-topos.

Diaconescu has proved a generalization of Giraud's theorem which states that an $\underline{E}$-topos $\underline{F} \xrightarrow{\ f\ } \underline{E}$ is bounded if and only if there exists a generating object in $\underline{F}$. This means that there exists an object G in $\underline{F}$ such that for every object X of $\underline{F}$, the natural map

$$f^* f_*(\tilde{X}^G) \times G \longrightarrow \tilde{X}$$

obtained by using the end adjunction and evaluation, is epic. Here $\tilde{X}$ denotes the classifier of partial maps into X, and is defined as the equalizer of

$$\Omega^X \underset{\xi}{\overset{1}{\rightrightarrows}} \Omega^X$$

where ξ is exponentially adjoint to the classifier

$$\Omega^X \times X \longrightarrow \Omega$$

of $X \xrightarrow{<\{\cdot\},\, 1>} \Omega^X \times X.$

As a corollary of this theorem, Diaconescu proves

<u>Proposition 9.4</u> Let the composite

$$\underline{E} \longrightarrow \underline{F} \longrightarrow \underline{G}$$

be bounded. Then $\underline{E} \longrightarrow \underline{F}$ is bounded.

We also have as a consequence of Diaconescu's work that pullbacks along bounded geometric morphisms exist and preserve bounded geometric morphisms.

<u>Proposition 9.5</u> A composite of bounded geometric morphisms is bounded.

Proof : Consider the diagram

$$\begin{array}{ccccccccc}
\underline{H} & \xrightarrow{\lambda} & \underline{F}^{\underline{B}^0} & \longrightarrow & \underline{F} & \xrightarrow{i} & \underline{E}^{\underline{A}^0} & \longrightarrow & \underline{E} \\
 & & & \searrow^{\sim} & \uparrow & & \uparrow & & \uparrow \\
 & & & & \underline{F}^{i^* i_* \underline{B}^0} & \xrightarrow{i'} & (\underline{E}^{\underline{A}^0})^{i_*(\underline{B})^0} & \xrightarrow{\sim} & \underline{E}^{\exists_{\underline{A}}(i_* \underline{B})^0}
\end{array}$$

where λ_*, i_* are full and faithful. We need the fact that i'_* is full and faithful because the centre square is a pullback.

§ 10. Local equivalence

Let $\underline{F} \xrightarrow{p} \underline{E}$ be a bounded geometric morphism. Pullback along p defines a functor

$$\mathrm{Top}_{\underline{E}} \longrightarrow \mathrm{Top}_{\underline{F}} \ .$$

We say that two $\underline{E}$-topoi are locally equivalent if there exists $K \in \underline{E}$, with $K \longrightarrow 1$ epic, such that under pullback along

$$\underline{E}/_K \longrightarrow \underline{E}$$

the two become equivalent $\underline{E}/_K$ - topoi.

Proposition 10.1

Let $\underline{F}_1 \xrightarrow{f} \underline{F}_2$ be a morphism of $\underline{E}$-topoi. If there exists $K \in \underline{E}$, with $K \longrightarrow 1$ epic, such that under pullback along $\underline{E}/_K \longrightarrow \underline{E}$, f becomes an equivalence, then f is already an equivalence.

Proof. Let $\underline{F}_i \xrightarrow{p_i} \underline{E}$ $(i = 1, 2)$ be the structural morphisms. The pullback of f is

$$\underline{F}_1/_{p_1^*(K)} \xrightarrow{f'} \underline{F}_2/_{p_2^*(K)} \ .$$

Consider the front and end adjunctions of f. Under pullback along $p_2^*(K) \longrightarrow 1$ and $p_1^*(K) \longrightarrow 1$ respectively they become isomorphisms. By proposition 2.12, they are isomorphisms to begin with.

The same argument applied only to the end adjunction shows that a morphism of $\underline{E}$-topoi which is locally a sheaf-inclusion is a sheaf-inclusion.

<u>Proposition 10. 2</u> An $\underline{E}$-topos locally equivalent to $sh_j(\underline{E})$ is equivalent to $sh_j(\underline{E})$.

Proof : Let $\underline{F} \xrightarrow{p} \underline{E}$ be locally equivalent to $sh_j(\underline{E}) \xrightarrow{i} \underline{E}$. Form the pullback

$$\begin{array}{ccc} sh_{j'}(\underline{F}) & \xrightarrow{p'} & sh_j(\underline{E}) \\ {\scriptstyle i'}\downarrow & & \downarrow{\scriptstyle i} \\ \underline{F} & \xrightarrow{p} & \underline{E} \end{array} .$$

But i' and p' are locally equivalences, i.e. there exists $K \in \underline{E}$, with $K \longrightarrow 1$ epic, such that pullback along $\underline{E}/_K \longrightarrow \underline{E}$ takes p and i into the same geometric morphism. Since the pullback of a sheaf-inclusion along itself is an identity morphism we get that i' and p' are identity morphisms.

<u>Proposition 10. 3</u> Let $\underline{A} \xrightarrow{F} \underline{B}$ be an internal functor in $\underline{E}$ which is locally an equivalence of internal categories. Then F is full and faithful and

$$\underline{E}^{\underline{A}^0} \xrightarrow{F} \underline{E}^{\underline{B}^0}$$

is equivalence of $\underline{E}$ - topoi.

Proof We say that an internal functor F is full and faithful if

$$\begin{array}{ccc} A_1 & \xrightarrow{F_1} & B_1 \\ {\scriptstyle \langle dom, cod \rangle}\downarrow & & \downarrow{\scriptstyle \langle dom, cod \rangle} \\ A_0 \times A_0 & \xrightarrow{F_0 \times F_0} & B_0 \times B_0 \end{array}$$

is a pullback diagram. Pullback along epics reflects pullback diagrams. Proposition 10. 1 proves the last part.

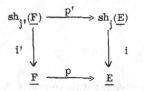

If K is an object of a spatial topos such that $K \longrightarrow 1$ is epic, then there is an open covering $\{U_i\}_{i \in I}$ of the space and an epic map

$$\coprod_{i \in I} U_i \longrightarrow K$$

so that for spatial topoi the phrase "locally" has its usual meaning, i.e. "on some open cover".

We call two objects X_1, X_2 of $\underline{E}$ locally isomorphic if $\underline{E}/X_1 \longrightarrow \underline{E}$ and $\underline{E}/X_2 \longrightarrow \underline{E}$ are locally equivalent. For example, any two vector bundles on a topological space, of the same dimension, are locally isomorphic (that is to say, their sheaves of sections are locally isomorphic).

Definition 10.4 Let G be a group object in an elementary topos $\underline{E}$. Let M be a right G-object with action $M \times G \xrightarrow{\mu} M$. Then M is a right G-torsor if

i) $M \longrightarrow 1$ is epic,

ii) $M \times G \xrightarrow{\langle p_1, \mu \rangle} M \times M$ is an isomorphism.

Proposition 10.5 A functor $\underline{G} \longrightarrow \underline{E}$ is flat if and only if the right G-object it determines is a G-torsor.

It follows that $\underline{E}^{\underline{G}^0}$ classifies right G-torsors. If

$$\underline{E}^{\underline{G}^0} \xrightarrow{c} \underline{E}$$

is the structural morphism, $c^*(G)$ is G with trivial G-action. G considered as a left G-object via multiplication is a right $c^*(G)$-torsor, the universal one.

Examples

i) Let S^1 denote a circle. In Top(S^1) let Z_2 denote the constant sheaf on the cyclic group of order 2. In pictures

Z_2 =

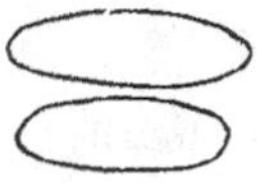

Then the double covering

is a Z_2-torsor.

ii) Let X be a topological space with a universal covering space $\tilde{X} \longrightarrow X$. Let X also be connected, and let $\pi_1(X)$ denote the constant sheaf on the fundamental group of X. Then $\tilde{X}$ is a $\pi_1(X)$ - torsor.

For any group G, G itself, with right G-action, is a right G-torsor. We call it the trivial G-torsor. From the definition of G-torsor it follows that any G-torsor is locally isomorphic as a G-object to the trivial G-torsor. Hence, any two G-torsors are locally isomorphic.

It may easily be verified that in S, any G-torsor is isomorphic as a G-object to the trivial G-torsor.

Proposition 10.6 Mod($\underline{G}$, $\underline{E}$) is a groupoid.

Proof: Considering G as a right G-object, let

$$G \xrightarrow{\phi} G$$

be a map of right G-objects. If $1 \xrightarrow{e} G$ is the unit element of G, let

$$1 \xrightarrow{g} G = 1 \xrightarrow{e} G \xrightarrow{\phi} G$$

and let

$$G \xrightarrow{\psi} G = G \xrightarrow{g^{-1} \times 1} G \times G \xrightarrow{\mu} G .$$

Then ϕ and ψ are inverse isomorphisms. Hence, any G-endomorphism of the trivial G-torsor is an automorphism. Any two G-torsors are locally trivial, and any map which is locally an isomorphism is an isomorphism.

Proposition 10.7 A torsor with an element is trivial.

Proof: Let M be a right G-torsor, and let

$$1 \xrightarrow{u} M$$

be a map. Then

$$G \xrightarrow{u \times 1} M \times G \xrightarrow{\mu} M$$

is a G-map, and so is an isomorphism.

If $\underline{A}$, $\underline{B} \in \mathrm{Cat}(\underline{E})$ then corollary 9.2. tells us that $\underline{E}^{\underline{A}^0} \times \underline{E}^{\underline{B}^0}$ is the product of $\underline{E}^{\underline{A}^0}$ and $\underline{E}^{\underline{B}^0}$ in the category of $\underline{E}$-topoi. It follows from Diaconescu's theorem that

$$G \longmapsto \mathrm{Mod}(\underline{G}, \underline{E})$$

is a product preserving functor from groups in $\underline{E}$ to groupoids.

We denote by $H^1(\underline{E}, G)$ the class of components of $Mod(\underline{G}, \underline{E})$. If G is an abelian group, then it is an abelian group object in the category of groups in $\underline{E}$, so, as $H^1(\underline{E}, -)$ preserves products, $H^1(\underline{E}, G)$ has an abelian group structure. The trivial G-torsor acts as unit element.

<u>Proposition 10.8</u> For any $\alpha \in H^1(\underline{E}, G)$, there is a monomorphism of groups $G \xrightarrow{\phi} H$ such that $H^1(\underline{E}, \phi)$ takes α to zero.

Proof: Suppose α is represented by the right G-torsor M. Take $H = G^M$ with ϕ induced by $M \longrightarrow 1$.

It is conventional to denote $Hom_{\underline{E}}(1, X)$ by $H^0(\underline{E}, X)$. In this way we can extend the definition of Grothendieck cohomology to arbitrary coefficient objects in dimension zero, and group coefficient objects in dimension one. This is suggestive of the definition of homotopy groups, where the same phenomenon occurs.

If $0 \longrightarrow A \longrightarrow B \longrightarrow C \longrightarrow 0$ is a short exact sequence of abelian groups in $\underline{E}$ it is instructive to see how the connecting map

$$\delta: H^0(\underline{E}, C) \longrightarrow H^1(\underline{E}, A)$$

is defined. Given $1 \xrightarrow{c} C$, form the pullback

$$\begin{array}{ccc} P & \longrightarrow & 1 \\ \downarrow & & \downarrow c \\ B & \longrightarrow & C \end{array} \quad .$$

We may prove that $P \times A \rightarrowtail B \times B \longrightarrow B$ factors through $P \rightarrowtail B$, making P into an A-object. Then we show that P is acturally an A-torsor, whose class defines $\delta(c)$.

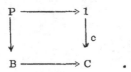

In $\mathrm{Top}(S^1)$, for example, consider the following extension of Z_2 by itself: -

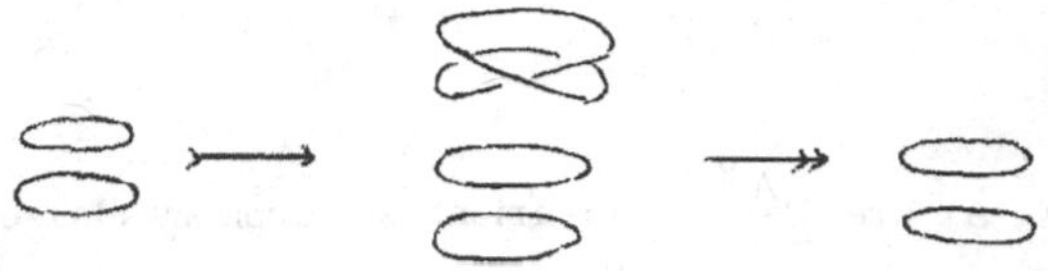

where the middle group is a "twisted" 4-group.

If A is an abelian group in $\underline{E}$, then $\underline{E}^{\underline{A}^0}$ is an abelian group object in $\mathrm{Top}_{\underline{E}}$. We may consider $\underline{E}$-topoi $\underline{F}$ which have an $\underline{E}^{\underline{A}^0}$ - action, i.e. a geometric morphism

$$\underline{F} \times_{\underline{E}} \underline{E}^{\underline{A}^0} \xrightarrow{u} \underline{F}$$

satisfying the usual requirements. If $\underline{F} \xrightarrow{p} \underline{E}$ is the structural morphism of $\underline{F}$, then

$$\underline{F} \times_{\underline{E}} \underline{E}^{\underline{A}^0} \simeq \underline{F}^{p^*(A)^0} .$$

Thus, for any object X of $\underline{F}$, $u^*(X)$ makes X into a $p^*(A)$-object (the condition for the unit ensures that $u^*(X)$ has X for its underlying object). In this way, we see that an $\underline{E}^{\underline{A}^0}$ - action on $\underline{F}$ is equivalent to giving every object of $\underline{F}$ a $p^*(A)$-action for which the maps of $\underline{F}$ are equivariant. The trivial action corresponds to the projection $\underline{F} \times_{\underline{E}} \underline{E}^{\underline{A}^0} \xrightarrow{p_1} \underline{F}$.

We may go on to consider $\underline{E}^{\underline{A}^0}$ - equivariant geometric morphisms between $\underline{E}$-topoi with $\underline{E}^{\underline{A}^0}$ -action, and so on.

Now Giraud, in his book "Cohomologie non-abelienne", has a description of $H^2(\underline{E}, A)$, for A an abelian group object in $\underline{E}$, which I think I have understood to be as follows: -

The elements of $H^2(\underline{E}, A)$ are $\underline{E}^{\underline{A}^0}$ - equivariant isomorphism classes of $\underline{E}$-topoi with $\underline{E}^{\underline{A}^0}$-action, which are locally $\underline{E}^{\underline{A}^0}$ - equivariantly equivalent to $\underline{E}^{\underline{A}^0}$.

The analogy with torsors, is quite striking. Let us call an $\underline{E}$-topos with $\underline{E}^{\underline{A}^0}$-action which is locally $\underline{E}^{\underline{A}^0}$ - equivariantly equivalent to $\underline{E}^{\underline{A}^0}$ an extension of $\underline{E}$ by A, following Giraud. Then, as for torsors, any two extensions are locally isomorphic. Any $\underline{E}^{\underline{A}^0}$-equivariant morphism of $\underline{E}$-topoi between two extensions of $\underline{E}$ by A is an equivalence. Any extension of $\underline{E}$ by A which has a section is equivalent to the trivial extension, i.e. $\underline{E}^{\underline{A}^0}$ itself.

Let us see how the connecting map

$$\delta : H^1(\underline{E}, C) \longrightarrow H^2(\underline{E}, A)$$

for a short exact sequence

$$0 \longrightarrow A \longrightarrow B \longrightarrow C \longrightarrow 0$$

of abelian groups in $\underline{E}$, works. First note that we have a pullback diagram of $\underline{E}$-topoi

$$\begin{array}{ccc} \underline{E}^{\underline{A}^0} & \longrightarrow & \underline{E}^{\underline{B}^0} \\ \downarrow & & \downarrow \\ \underline{E} & \longrightarrow & \underline{E}^{\underline{C}^0} \end{array}$$

where 0 is induced by $0 \longrightarrow C$ and represents the trivial C-torsor. An element $x \in H^1(\underline{E}, C)$ is represented by a morphism

$$\underline{E} \xrightarrow{\;x\;} \underline{E}^{\underline{C}^0}$$

of $\underline{E}$-topoi. Form the pullback diagram

$$\begin{array}{ccc} \underline{F} & \longrightarrow & \underline{E}^{\underline{B}^0} \\ \downarrow & & \downarrow \\ \underline{E} & \xrightarrow{x} & \underline{E}^{\underline{C}^0} \end{array} .$$

Since x and 0 are locally isomorphic, $\underline{F}$ and $\underline{E}^{\underline{A}^0}$ are locally equivalent. We may show that

$$\underline{F} \times_{\underline{E}} \underline{E}^{\underline{A}^0} \longrightarrow \underline{E}^{\underline{B}^0} \times_{\underline{E}} \underline{E}^{\underline{B}^0} \longrightarrow \underline{E}^{\underline{B}^0}$$

factors through $\underline{F} \longrightarrow \underline{E}^{\underline{B}^0}$, so that $\underline{F}$ has an $\underline{E}^{\underline{A}^0}$-action. In this way we get an element $\delta(x) \in H^2(\underline{E}, A)$ represented by $\underline{F}$.

It is a straightforward matter to check the exactness of the sequence

$$0 \longrightarrow H^0(\underline{E}, A) \longrightarrow \ldots\ldots\ldots \longrightarrow H^2(\underline{E}, C) .$$

As Giraud has pointed out, the beauty of the above description of $H^2(\underline{E}, A)$ is how it ties up with the description known for the cohomology of groups.

If $\underline{E}$ = G-sets, for G a group, then $H^0(\underline{E}, X)$ is simply the fixed point set of the G-set X. It follows that $H^n(\underline{E}, A)$ is simply the nth cohomology of G with coefficients in the G-module A. It is well known that the elements of $H^2(\underline{E}, A)$ correspond to isomorphism classes of extensions of G by A. If $x \in H^2(\underline{E}, A)$ corresponds to the extension

$$0 \longrightarrow A \longrightarrow F \longrightarrow G \longrightarrow 1$$

then we find that $\underline{F}$ = F-sets is the topos extension of $\underline{E}$ corresponding to x.

What has made much of the analysis above possible is the fact that an abelian group, considered as a category, is a group object in Cat. It may be shown that the underlying category of a group object in Cat is always a groupoid - such objects are generally known as crossed groups. Now, if A is a crossed group in an elementary topos $\underline{E}$, i.e. an object of $Gp(Cat(\underline{E}))$, then the $\underline{E}$-topos $\underline{E}^{\underline{A}^0}$ is a group object in the category of $\underline{E}$-topoi. Perhaps this observation may explain why crossed groups occur in non-abelian cohomology.

[The search for group objects in the category of topoi seems rather interesting; if G is a topological group, Top(G) is not necessarily a group object in Top_S, because the functor Top does not preserve products.]

<u>Definition 10.9</u> A morphism of S-topoi

$$\underline{E} \xrightarrow{f} \underline{F}$$

is a <u>weak homotopy equivalence</u>, if for every locally constant object (group for $n = 1$, abelian group for $n > 1$) A,

$$H^n(f,A) : H^n(\underline{F}, A) \longrightarrow H^n(\underline{E}, f^*(A))$$

is an isomorphism (see the end of § 4).

An object of an S-topos is locally constant if it is locally isomorphic to a constant object, i.e. to a coproduct of the terminal object.

A modification due to D. Quillen of a theorem of Whitehead asserts that the above definition of weak homotopy equivalence agrees with the usual one for spatial topoi. In order not to lose the "information" given by the fundamental group, it is essential to allow non-abelian coefficient groups in dimension one.

As an application, let G be a discrete group, and X a G-space. We may consider X as a topological space internal to G-sets. Forming the topos of sheaves on X internally to S^{G^0} gives an S^{G^0}-topos, $\mathrm{Top}(X, G)$, of sheaves on X with a compatible G-action. We get a geometric morphism

$$\mathrm{Top}(X, G) \xrightarrow{f} S^{G^0} .$$

Let $S \xrightarrow{u} S^{G^0}$ be induced by the unit $1 \longrightarrow G$. We get a pullback diagram

$$\begin{array}{ccc} \mathrm{Top}(X) & \xrightarrow{u'} & \mathrm{Top}(X, G) \\ {\scriptstyle f'}\downarrow & & \downarrow{\scriptstyle f} \\ S & \xrightarrow{u} & S^{G^0} \end{array}$$

where u^*, u'^* are functors which forget G-action. In pullback diagrams of this kind, the "Beck condition"

$$f'_* u'^* \simeq u^* f_*$$

holds. This tells us that for $A \in \mathrm{Top}(X, G)$, $f_*(A)$ is the G-set of global sections of A. Hence $R^n f_*(A)$ is the G-set $H^n(X, A)$. The Leray spectral sequence of f gives a spectral sequence

$$H^p(G, H^q(X, A)) \Longrightarrow H^n(\mathrm{Top}(X, G), A) .$$

The G-action is <u>good</u> if for all $g \neq 1$ in G and $x \in X$ there exists an open neighbourhood U of x such that $gU \cap U = \phi$.

If X/G denotes the space of orbits under G, then the projection map $X \longrightarrow X/G$ is a local homeomorphism if the action is good. We may also prove that if the action is good there is an equivalence of categories

$$\mathrm{Top}(X,\ G) \simeq \mathrm{Top}(X/G)$$

making the diagram

$$\begin{array}{ccc} \mathrm{Top}(X,\ G) & \xrightarrow{\ \sim\ } & \mathrm{Top}(X/G) \\ {\scriptstyle u'}\nwarrow & & \nearrow \\ & \mathrm{Top}(X) & \end{array}$$

commute.

Suppose that $EG \longrightarrow BG$ is a universal principal G-bundle. Then $X \times EG \xrightarrow{pr_1} X$ is a G-equivariant map whose underlying map is a homotopy equivalence. Also, $X \times EG$ has a good G-action, so that

$$\mathrm{Top}(X \times EG,\ G) \simeq \mathrm{Top}(X \times EG/G)\ .$$

From the Leray spectral sequence of $pr_1 : X \times EG \longrightarrow X$ we get immediately an isomorphism for the E_2-term

$$H^p(G,\ H^q(X,\ A)) \longrightarrow H^p(G,\ H^q(X \times EG,\ A))$$

and so we deduce that

$$\mathrm{Top}(X \times EG,\ G) \longrightarrow \mathrm{Top}(X,\ G)$$

is a weak homotopy equivalence. Hence we get a weak homotopy equivalence between Top(X, G) and Top(X x EG/G), showing that

$$H^n(\mathrm{Top}(X,\ G),\ A) \simeq H^n_G(X,\ A)$$

where the right hand side stands for G-equivariant cohomology.

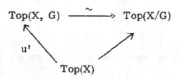

Bibliography

1. M. Artin, Grothendieck Topologies, Harvard University Press (1962).

2. M. Artin, A. Grothendieck, J. Verdier, Theorie des Topos et Cohomologie Etale des Schemas, Springer Lecture Notes Nos. 269 and 270 (revised version of SGA4 1963/64).

3. M. Barr, Toposes without points, Preprint. McGill University (1973).

4. J. C. Cole, Categories of sets and models of set theory, Thesis (University of Sussex 1972). Aarhus preprint No. 52 (1971).

5. Diaconescu, Thesis (Rutgers University 1973).

6. P. Freyd, Aspects of Topoi, Bull. Australian Math. Soc. (1972) Vol. 7. pp. 1 -76.

7. P. Freyd, On the logic of Topoi, Preprint. University of Pennsylvania, (1973).

8. J. Giraud, Analysis Situs. Sem. Bourbaki (1962/63).

9. J. Giraud, Methode de la descente. Mem. Soc. Math. France (1964).

10. J. Giraud, Cohomologie non-abelienne. Springer. Grundlehren Band 179. (1971).

11. M. Hakim, Topos Anneles et Schemas relatifs, Springer. Ergebuisse band 64. (1972).

12. L. Illusie, Complexe Cotangent et Deformations. Springer Lecture Notes Nos. 239 and 283.

13. G. M. Kelly and R. Street, Abstracts of the Sydney Category Theory Seminar (1972). University of New South Wales.

14. A. Kock, On a theorem of Läuchli concerning proof bundles. Preprint. Aarhus (1970).

15. A. Kock and Chr. Juul Mikkelsen, Non-standard Extensions in the Theory of Toposes. Aarhus Preprint series No. 25 (1971/72).

16. A. Kock and Chr. Juul Mikkelsen, Topos-Theoretic Factorization of Non-standard Extensions. Preprint. Aarhus (1972).

17. A. Kock and G. Wraith. Elementary Toposes. Aarhus Lecture Notes No. 30 (1970/71).

18. F. W. Lawvere, An elementary theory of the category of sets. Proc. Nat. Acad. Sci. 52 (1964) pp. 1506-1511.

19. F.W. Lawvere, Adjointness in Foundations. Dialectica. 23 (1969) pp. 281-296.

20. F. W. Lawvere, Equality in hyperdoctrines and comprehension as an adjoint functor. Symposia in Pure Maths. Vol. XVII A. M. S. (1970).

21. F. W. Lawvere, Quantifiers and sheaves. Actes du Congres International des Mathematiciens. Nice. (1970). pp. 329-334.

22. F. W. Lawvere and M. Tierney, Summary by J. Gray. Springer Lecture Notes No. 195 (1971).

23. F. W. Lawvere, Toposes, Algebraic Geometry and Logic. Springer Lecture Notes. No. 274. (1972).

24. W. Mitchell, On Topoi as closed Categories. J. Pure and Applied Algebra. Vol. 3 No. 2. 1973.

25. D. Mumford, Picard group of moduli problems. Proc. Conference on arithmetical algebraic geometry at Purdue. (1963).

26. R. Paré. Colimits in Topoi, Preprint. Dalhousie University (1973).

27. D. Quillen, Homotopical Algebra. Springer Lecture Notes. No. 43 (1967).

28. M. Tierney, Axiomatic Sheaf Theory. C. I. M. E. conference on Categories and Commutative Algebra. Varenna (1971). pp. 249-326. Edizioni Cremonese. Rome (1973).

School of Mathematical and Physical Sciences
The University of Sussex.

PART III

(presented at a conference in Berlin, October 1973, organized by Ch. Maurer; manuscripts received by the editors in June 1974)

SOME TOPOS THEORETIC CONCEPTS OF FINITENESS

A. Kock, P. Lecouturier, and C. J. Mikkelsen

In an elementary topos $\underline{E}$, there are many different notions of "finiteness" which specialize to the same "usual" finiteness notion when $\underline{E}$ is the category of sets. The notion we study here comes about by extending Kuratowski's description of finite sets, [10], [18], to an arbitrary elementary topos: a set A is (Kuratowski-) finite if $A \in K(A)$, where $K(A)$ ("the Kuratowski family") is the smallest family of subsets of A which contains $\emptyset$, contains all singletons $\{a\}$ (where $a \in A$), and is closed under binary union. (K is actually a submonad of the power set monad, as described in, say, [8] or [12].)

This finiteness notion is proved equivalent to two other finiteness notions; one which we essentially learned from Joyal, and which in some sense goes back to Birkhoff and Frink [4], who defined the notion of inaccessible element in a lattice. A set is finite if and only if the maximal element in its lattice of subsets is inaccessible. The third finiteness notion we study is of more category theoretic nature, hinging on the notion of cofinality. We call the three notions K-finite (K for Kuratowski), J-finite, and D-finite. Whereas K-finiteness is essentially impredicative, J-finiteness is predicative in a certain sense which allows us to prove that finiteness is preserved under logical functors (it is also preserved under "inverse image functors" for geometric morphisms (see e.g. [9]) of elementary toposes - this has been proved by Mikkelsen). Finally,

"D-finite" is the simplest of the three notions to state: The class of D-finite objects is the Galois-closure of the class consisting of the two objects 0 and $2 = 1+1$, under a certain simple Galois-correspondence derived from the notion of cofinality of maps into ordered objects.

Motivating this research is of course the line of thought that "an important technique is to lift constructions first understood for "the" category $\mathbb{S}$ of abstract sets to an arbitrary topos" (Lawvere), and then to apply the lifted construction to some specific topos, like sheaves on a space, or group representations, to get new knowledge of the space or the group. To this step of reapplying, we have not contributed anything yet. For lifting some useful standard algebra (like linear algebra) into a topos, one of course needs a notion of finiteness. Mulvey and Tierney have done that successfully (they reapplied it) by means of a more restricted finiteness-notion "cardinal-finite" (it is in fact possible to prove by induction that cardinal-finiteness implies our kind of finiteness. The converse is false). Our more general finiteness notion seems more to be fit for fitting lattice theoretic ideas into an arbitrary topos (this viewpoint is illustrated in Section 4).

There are four sections. In the first we give some general remarks about the method used, and, in particular, state and prove some useful principles concerning the internal power-set functor. We use extensively a method which we essentially learned from Joyal for "working with elements in a category $\underline{E}$" where $\underline{E}$ is any suitably good category (say, a regular category [2]; in particular, an elementary topos is "suitably good"). In Section 2, we define the three finiteness notions J, K and D, and prove them equivalent by proving $K \Rightarrow J \Rightarrow D \Rightarrow K$ (Theorem at the end of the section).

Section 3 is devoted to examples and to give some hereditary properties which J-finiteness has, like being closed under finite products. Of the more surprising things is that a subobject of a J-finite object need not be J-finite (although it is if it is detachable). In some concrete toposes, we describe completely what J-finite objects are; in spatial toposes sh(X), we can only give some necessary conditions: a J-finite sheaf must be flabby, have finite stalks, and a finite set of cross-sections over each open set. If X is the Sierpinski two-point space, these conditions are sufficient.

The notation employed is mostly standard. We sometimes write $A \pitchfork B$ for B^A (the exponential object). We use 1_A as well as id_A (or just 1 or id) for the identity map of A. We denote by ω or ω_A the unique map $A \to 1$.

§1. Some preliminary remarks on methods used

As mentioned in the introduction, we do a good deal of reasoning in the given elementary topos $\underline{E}$ in terms of elements. If $B \in \underline{E}$ is an object in $\underline{E}$ under consideration, an element of B is here by definition an arbitrary map with codomain B,

$$b: X \to B.$$

We usually denote objects which occur as domain of elements by capital letters near the end of the alphabet: $X, Y, Z, Z', \ldots$. An important feature in the elementwise method is the "change of domain of elements", i.e. maps in $\underline{E}$ between domains of elements. These are usually denoted by lower case greek letters like $\alpha: Y \to X$. They are related to the "change of time" occurring in Kripke's semantics for intuitionistic logic. The philosophy of this method is explained in Lawvere's [11]. The reader may reconstruct usual set theoretic ideas and arguments from the element-wise ideas and arguments used here, by putting $X = Y = .. = 1$.

Of course, the "power-objects" Ω^A in the elementary topos under consideration are going to play an important role. Since they here occur highly iterated, we sometimes use the on-line notation $\Omega^A = A \pitchfork \Omega$, and, more generally,

$$B^A = A \pitchfork B$$

(read "A hom B").

If $A': X \to A \pitchfork \Omega$ and $a: X \to A$ are elements (with same domain X), we shall write

$$a \underline{\in} A' \tag{1.1}$$

as an abbreviation for

$$(X \xrightarrow{\langle A', a\rangle} (A \pitchfork \Omega) \times A \xrightarrow{ev} \Omega) = (X \to 1 \xrightarrow{true} \Omega),$$

(which in turn is equivalent to

$$\text{"}\langle A',a\rangle \text{ factors through } \in_A \rightarrowtail (A \pitchfork \Omega) \times A\text{"},$$

whence the notation).

Clearly, the relation $\underline{\in}$ in (1.1) is <u>stable</u> in the sense that for any $\alpha\colon Y \to X$

$$a \underline{\in} A' \quad \text{implies} \quad \alpha.a \underline{\in} \alpha.A'.$$

The relation $\underline{\in}$ compares as follows with $A \pitchfork \Omega$ viewed as a contravariant functor in the first variable:

1.1 <u>Pull-back Principle</u>. Given the situation

$$\begin{array}{ccc} X & \xrightarrow{\quad G \quad} & B \pitchfork \Omega \\ & {}_{a}\searrow & \\ & & A \xrightarrow[f]{} B \end{array}$$

Then

$$a.f \underline{\in} G \quad \text{iff} \quad a \underline{\in} G.f \pitchfork id_\Omega .$$

<u>Proof</u>. This is a version of the fact that $f \pitchfork id_\Omega\colon B \pitchfork \Omega \to A \pitchfork \Omega$ represents pulling back subobjects along f. A more formal proof goes like this:

$$\begin{array}{ll} & a.f \underline{\in} G \\ \text{iff} & \langle G,a\rangle.id_{B \pitchfork \Omega} \times f.ev_B = true_X \\ \text{iff} & \langle G,a\rangle.(f \pitchfork id_\Omega) \times id_A.ev_A = true_X \qquad \text{(naturality of ev)} \\ \text{iff} & \langle G.f \pitchfork id_\Omega, a\rangle.ev_A = true_X \\ \text{iff} & a \underline{\in} G.f \pitchfork id_\Omega . \end{array}$$

Suppose that an object $B \in \underline{E}$ is given together with a subobject $\textcircled{\le}$ of $B \times B$

(1.2) $$\textcircled{\le} \rightarrowtail B \times B.$$

Then we may say that two elements

$$b_1 : X \to B, \quad b_2 : X \to B$$

stand in the relation $\leq$ if $\langle b_1, b_2\rangle : X \to B\times B$ factors through $\textcircled{\leq}$; and (as usual), the subobject $\textcircled{\leq}$ is said to make B into a (partially) ordered object if for each X, the relation $\leq$ on the set $\hom_{\underline{E}}(X,B)$, which we just defined, makes $\hom_{\underline{E}}(X,B)$ into a partially ordered set. Clearly, $\leq$ is a stable relation in the same sense as $\underline{\in}$ was: for $\alpha: Y \to X$ arbitrary

$$b_1 \leq b_2 \quad \text{implies} \quad \alpha.b_1 \leq \alpha.b_2 .$$

Every power object $\Omega^A = A \pitchfork \Omega$ carries a canonical order-relation (being in the set case the inclusion ordering of subsets of A); see e.g. [9], p.34. We shall mainly use the following criterion for that relation:

1.2 Extensionality Principle. Let A_1 and A_2 be two elements of $A \pitchfork \Omega$ with same domain X, i.e. $A_i: X \to A \pitchfork \Omega$ $(i = 1,2)$. Then $A_1 \leqq A_2$ if and only if for every $\alpha: Y \to X$ and every element $a: Y \to A$,

$$a \mathrel{\underline{\in}} \alpha.A_1 \quad \text{implies} \quad a \mathrel{\underline{\in}} \alpha.A_2 .$$

This is just a slight generalization of Proposition 1.6 in [9].

Recall the Singleton map $\{\cdot\}_A : A \to A \pitchfork \Omega$ (e.g. [9], p.10). We have

1.3. Singleton Principle. Let $a: X \to A$ and $F: X \to A \pitchfork \Omega$ be arbitrary. Then

$$a \mathrel{\underline{\in}} F \quad \text{iff} \quad a.\{\cdot\}_A \leq F.$$

Proof. This follows from the extensionality principle and the fact that $b \mathrel{\underline{\in}} \alpha.a.\{\cdot\}$ implies $b = \alpha.a$ (which stems directly from the construction of $\{\cdot\}$).

A map $f: B \to C$ between ordered objects is called <u>monotone</u> if

$$b_1 \leq b_2 \quad \text{implies} \quad b_1.f \leq b_2.f$$

for an arbitrary pair of elements $b_i: X \to B$ in B, $(i=1,2)$. If g is a monotone map in the other direction, $g: C \to B$, then f is said to be <u>left adjoint</u> to g (and g <u>right adjoint</u> to f), in symbols $f \dashv g$, if

$$id_B \leq f.g \quad \text{and} \quad g.f \leq id_C.$$

This is equivalent to

$$b.f \leq c \quad \text{iff} \quad b \leq c.g$$

for every pair of elements $b: X \to B$ and $c: X \to C$.

If B is an ordered object, one produces a monotone map

$$\downarrow seg_B: B \to B \pitchfork \Omega$$

as exponential adjoint of the charateristic map $B \times B \to \Omega$ of the (twisted) order-relation $\textcircled{\geq}$. It is characterized by the property that for any pair b_1, b_2 of elements in B (with same domain),

$$b_1 \leq b_2 \quad \text{iff} \quad b_1 \subseteq b_2.\downarrow seg_B.$$

The anti-symmetry of $\leq$ can also be phrased: $\downarrow seg_B$ is a monic map.

Furthermore, $\downarrow seg_B: B \to B \pitchfork \Omega$ is <u>full</u> in the sense that not only does it preserve order-relations, but is also reflects them: for $b_i: X \to B$ $(i=1,2)$, we have

$$b_1.\downarrow seg_B \leq b_2.\downarrow seg$$

iff

$$b_1 \leq b_2.$$

This follows from the extensionality principle and from the above characteristic property for $\downarrow seg$, using the reflexity of the order-relation which implies $b_1 \subseteq b_1.\downarrow seg$.

We call B a <u>complete</u> ordered object provided $\downarrow seg_B$ has a left adjoint $sup: B \pitchfork \Omega \to B$. (This property of B, being "of the nature of second order logic" does <u>not</u> reflect itself down to properties on the hom-sets $hom_{\underline{E}}(X,B)$.)

From the fullness of $\downarrow seg_B$ it follows that the end-adjunction for $sup \dashv seg_B$ is an identity, so that we have

$$* \qquad \downarrow seg_B . sup = id_B$$

and

$$id_{B \pitchfork \Omega} \leqq sup . \downarrow seg_B .$$

From the latter inequality, we get by multiplication on the left that

$$\{\cdot\}_B \leqq \{\cdot\}_B . sup . \downarrow seg,$$

and since $id_B \underline{\in} \{\cdot\}_B$ by Singleton Principle, the extensionality principle yields

$$id_B \underline{\in} \{\cdot\}_B . sup . \downarrow seg,$$

that is

$$** \qquad id_B \leqq \{\cdot\}_B . sup .$$

On the other hand, the extensionality principle and the reflexitivity of the order relation yields

$$\{\cdot\}_B \leqq \downarrow seg_B ,$$

and multiplying on the right by the monotone map sup, we get

$$\{\cdot\}_B . sup \leqq \downarrow seg_B . sup = id_B$$

using *. Combining * and **, we get

$$\{\cdot\}_B . sup = id_B .$$

It may also at this point be worthwhile to record explicitly in which sense left adjoints preserve sup's. Let A and B be complete ordered objects, and let $f: A \to B$ be left adjoint to $g: B \to A$. Then f preserves sup's in the sense that

$$\begin{array}{ccc} A \pitchfork \Omega & \xrightarrow{\exists_f} & B \pitchfork \Omega \\ {\scriptstyle sup_A}\downarrow & & \downarrow{\scriptstyle sup_B} \\ A & \xrightarrow[f]{} & B \end{array}$$

commutes. To prove this, take right adjoints of all four maps: it then amounts to proving the following diagram commutative

$$\begin{array}{ccc} A \pitchfork \Omega & \xleftarrow{f \pitchfork 1} & B \pitchfork \Omega \\ {\scriptstyle \downarrow seg_A}\uparrow & & \uparrow{\scriptstyle \downarrow seg_B} \\ A & \xleftarrow[g]{} & B \end{array}$$

This can easily be done by the extensionality principle: let $b\colon X \to B$ and $a\colon X \to A$ be arbitrary. Then

$$a \mathrel{\underline{\in}} b.g.{\downarrow}seg_A$$

iff

$$a \mathrel{\underline{\leq}} b.g$$

iff

$$a.f \mathrel{\underline{\leq}} b, \qquad \text{by adjointness } f \dashv g$$

iff

$$a.f \mathrel{\underline{\in}} b.{\downarrow}seg_B$$

iff

$$a \mathrel{\underline{\in}} b.{\downarrow}seg_B.f \pitchfork 1, \text{ by pull-back principle.}$$

(Further information about ordered objects in an elementary topos may be found in [13].)

Power objects $A \pitchfork \Omega$ are, with the canonical ordering, complete ordered objects. The functorial properties of the power object formation has been studied by several people. We here summarize what we need about this:

<u>Summary of facts about the "power-set" functor in a topos $\underline{E}$</u>

For each map $f\colon A \to B$ in $\underline{E}$, we have, for pure closed category reasons, a map

$$B \pitchfork \Omega \xrightarrow{f \pitchfork id} A \pitchfork \Omega .$$

(It "externalizes" to "pull-back along f".) Now it is a well known fact (first realized by Lawvere and Tierney) that f id is monotone and has adjoints on both sides

$$\exists_f : A \pitchfork \Omega \to B \pitchfork \Omega, \qquad \text{left adjoint to } f \pitchfork \mathrm{id}$$
$$\forall_f : A \pitchfork \Omega \to B \pitchfork \Omega, \qquad \text{right adjoint to } f \pitchfork \mathrm{id}.$$

We shall in this paper be concerned in particular with $\exists_F$. It makes $- \pitchfork \Omega$ into a covariant functor $\underline{E} \to \underline{E}$, which in fact is known to be the functor part of a symmetric monoidal monad on $\underline{E}$ (see for instance [8]), and therefore, by [8], also a strong (and commutative) monad on $\underline{E}$. We shall need to make explicit this description of the strength (denoted $\exists\!\!\exists$) of the covariant power-set functor. It is a map

$$\exists\!\!\exists_{A,B} : A \pitchfork B \to (A \pitchfork \Omega) \pitchfork (B \pitchfork \Omega)$$

(which "externalizes" to the map which sends $f: A \to B$ to $\exists_f : A \pitchfork \Omega \to B \pitchfork \Omega$). We shall also make explicit a construction derived from this strength,

$$\mathrm{Im}: A \pitchfork B \to B \pitchfork \Omega$$

(which set theoretically, to $f: A \to B$ associates (the characteristic function of) $f(A) \subseteq B$).

We begin, by giving the following element-wise description of $\exists_f : A \pitchfork \Omega \to B \pitchfork \Omega$ (where $f: A \to B$). ("Element-wise" in the sense used here: "elements with arbitrary domain"):

1.4 <u>Existence Principle</u>. Let $A' : X \to A \pitchfork \Omega$ and $b: X \to B$ be arbitrary maps with same domain X. Then

$$b \mathrel{\underline{\in}} A'.\exists_f \tag{1.3}$$

if and only if

(1.4) there is an epic $\beta: Y \twoheadrightarrow X$ and a map $a: Y \to A$ such that
$$a \mathrel{\underline{\in}} \beta.A' \quad \text{and} \quad \beta.b = a.f.$$

<u>Proof</u>. We shall take as known the fact that $\exists_f : A \pitchfork \Omega \to B \pitchfork \Omega$ may be described as the map whose exponential adjoint $\check{\exists}_f : (A \pitchfork \Omega) \times B \to \Omega$ is the characteristic map of the image F of the composite

$$\epsilon_A \rightarrowtail (A \pitchfork \Omega) \times A \xrightarrow[1 \times f]{} (A \pitchfork \Omega) \times B.$$

The implication $\Rightarrow$ is now proved as follows. First

$$b \in A' . \exists_f \quad \text{iff} \quad \langle A', b \rangle \text{ factors through } F;$$

then we can define Y and β by the pull-back diagram $*$

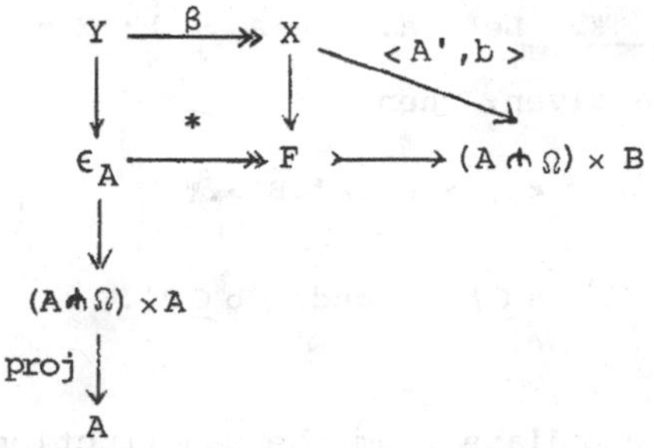

and let the vertical column at the left be a. Then it is immediate to see that (1.4) holds.

Conversely, if there exist some $\beta : Y \twoheadrightarrow X$ and $a : Y \to A$ which satisfy (1.4), then

$$\beta . \langle A', b \rangle = \langle \beta . A' . a \rangle . 1 \times f,$$

but $\langle \beta . A', a \rangle$ factors across ϵ_A by assumption. Hence $\beta . \langle A', b \rangle$ factors across ϵ_A, and thus the image of $\beta . \langle A', b \rangle$ is contained in F. Since β is epic, the image of $\langle A', b \rangle$ is also contained in F, so since $ch(F) = \check{\exists}_f$, we have $b \in A' . \exists_f$. This proves the principle.

Let us next recall the monoidal structure Ψ on the covariant power-"set" functor

$$\Psi_{A,B} : (A \pitchfork \Omega) \times (B \pitchfork \Omega) \to (A \times B) \pitchfork \Omega .$$

Set-theoretically, it associates to $A' \subseteq A$ and $B' \subseteq B$ the subset $A' \times B' \subseteq A \times B$. In the general setting here, it can be described as the composite

$$(A \pitchfork \Omega) \times (B \pitchfork \Omega) \xrightarrow{(proj_1 \pitchfork 1) \times (proj_2 \pitchfork 1)} \left[(A \times B) \pitchfork \Omega\right] \times \left[(A \times B) \pitchfork \Omega\right] \xrightarrow{\wedge} (A \times B) \pitchfork \Omega,$$

using the lower semi-lattice structure $\wedge$ ("intersection") on objects of form $X \pitchfork \Omega$. We can describe the monoidal structure in elementwise terms:

1.5. <u>Principle Ψ</u>. Let $a: X \to A$, $b: X \to B$, $A': X \to A \pitchfork \Omega$, and $B': X \to B \pitchfork \Omega$ be given. Then

$$\langle a,b \rangle \underline{\in} \langle A',B' \rangle . \Psi$$

iff

$$a \underline{\in} A' \quad \text{and} \quad b \underline{\in} B'.$$

<u>Proof</u>. Almost immediate from the construction of Ψ in terms of $\wedge$.

We shall next analyse the strength derived from Ψ, that is $\exists_{A,B}: A \pitchfork B \to (A \pitchfork \Omega) \pitchfork (B \pitchfork \Omega)$, in elementwise terms. Rather, we analyse its exponential adjoint

$$\check{\exists}_{A,B}: (A \pitchfork B) \times (A \pitchfork \Omega) \to B \pitchfork \Omega.$$

According to the procedure of [8], $\check{\exists}_{A,B}$ is the composite

$$(A \pitchfork B) \times (A \pitchfork \Omega) \xrightarrow{\{\cdot\} \times 1} ((A \pitchfork B) \pitchfork \Omega) \times (A \pitchfork \Omega) \to$$

$$\xrightarrow{\Psi} ((A \pitchfork B) \times A) \pitchfork \Omega \xrightarrow{\exists_{ev}} B \pitchfork \Omega$$

(using that $\{\cdot\}$ is the unit for the power-"set" monad, and $\exists$ its functor part on maps). Then

1.6 <u>Principle of strength-of-existence</u>. Let $f: X \to A \pitchfork B$, $A': X \to A \pitchfork \Omega$, and $b: X \to B$ be given. Then

(1.5) $b \underline{\in} \langle f, A' \rangle . \check{\exists}$

if and only if

(1.6) $\begin{cases} \text{there is an epic } \beta: Y \twoheadrightarrow X \text{ and a map } a: Y \to A, \text{ with} \\ \text{(i)} \quad a \underline{\in} \beta . A', \quad \text{and} \\ \text{(ii)} \quad \langle \beta . f, a \rangle . ev = \beta . b \end{cases}$

Proof. By the existence-principle and the construction of $\check{\exists}$ in terms of $\exists_{ev}$, we see that (1.5) holds if and only if

(1.7) $\begin{cases} \text{there is an epic } \beta: Y \twoheadrightarrow X \text{ and } \langle g, a \rangle: Y \to (A \pitchfork B) \times A, \\ \text{so that} \\ \text{(iii)} \quad \langle g, a \rangle . ev = \beta . b \quad \text{and} \\ \text{(iv)} \quad \langle g, a \rangle \underline{\in} \beta . \langle f, A' \rangle . \{\cdot\} \times 1 . \Psi \\ \qquad\qquad = \langle \beta . f . \{\cdot\}, \beta . A' \rangle . \Psi \end{cases}$

Condition (iv), however, is equivalent, by Principle Ψ, to the conjunction of (v) and (vi):

(v) $g \underline{\in} \beta . f . \{\cdot\}$ (i.e. $g = \beta . f$)

(vi) $a \underline{\in} \beta . A'$,

and therefore the condition (iii) is equivalent (in presence of (iv)) to the condition

(vii) $\langle \beta . f, a \rangle . ev = \beta . b$.

So (1.5) is equivalent to the existence of an epic β and an a so that (vi) and (vii) hold. But (vi) and (vii) are the same as (i) and (iii), respectively.

A certain construction derived from $\check{\exists}$ deserves a special name, namely the map $Im: A \pitchfork B \to B \pitchfork \Omega$, whose behaviour in the set-case was described above. It can be constructed as the composite

$$Im = A \pitchfork B \xrightarrow{\langle id, \omega . \ulcorner true_A \urcorner \rangle} (A \pitchfork B) \times (A \pitchfork \Omega) \xrightarrow{\check{\exists}} B \pitchfork \Omega .$$

As a corollary of the Principle 1.6 above, we then get, by putting $A': X \to A \pitchfork \Omega$ equal to $\omega_X . \ulcorner true_A \urcorner$:

1.7 <u>Image-Principle</u>. Let $b: X \to B$ and $f: X \to A \pitchfork B$ be given. Then

$$b \underline{\in} f.\mathrm{Im}$$

iff

there is an epic $\beta: Y \twoheadrightarrow X$ and a map $a: Y \to A$ such that $\langle \beta.f, a \rangle.ev = \beta.b$.

The left adjoint of $\downarrow seg_{A\pitchfork\Omega}: A \pitchfork \Omega \to (A \pitchfork \Omega) \pitchfork \Omega$ is denoted "union":

$$\bigcup : (A \pitchfork \Omega) \pitchfork \Omega \to A \pitchfork \Omega .$$

Using the explicit construction of $\bigcup$ (see e.g. [9], p. 111, or [12], p. 51), the reader will be able to prove

1.8. <u>Union Principle</u>. Let $F: X \to (A \pitchfork \Omega) \pitchfork \Omega$ and $a: X \to A$ be given. Then

$$a \underline{\in} F.\bigcup$$

iff

there is an epic $\beta: Y \twoheadrightarrow X$ and an $A': Y \to A \pitchfork \Omega$ such that $\beta.a \underline{\in} A'$ and $A' \underline{\in} \beta.F$.

If $B, \leq$ is an ordered object, we define

$$\downarrow cl: B \pitchfork \Omega \to B \pitchfork \Omega$$

to be the composite

$$B \pitchfork \Omega \xrightarrow{\exists_{\downarrow seg}} (B \pitchfork \Omega) \pitchfork \Omega \xrightarrow{\bigcup} B \pitchfork \Omega .$$

In the set case, it associates to a subset B' of B the set of all elements of B which are dominated by some element in B' (i.e. "the downward closure of B' "). We leave to the reader to use the existence- and union-principles to prove

1.9 <u>↓cl-Principle</u>. Let $B, \leq$ be an ordered object, and let $B'\colon X \to B \pitchfork \Omega$ and $b\colon X \to B$ be given. Then

$$b \mathrel{\underline{\in}} B'.{\downarrow}cl$$

iff

> there is an epic $\beta\colon Y \twoheadrightarrow X$ and an $c\colon Y \to B$ with $c \mathrel{\underline{\in}} \beta.B'$ and $\beta.b \leqq c$.

§ 2. The finiteness notions

We begin by lifting some well-known lattice theoretic notions into an arbitrary elementary topos $\underline{E}$. This lifting is actually canonical, once one writes down the notions in first order language and then translates them into W. Mitchell's or J. Benabou's language $L(\underline{E})$ (see [15] for an (incomplete) account of this). In order not to involve ourselves in syntax, we instead describe the lifted notions one by one.

Classically, a directed ordered set is an ordered set which is (i) non-empty, and (ii) has the property that any two elements of it has a common upper bound. This lifts as follows:

Let $B, \leq$ be an ordered object in $\underline{E}$. It is called directed provided (i) $\omega_B: B \to 1$ is epic, and (ii) for every pair b_1, b_2 of elements of B with same domain X, $b_i: X \to B$ $(i=1,2)$, there is an epic $\beta: Y \twoheadrightarrow X$ and an element $b: Y \to B$ with $\beta.b_1 \leq b$ and $\beta.b_2 \leq b$. (These two conditions, individually, may be called: "B is 0-directed" and "B is 2-directed", respectively).

An equivalent definition is given later. We started out with this one because it is typical for the lifting method. - One can also talk about when arbitrary subsets F of an ordered set are directed subsets. This lifts as follows (thinking of a subset of B as an element of $B \pitchfork \Omega$):

Let $B, \leq$ be an ordered object in $\underline{E}$. Then we say that an element

$$F: \quad X \to B \pitchfork \Omega$$

is directed, or a directed family, provided

(i) $\begin{cases} \text{there is an epic } \beta\colon Y \twoheadrightarrow X \text{ and a map } b\colon Y \to B \\ \text{so that } b \,\underline{\in}\, \beta.F, \text{ and} \end{cases}$

(ii) for every $\alpha\colon Z \to X$ and every pair

$$b_1\colon Z \to B, \quad b_2\colon Z \to B,$$

with

(2.1) $$b_1 \,\underline{\in}\, \alpha.F \quad \text{and} \quad b_2 \,\underline{\in}\, \alpha.F,$$

there is an epic $\beta\colon Z' \twoheadrightarrow Z$ and a $b_3\colon Z' \to B$ with

(2.2) $$b_3 \,\underline{\in}\, \beta.\alpha.F \quad \text{and with} \quad \beta.b_i \leq b_3 \quad (i=1,2).$$

Let B be a complete ordered object. An element $b\colon X \to B$ is called <u>intranscessible</u> (Diener [6]) provided it satisfies:

$$\begin{cases} \text{for every } \alpha\colon Y \to X \text{ and every } F\colon Y \to B \pitchfork \Omega \text{ with} \\ F \text{ directed and with } F.\sup_B \geq \alpha.b, \text{ there exists} \\ \text{an epic } \beta\colon Z \twoheadrightarrow Y \text{ and a } d\colon Z \to B \text{ with} \\ \qquad \beta.\alpha.b \leq d \text{ and } d \,\underline{\in}\, \beta.F. \end{cases}$$

Set-theoretically, $b \in B$ is intranscessible if every directed family F with $\sup(F) \geq b$ has a member $d \in F$ with $b \leq d$. If B satisfies "AB5" ("finite meets distribute over directed sup's") then $b \in B$ is intranscessible if and only if it is inaccessible in the sense of Birkhoff and Frink [4], or compact in the sense of Nachbin.

It is a consequence of a general method of carving out subobjects of objects in $\underline{E}$ by means of statements in $L(\underline{E})$ (W. Mitchell, J. Bénabou, G. Osius) that there is a subobject $S(B) \to B$, such that $b\colon X \to B$ is intranscessible if and only if b factors through $S(B)$. If the reader insists, he can construct S(B) (without reference to L($\underline{E}$)); but all information of S(B) needed is contained in the universal property: that it is the smallest subobject of B through which the intranscessible elements factor.

We now consider the upper-semilattice structure $(\vee,0)$ on the complete ordered object B

$$\vee: B \times B \to B, \qquad 0: 1 \to B,$$

(left adjoint to $\Delta: B \to B \times B$ and $\omega: B \to 1$, respectively). For each X, an upper-semilattice structure is induced on the hom-set $hom_{\underline{E}}(X,B)$, which we denote also $(\vee,0)$.

<u>2.1 Proposition</u>. The set of intranscessible elements $X \to B$ form a sub-semilattice of $hom_{\underline{E}}(X,B)$. (Alternatively, $S(B) \to B$ is a sub-semilattice-object of B.)

<u>Proof</u>. The fact that $0: X \to B$ is intranscessible is an immediate consequence of the requirement that a directed $F: Y \to B \pitchfork \Omega$ is "non-empty" (axiom (i) in the definition). Assume now that b_1 and b_2 are intranscessible elements $X \to B$. To prove that $b_1 \vee b_2$ is intranscessible, let $\alpha: Y \to X$ and $F: Y \to B \pitchfork \Omega$ be given, with F directed and with $F.sup_B \geq \alpha.(b_1 \vee b_2)$. Then, for $i = 1,2$,

$$F.sup \geq \alpha.b_i$$

and since b_i (and thus $\alpha.b_i$) is intranscessible, we can find epics $\beta_i: Z_i \twoheadrightarrow Y$ and maps $d_i: Z_i \to B$ with

(2.3) $\beta_i.\alpha.b_i \leq d_i \in \beta_i.F$ $i = 1,2.$

Let Z be formed as the pull-back

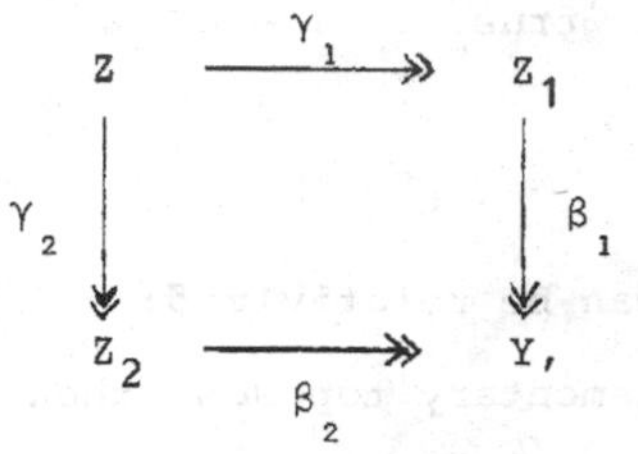

and let $\beta\colon Z \to Y$ be the diagonal map of this. It is again epic. Also, let

$$d_i'\colon \quad Z \to B$$

be defined as the composite $\gamma_i.d_i$. Then, by (2.3),

$$\gamma_i.\beta_i.\alpha.b_i \leq \gamma_i.d_i \in \gamma_i.\beta_i.F, \quad i = 1,2$$

so that

(2.4) $\beta.\alpha.b_i \leq d_i' \in \beta.F,$ $i = 1,2.$

Since $\beta.F$ is directed, we can find yet another epic $\delta\colon Z' \twoheadrightarrow Z$ and $d\colon Z' \to B$ with

(2.5) $\delta.d_i' \leq d \in \delta.\beta.F,$ $i = 1,2.$

Combining (2.4) and (2.5), we get

$$(\delta.\beta.\alpha.b_1) \vee (\delta.\beta.\alpha.b_2) \leq \delta.d_1' \vee \delta.d_2' \leq d \in \delta.\beta.F.$$

The left-hand side here is $\delta.\beta.\alpha.(b_1 \vee b_2)$, so the epic $\delta.\beta\colon Z' \to Y$ and the element $d\colon Z' \to B$ now are the maps we were required to produce to prove $b_1 \vee b_2$ intranscessible.

2.2 Definition. An object $A \in |\underline{E}|$ is called J-finite provided the complete ordered object $A \pitchfork \Omega$ ("the power set of A") has the property that its maximal element

$$(2.6) \qquad \ulcorner true_A \urcorner : 1 \to A \pitchfork \Omega$$

is intranscessible.

(This notion can be relativized; if $\Gamma: \underline{E} \to \underline{S}$ is a geometric morphism between elementary toposes, then $A \in \underline{E}$ might be called J-finite relative to $\underline{S}$ if $\Gamma_*: \underline{E} \to \underline{S}$ takes (2.6) into an intranscessible element in $\Gamma_*(A \pitchfork \Omega)$. Specializing to the case where $\underline{E}$ is a Grothendieck topos and $\underline{S}$ the category of sets, $A \in \underline{E}$ is J-finite relative to $\underline{S}$ if and only if A is quasi-compact in the sense of SGA 4, Exposé 6 [1]. This follows from loc.cit., 1.5.2).

Let us remark that "A is J-finite" also can be formulated as follows: If $F: X \to (A \pitchfork \Omega) \pitchfork \Omega$ is directed, and covering (meaning $F.\cup = \omega_X . \ulcorner true_A \urcorner$), then F is "trivially covering" in the sense that $\omega_X . \ulcorner true_A \urcorner \underline{\in} F$.

We now turn to the next finitness-notion. First, it is a straightforward consequence of the Dedekind-Tarski fixpoint theorem for elementary toposes, as proved by Mikkelsen [14], that, for each object B equipped with, say, one binary operation $*: B \times B \to B$ and, say, one nullary operation $0: 1 \to B$, one can, for any subobject $A \rightarrowtail B$, form the smallest subobject $\bar{A} \rightarrowtail B$ which contains A and is closed under the operations * and 0. (Apply the fixpoint theorem to the complete ordered object $B \pitchfork \Omega$).

In particular, if B is an upper-semi-lattice by means of $v: B \times B \to B$ and $0: 1 \to B$, we can form, for any $A \rightarrowtail B$, the subsemilattice $\bar{A} \rightarrowtail B$ generated by A.

Now let A be any object in $\underline{E}$. The power object $A \pitchfork \Omega$ has a canonical upper-semi-lattice structure. Thus, we can form the smallest sub-semi-lattice containing $\{\cdot\}_A: A \to A \pitchfork \Omega$. Set-theoretically, this sub-semi-lattice is the smallest subset of the powerset of A which contains $\emptyset$, all singletons $\{a\}$ (with $a \in A$), and which is closed under binary union. This is the family of sets considered by Kuratowski in his investigations on finiteness [10], or [18] §22:

2.3 Definition. Let $A \in \underline{E}$. The Kuratowski-object for A, $K(A) \rightarrowtail A \pitchfork \Omega$, is the smallest sub-(upper-)semi-lattice of $A \pitchfork \Omega$ which contains $\{\cdot\}_A: A \to A \pitchfork \Omega$.

(It actually can be proved, as a Corollary of the results of the present section, that $K(A) = S(A \pitchfork \Omega)$).

2.4 Definition. Let $A \in \underline{E}$. Then A is called K-finite provided $\ulcorner \mathrm{true}_A \urcorner: 1 \to A \pitchfork \Omega$ factors through $K(A)$.

2.5 Proposition. Any object A which is K-finite is also J-finite.

Proof. It suffices to prove $K(A) \subseteq S(A \pitchfork \Omega)$. For this, it suffices to prove that $S(A \pitchfork \Omega)$ is a sub-semilattice of $A \pitchfork \Omega$ containing $\{\cdot\}_A: A \to A \pitchfork \Omega$. By Proposition 2.1, we know already that it is a sub-semilattice, so we just have to prove that $\{\cdot\}_A$ factors through $S(A \pitchfork \Omega)$, i.e. that "singletons are intranscessible". This is Lemma 2.7 below. We now start the preparations for that Lemma:

Recall the map $\downarrow cl\colon B \pitchfork \Omega \to B \pitchfork \Omega$ from Principle 1.7, where B is an arbitrary ordered object. We then have

<u>2.6 Lemma</u>. For any object A, the following diagram commutes

$$\begin{array}{ccc} (A \pitchfork \Omega) \pitchfork \Omega & \xrightarrow{\ \downarrow cl_{A \pitchfork \Omega}\ } & (A \pitchfork \Omega) \pitchfork \Omega \\ & \searrow^{\bigcup} & \downarrow {\scriptstyle \{\cdot\} \pitchfork 1} \\ & & A \pitchfork \Omega . \end{array}$$

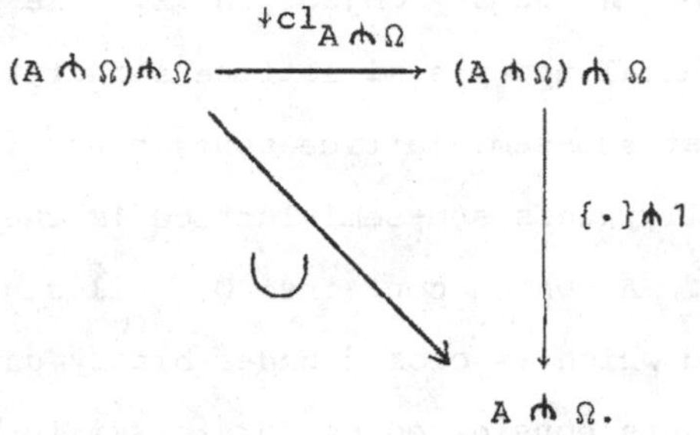

<u>Proof</u>. We have

$$\begin{aligned} \downarrow cl_{A \pitchfork \Omega} . \{\cdot\} \pitchfork 1 &= \exists(\downarrow seg_{A \pitchfork \Omega}) . \textstyle\bigcup . \{\cdot\} \pitchfork 1 \\ &= \exists(\downarrow seg_{A \pitchfork \Omega}) . \exists_{\{\cdot\} \pitchfork 1} . \textstyle\bigcup \\ &= \exists(\downarrow seg_{A \pitchfork \Omega} . \{\cdot\} \pitchfork 1) . \textstyle\bigcup . \end{aligned}$$

Only the second equality needs explanation: it is a consequence of the fact that $\{\cdot\} \pitchfork 1$ is a homomorphism of complete upper semi-lattices. (More generally, if $f\colon C \to D$, then

$$\textstyle\bigcup_D . f \pitchfork 1 = \exists_{f \pitchfork 1} . \bigcup_C ,$$

due to the fact that $f \pitchfork 1$ has a right adjoint $\forall_f$). Now we just have to prove

$$A \pitchfork \Omega \xrightarrow{\ \downarrow seg\ } (A \pitchfork \Omega) \pitchfork \Omega \xrightarrow{\ \{\cdot\} \pitchfork 1\ } A \pitchfork \Omega$$

is the identity map of $A \pitchfork \Omega$. Both maps in this composite have left adjoints:

$$A \pitchfork \Omega \xleftarrow{\ \bigcup\ } (A \pitchfork \Omega) \pitchfork \Omega \xleftarrow{\ \exists_{\{\cdot\}}\ } A \pitchfork \Omega ,$$

but this composite is the identity map (by one of the monad laws for the power-"set" monad). This proves the Lemma.

Now we can easily prove the following Lemman, and thereby, as we have seen, the Proposition 2.5.

2.7 Lemma. The map

$$\{\cdot\}_A : A \to A \pitchfork \Omega$$

is intranscessible.

Proof. Suppose given Y, α, and F as displayed in

$$\begin{array}{ccc} Y & \xrightarrow{\ \alpha\ } & A \\ {\scriptstyle F}\big\downarrow & \geq & \big\downarrow{\scriptstyle \{\cdot\}_A} \\ (A \pitchfork \Omega) \pitchfork \Omega & \xrightarrow[\ \bigcup\ ]{} & A \pitchfork \Omega \end{array}$$

with

$$\text{(2.7} \qquad F.\bigcup \;\geq\; \alpha.\{\cdot\}_A$$

and with F directed (actually directedness is not needed here). Using Lemma 2.6, the assumed inequality (2.7) may be written

$$\alpha.\{\cdot\} \leq F.\!\downarrow\! cl.\ \{\cdot\} \pitchfork 1.$$

By adjointness, this is equivalent to

$$\text{(2.8)} \qquad \alpha.\{\cdot\}.\ \exists_{\{\cdot\}} \leq F.\!\downarrow\! cl.$$

Clearly, we have (by Existence Principle and Singleton Principle)

(2.9) $\alpha.\{\cdot\} \underline{\in} \alpha.\{\cdot\}\ \exists_{\{\cdot\}}$,

and then the inequality (2.8) together with (2.9) gives

$$\alpha.\{\cdot\} \underline{\in} F.{\downarrow}cl.$$

By the ${\downarrow}cl$-principle 1.9, we get the existence of an epic $\beta: Z \twoheadrightarrow Y$ and $A': Z \to A \pitchfork \Omega$ with $\beta.\alpha.\{\cdot\} \leq A'$ and $A' \underline{\in} \beta.F$, so that A' and β witness the intranscessibility of $\{\cdot\}$.

We now turn to the third finiteness-notion. This hinges on a certain Galois-connection arising out of the notion of <u>cofinality</u>. If C is a partially ordered object in $\underline{E}$, and $g: D \to C$ a map from an arbitrary object into C, we say that g is <u>cofinal</u> if for every $c: X \to C$, there is an epic $\beta: Z \twoheadrightarrow X$ and a $d: Z \to C$ so that

$$d.g \geq \beta.c.$$

If A is an arbitrary object and B a partially ordered object, we say that B is <u>A-directed</u> if the diagonal

$$B \xrightarrow{\Delta} B^A$$

is cofinal (the diagonal is the exponential adjoint $\widehat{proj}$ of $proj: B \times A \to B$, and the order-relation on B^A is induced canonically from that of B). We write (temporarily)

$$\mathfrak{D}(A,B)$$

for "B is A-directed". Then $\mathfrak{D}$ establishes a relation between objects in $\underline{E}$ and partially ordered objects in $\underline{E}$, and thus gives rise to a Galois correspondence

$$\text{classes of objects in } \underline{E} \underset{\psi}{\overset{\Phi}{\rightleftarrows}} \text{ classes of ordered objects in } \underline{E}.$$

The class of D-finite objects is defined to be the Galois closure of the class consisting of 0 and $2 = 1 + 1$:

$$\text{class of D-finite objects} = \psi(\Phi(\{0,2\})).$$

In slightly more simple-minded terms, observe that $\Phi(\{0,2\})$ is the class of directed ordered objects, as defined right at the beginning of §2; for, the conditions (i) and (ii) there are equivalent, respectively, to the requirements that

$$B \overset{\omega}{\to} B^{0} = 1$$

and

$$B \overset{\Delta}{\to} B^{2} = B \times B$$

are cofinal. So the more simple-minded definition of D-finiteness now goes

2.8 Definition. An object $A \in \underline{E}$ is D-finite, provided for every directed ordered object B,

$$B \overset{\Delta}{\to} B^{A}$$

is cofinal.

In set-theoretic terms, A is D-finite if any map $A \to B$ into a directed ordered object B can be uniformly dominated by some $b \in B$.

In order to connect this with the previous two finiteness notions "J and K", we must study directed ordered objects more closely.

<u>2.9 Proposition</u>. The ordered object $(B,\leq)$ is directed if and only if the map bd defined as the composite

$$bd = (B \pitchfork \Omega)^{op} \xrightarrow{\uparrow seg} (B \pitchfork \Omega) \pitchfork \Omega \xrightarrow{\downarrow seg \pitchfork 1} B \pitchfork \Omega \xrightarrow{\exists_B} \Omega$$

is a lower-semilattice map. (This map bd is precisely the characteristic function of the subobject

$$(2.10) \qquad Bd = Bd(B) \overset{j}{\rightarrowtail} B \pitchfork \Omega$$

of "bounded subsets").

The label 'op' in the domain of bd has only the effect of reversing the order-relation. Also $\exists_B$ denotes the composite

$$B \pitchfork \Omega \xrightarrow{\exists_\omega} 1 \pitchfork \Omega \simeq \Omega.$$

We note that $\{\cdot\}_B$ composed with bd yields $true_B$ ("any Singleton is bounded.") To see this formally, we prove that

$$id_B \in \{\cdot\}.\uparrow seg.\downarrow seg \pitchfork 1.$$

Namely, this is equivalent to

$$(2.11) \qquad \downarrow seg \in \{\cdot\}.\uparrow seg,$$

by the Pull-back Principle, and (2.11) in turn is equivalent to

$$\downarrow seg \geq \{\cdot\}$$

which is true by reflexivity, and by the extensionality principle.

<u>Proof</u> of Proposition 2.9. Letting $Bd(B) = Bd$ be the subobject of $B \pitchfork \Omega$ displayed in (2.10) and defined as the subobject whose characteristic map is bd, it is straightforward to use Existence- and Pull-back principle together with the definition of bd to prove that $B': X \to B \pitchfork \Omega$ factors through bd if and only

if there is an epic $\beta\colon Y \twoheadrightarrow X$ and a map $b\colon Y \to B$ with

(2.12) $$\beta.B' \leq b.{\downarrow}\mathrm{seg}.$$

Now, since $bd \times bd.\wedge$ is the characteristic map for

$$j \times j\colon Bd \times Bd \longrightarrow (B \pitchfork \Omega) \times (B \pitchfork \Omega),$$

we see, by passing to subobjects classified by the maps, that bd preserving 2-intersections $\wedge$ is equivalent to the existence of a map ℓ (necessarily unique) making the diagram

(2.13)
$$\begin{array}{ccc} Bd \times Bd & \xrightarrow{\ \ell\ } & Bd \\ {\scriptstyle j\times j}\downarrow & & \downarrow{\scriptstyle j} \\ (B \pitchfork \Omega)\times(B \pitchfork \Omega) & \xrightarrow[\vee_{B\pitchfork\Omega}]{} & B \pitchfork \Omega \end{array}$$

into a pull-back. We shall see that such ℓ exists if and only if B is 2-directed. Let us first assume that $j \times j.\ \vee_{B\pitchfork\Omega}$ factors across Bd by a map ℓ, as displayed in (2.13) (we shall actually not need that it is a pull-back). To prove B 2-directed, let b_1 and b_2 be morphisms $X \to B$. Now $\{\cdot\}_B$ factors through j

$$\{\cdot\}_B = a.j,$$

as we have seen above. Consequently

$$\begin{aligned} &\langle b_1.\{\cdot\}_B,\ b_2.\{\cdot\}_B\rangle.\vee_{B\pitchfork\Omega}.bd \\ &= \langle b_1.a,\ b_2.a\rangle.j \times j.\vee_{B\pitchfork\Omega}.bd \\ &= \langle b_1.a, b_2.a\rangle.\ell.j.bd = \mathrm{true}_X, \end{aligned}$$

so that we have an epic $\beta\colon Y \twoheadrightarrow X$ and a $b\colon Y \to B$ with

$$b.\!\downarrow\!\mathrm{seg} \geq \beta.\langle b_1.\{\cdot\}_B,\ b_2.\{\cdot\}_B\rangle.\vee_{B \pitchfork \Omega}$$

$$\geq \beta.b_i.\{\cdot\}_B, \qquad\qquad i = 1,2,$$

which means $\beta.b_i \underline{\in} b.\!\downarrow\!\mathrm{seg}$ (by singleton principle), or, equivalently, $\beta.b_i \leq b$, for $i = 1,2$. Thus β and b witness that B is 2-directed.

Assume conversely that B is 2-directed. Consider, for $i = 1,2$, the diagonal map in the commutative

$$\begin{array}{ccc}
Bd \times Bd & \xrightarrow{\ \mathrm{proj}_i\ } & Bd \\
\Big\downarrow{\scriptstyle j\times j} & & \Big\downarrow{\scriptstyle j} \\
(B \pitchfork \Omega)\times(B \pitchfork \Omega) & \xrightarrow[\ \mathrm{proj}_i\ ]{} & B \pitchfork \Omega \ ;
\end{array}$$

since it factors through Bd, we know by (2.12) that there exists an epic $\pi_i\colon X_i \to Bd \times Bd$ and a map $b_i'\colon X_i \to B$ with

$$\pi_i.\mathrm{proj}_i.j \leq b_i'.\!\downarrow\!\mathrm{seg}, \qquad\qquad i = 1,2.$$

Taking the fiber product

$$\begin{array}{ccc}
X & \xrightarrow{\ \pi_2'\ } & X_2 \\
\Big\downarrow{\scriptstyle \pi_1'} & & \Big\downarrow{\scriptstyle \pi_2} \\
X_1 & \xrightarrow[\ \pi_1\ ]{} & Bd \times Bd
\end{array}$$

and putting $\pi = \pi_i'.\pi_i$, $b_i = \pi_i'.b_i$, we get

(2.14) $\pi.j \times j.\mathrm{proj}_i = \pi.\mathrm{proj}_i.j \leq b_i.{\downarrow}\mathrm{seg}, \qquad i = 1,2.$

Now we have assumed that B is 2-directed; thus there is an epic $\beta: Y \twoheadrightarrow X$ and a $b: Y \to B$ with

$$\beta.b_i \leq b, \qquad i = 1,2.$$

Then, by (2.14),

$$\beta.\pi.j \times j.\mathrm{proj}_i \leq b.{\downarrow}\mathrm{seg}, \qquad i = 1,2,$$

hence

$$\beta.\pi.j \times j.\vee_{B \pitchfork \Omega} \leq b.{\downarrow}\mathrm{seg}$$

since $\vee_{B \pitchfork \Omega}$ gives the least upper bound of $\mathrm{proj}_1, \mathrm{proj}_2$. Thus $j \times j.\ \vee_{B \pitchfork \Omega}$ "has an upper bound" b, and factors through Bd by some map ℓ (necessarily unique). To prove that (2.13) is actually a pull-back, let

$$B': X \to \mathrm{Bd}, \quad B_i: X \to B \pitchfork \Omega, \qquad i = 1,2$$

be given with $\langle B_1, B_2\rangle.\vee_{B \pitchfork \Omega} = B'.j$. Then

$$B_i \leq B'.j, \qquad i = 1,2$$

and the argument "a subset of a bounded set is bounded" also applies here to give that B_i factors through j, for i = 1,2; formally, if $\beta: Y \twoheadrightarrow X$ and $b: Y \to B$ witness that $B'.j$ is bounded, $\beta.B'.j \leq b.{\downarrow}\mathrm{seg}$, then the same β and b will witness boundedness of B_i, i = 1,2.

We next have to prove that bd preserves the greatest element if and only if B is 0-directed (the greatest element of $(B \pitchfork \Omega)^{op}$ is $\ulcorner \mathrm{false}_B \urcorner$). We have that bd preserves the greatest element if and only if

$$\ulcorner \mathrm{false}_B \urcorner .\mathrm{bd} = \mathrm{true}.$$

But

$$\ulcorner \mathrm{false}_B \urcorner .\mathrm{bd} = \ulcorner \mathrm{false}_B \urcorner .\uparrow\mathrm{seg}_{B \pitchfork \Omega}.\downarrow\mathrm{seg}_B \pitchfork 1.\exists_B$$

$$= \ulcorner \mathrm{true}_{B \pitchfork \Omega} \urcorner .\downarrow\mathrm{seg}_B \pitchfork 1.\exists_B = \ulcorner \mathrm{true}_B \urcorner .\exists_B,$$

and $\ulcorner \mathrm{true}_B \urcorner .\exists_B = \mathrm{true}$ if and only if $\omega_B\colon B \to 1$ is epic, i.e. if and only if B is 0-directed. This proves Proposition 2.9.

If B and C are lower semilattices by means of $\wedge_B\colon B \times B \to B$, $t_B\colon 1 \to B$, $\wedge_C\colon C \times C \to C$, and $t_C\colon 1 \to C$, we have called a map $f\colon B \to C$ a <u>lower semilattice map</u> if it preserves $\wedge$ and t. (Then one easily sees that f is also order-preserving). It is easy to intersect two certain equalizers (which we display below) to get a subobject $B \pitchfork\!\!\pitchfork C \rightarrowtail B \pitchfork C$ of $B \pitchfork C$ "consisting of the lower semilattice maps", more precisely, with the property that $f\colon X \to B \pitchfork C$ factors through $B \pitchfork\!\!\pitchfork C$ if and only if the cartesian twisted version of f

$$B \to X \pitchfork C$$

is a lower semi-lattice map; alternatively, if for all $\alpha\colon Y \to X$, $b_1, b_2\colon Y \to B$

$$<b_1, b_2>.\wedge_B = <(b_1)\alpha.f,\ (b_2)\alpha.f>.\wedge_C$$

and

$$f.t_B \pitchfork 1 = \omega_X.t_C,$$

where $(b_i)\alpha.f$ denotes the composite

$$Y \xrightarrow{<\alpha.f,b_i>} (B \pitchfork C) \times B \xrightarrow{ev} C.$$

It is carved out of $B \pitchfork C$ by intersecting the equalizer of $\hat{u}_{B,C}$ and $\hat{v}_{B,C}$ with the equalizer of $r_{B,C}$ and $s_{B,C}$, where

$$(2.15) \qquad u_{B,C} = (B \pitchfork C) \times B \times B \xrightarrow{id \times \wedge} (B \pitchfork C) \times B \xrightarrow{ev} C$$

and

$$(2.16) \qquad v_{B,C} = (B \pitchfork C) \times B \times B \xrightarrow{<ev^{(1)},ev^{(2)}>} C \times C \xrightarrow{\wedge} C$$

($ev^{(i)}$ being $proj_i.ev^B_C$, $i = 1,2$); - and where

$$r_{B,C} = B \pitchfork C \xrightarrow{<id,\omega.t_B>} (B \pitchfork C) \times B \xrightarrow{ev} C$$

and

$$s_{B,C} = B \pitchfork C \xrightarrow{\omega} 1 \xrightarrow{t_C} C.$$

Let us remark that if $f\colon C \to D$ is a lower semi-lattice map, then $1 \pitchfork f\colon B \pitchfork C \to B \pitchfork D$ restricts to a map $1 \pitchfork\!\!\pitchfork f\colon B \pitchfork\!\!\pitchfork C \to B \pitchfork\!\!\pitchfork D$, where $B,C,$ and D are lower semilattices. This is straightforward to prove.

Now we shall see that

$$\exists : A \pitchfork B \to (A \pitchfork \Omega)^{op} \pitchfork (B \pitchfork \Omega)^{op}$$

factors through $(A \pitchfork \Omega)^{op} \pitchfork\!\!\pitchfork (B \pitchfork \Omega)^{op}$. Essentially, this is because $\exists_f$ is cocontinuous for all f, by $\exists_f \dashv f \pitchfork 1$. But internally, we must prove that $\exists$ equalizes $\hat{u}$ and $\hat{v}$ (as well as r and s) (where $u = u_{(A \pitchfork \Omega)^{op},\ (B \pitchfork \Omega)^{op}}$). This is equivalent to proving

$$\exists \times 1.u = \exists \times 1.v,$$

or again (the $\wedge$ on $(A \pitchfork \Omega)^{op}$ being denoted $\vee_{A \pitchfork \Omega}$)

$$(2.17) \qquad 1_{A \pitchfork B} \times \vee_{A \pitchfork \Omega} . \check{\exists} = (\exists \times 1) . \langle ev^{(1)}, ev^{(2)} \rangle . \vee_{B \pitchfork \Omega} .$$

From the following commutativity

$$\begin{array}{ccc}
(A \pitchfork \Omega) \times (A \pitchfork \Omega) \times A & \xrightarrow{\vee \times 1} & (A \pitchfork \Omega) \times A \\
\Big\downarrow {\scriptstyle \langle ev^{(1)}, ev^{(2)} \rangle} & & \Big\downarrow {\scriptstyle ev} \\
\Omega \times \Omega & \xrightarrow[\vee]{} & \Omega ,
\end{array}$$

we know that

$$a: X \to A \in \langle F_1, F_2 \rangle . \vee_{A \pitchfork \Omega}$$

(where $F_i: X \to A \pitchfork \Omega$) if and only if X is the union of two subobjects $j_i: X_i \rightarrowtail X$ such that $j_i . a \in j_i . F_i$ $(i=1,2)$. According to this fact and the fact that direct image preserves unions, we get the equality (2.17), by using extensionality principle and principle-of-strength-of-existence.

These two principles are also used in proving that $\exists$ equalizes r and s. We omit further details.

<u>2.10 Proposition</u>. If $(C, \leq)$ is an $(\vee\text{-}0)$-upper semilattice object, then any $F: X \to C \pitchfork \Omega$ which factors through $C^{op} \pitchfork \Omega$ is a directed family.

<u>Proof</u>. F is 0-directed, because

$$F . \langle 1, \omega_{C \pitchfork \Omega} . t_C \rangle . ev_C = F . \omega_{C \pitchfork \Omega} . true = true_X$$

so that $\omega_X . t_C \in F$ ($t_C: 1 \to C$ being 0, the maximal element in C^{op}). And F is 2-directed; for let α be a morphism $Y \to X$

and c_1, c_2 morphisms $Y \to C$ such that $c_i \in \alpha.F$ $(i=1,2)$. From $\hat{\alpha}.F.u = \hat{\alpha}.F.v$ (with u and v as in (2.15) and (2.16)) we get

$$\alpha.F \times 1.u = \alpha.F \times 1.v$$

and consequently

$$<1_Y, c_1, c_2>.\alpha F \times 1.1 \times v_C.ev_C$$

$$= <1_Y, c_1, c_2>.\alpha F \times 1.<ev^{(1)}, ev^{(2)}>.\wedge \quad ,$$

that is, letting $c_1 \vee c_2$ denote the composite $<c_1, c_2>.v_C$,

$$<\alpha F, c_1 \vee c_2>.ev_C = <<\alpha.F, c_1>.ev_C, <\alpha.F, c_2>.ev_C>.\wedge$$
$$= <true_Y, true_Y>.\wedge \; = true_Y,$$

so that $c_1 \vee c_2 \in \alpha.F$, as desired.

Now assume that $(B, \leq)$ is a directed ordered object. Consider, for any $A \in \underline{E}$ the composite

(2.18) $$k_{A,B} = A \pitchfork B \xrightarrow{\exists} (A \pitchfork \Omega)^{op} \pitchfork (B \pitchfork \Omega)^{op} \xrightarrow{1 \pitchfork bd} (A \pitchfork \Omega)^{op} \pitchfork \Omega .$$

Set-theoretically, it associates to $f: A \to B$ the family of those subsets $A' \subseteq A$ such that $f(A')$ is a <u>bounded</u> subset of B. - We have seen above that $\exists$ factors through the subobject of lower semilattice maps, and, if further B is directed, bd is a lower semilattice map, so that (2.18) factors as displayed in the diagram

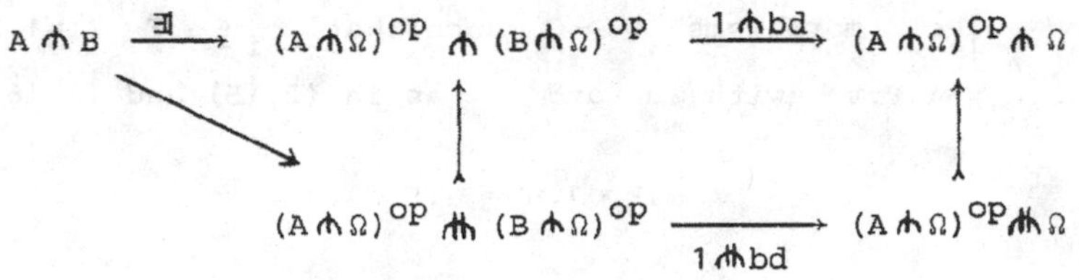

$$A \pitchfork B \xrightarrow{\exists} (A \pitchfork \Omega)^{op} \pitchfork (B \pitchfork \Omega)^{op} \xrightarrow{1 \pitchfork bd} (A \pitchfork \Omega)^{op} \pitchfork \Omega$$

$$(A \pitchfork \Omega)^{op} \pitchfork (B \pitchfork \Omega)^{op} \xrightarrow[1 \pitchfork bd]{} (A \pitchfork \Omega)^{op} \pitchfork \Omega$$

(note that $1 \pitchfork bd$ makes sense only because bd is a lower semi-lattice map). Hence, by Proposition 2.10, (2.18) is a directed family (whenever B is a directed ordered object).

We next prove that the family (2.18) is covering (whether or not B is directed). Consider the diagram

$$A \pitchfork B \xrightarrow{\exists} (A \pitchfork \Omega) \pitchfork (B \pitchfork \Omega) \xrightarrow{1 \pitchfork bd} (A \pitchfork \Omega) \pitchfork \Omega$$

with maps $1 \pitchfork \{\cdot\}$, $\{\cdot\} \pitchfork 1$, $\{\cdot\} \pitchfork 1$, $\exists_{\{\cdot\}}$

$$A \pitchfork (B \pitchfork \Omega) \xrightarrow[1 \pitchfork bd]{} A \pitchfork \Omega \xrightarrow[id]{} A \pitchfork \Omega.$$

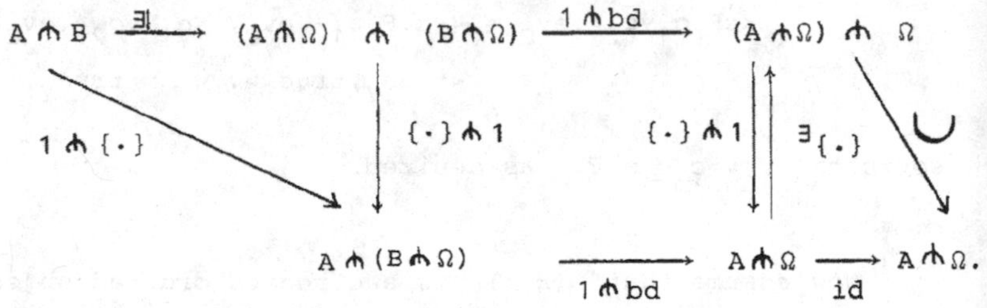

By adjointness $\exists_{\{\cdot\}} \dashv \{\cdot\} \pitchfork 1$, we get the first inequality in the string

$$\exists\!\mid .1 \pitchfork bd.\bigcup \geq \exists\!\mid .1 \pitchfork bd.\{\cdot\} \pitchfork 1.\exists_{\{\cdot\}}.\bigcup$$

$$= \exists\!\mid .1 \pitchfork bd.\{\cdot\} \pitchfork 1$$

$$= \exists\!\mid .\{\cdot\} \pitchfork 1.1 \pitchfork bd$$

$$= 1 \pitchfork \{\cdot\}.1 \pitchfork bd$$

$$= 1 \pitchfork (\{\cdot\}.bd).$$

But now $\{\cdot\}.bd = true_B$, as we have observed. So the right-hand side of the inequality is

$$1 \pitchfork true_B = \omega_{A \pitchfork B}.\ulcorner true_A \urcorner ,$$

which proves that the family (2.18) is covering.

Putting things together, we can now prove

2.11 Proposition. If A is J-finite, it is D-finite.

Proof. Let B be a directed object. We have just seen that in this case the map $k_{A,B}$ constructed in (2.18), $k_{A,B}: A \pitchfork B \to (A \pitchfork \Omega) \pitchfork \Omega$ is directed, and also covering. Since A is J-finite, it is "trivially-covering", that is,

$$\begin{array}{ccc} A \pitchfork B & \xrightarrow{\quad k_{A,B} \quad} & (A \pitchfork \Omega) \pitchfork \Omega \\ {\scriptstyle \omega_{A \pitchfork B}} \big\downarrow & \underline{\in} & \\ 1 & \xrightarrow[\ulcorner true_A \urcorner]{\qquad} & A \pitchfork \Omega \end{array} .$$

Now $\omega_{A \pitchfork B} \cdot \ulcorner true_A \urcorner \,\underline{\in}\, k_{A,B}$ is equivalent to

$$<1_{A \pitchfork B}, \omega_{A \pitchfork B} \cdot \ulcorner true_A \urcorner > . \check{k} = \omega_{A \pitchfork B} . true \quad (= true_{A \pitchfork B}),$$

and since $k = \exists . 1 \pitchfork bd$, we get that the following composite equals $true_{A \pitchfork B}$

$$\begin{array}{ccccccc} A \pitchfork B & \xrightarrow{<1,\omega.true_A>} & (A \pitchfork B) \times (A \pitchfork \Omega) & \xrightarrow{\exists \times 1} & ((A \pitchfork \Omega) \pitchfork (B \pitchfork \Omega)) \times (A \pitchfork \Omega) & \xrightarrow{ev} & B \pitchfork \Omega \\ & & & \searrow^{\check{k}} & & & \big\downarrow {\scriptstyle bd} \\ & & & & & & \Omega \end{array} . \qquad (2.19)$$

The horizontal composite here associates, in the set case, to $f \in A \pitchfork B$ its image (which is a subset of B). In fact, it is the map Im which occurs in Im-princple 1.7. So

$$Im.bd = true_{A \pitchfork B},$$

or, taking bd apart in its constituents,

$$\mathrm{Im}.\uparrow\mathrm{seg}.\downarrow\mathrm{seg} \pitchfork 1.\exists_B = \mathrm{true}_{A \pitchfork B}\colon A \pitchfork B \to \Omega,$$

so by Existence principle, there is an epic $\beta\colon Y \twoheadrightarrow A \pitchfork B$ and a $b\colon Y \to B$ so that

$$b \underline{\in} \beta.\mathrm{Im}.\uparrow\mathrm{seg}.\downarrow\mathrm{seg} \pitchfork 1$$

which is equivalent to

$$b.\downarrow\mathrm{seg} \underline{\in} \beta.\mathrm{Im}.\uparrow\mathrm{seg}$$

(by pull-back principle), which in turn is equivalent to

$$b.\downarrow\mathrm{seg} \geq \beta.\mathrm{Im};$$

in display

(2.20)
$$\begin{array}{ccc} Y & \xrightarrow{\;b\;} & B \\ {\scriptstyle\beta}\Big\downarrow & \leq & \Big\downarrow{\scriptstyle\downarrow\mathrm{seg}} \\ A \pitchfork B & \xrightarrow[\mathrm{Im}]{} & B \pitchfork \Omega. \end{array}$$

We shall from this prove the following inequality (which by passing to exponential adjoints gives the inequality guaranteeing the cofinality desired):

(2.21)
$$\begin{array}{ccc} Y & \xrightarrow{\;b\;} & B \\ {\scriptstyle\mathrm{proj}_Y}\Big\uparrow & \geq & \Big\uparrow{\scriptstyle\mathrm{ev}} \\ Y \times A & \xrightarrow[\beta \times 1]{} & (A \pitchfork B) \times A. \end{array}$$

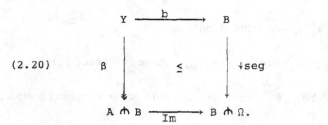

The inequality in (2.21) is equivalent to

(2.22) $$\beta \times 1.\mathrm{ev} \underline{\in} \mathrm{proj}_Y.b.{\downarrow}\mathrm{seg}.$$

To prove (2,22) it suffices, by (2.20) , to prove the $\underline{\in}$-sign in

$$\beta \times 1.\mathrm{ev} \underline{\in} \mathrm{proj}_Y.\beta.\mathrm{Im} = \beta \times 1.\mathrm{proj}_{A \pitchfork B}.\mathrm{Im},$$

which follows immediately from the following general fact about the strength of the "power set monad":

2.12 Lemma. We have $\mathrm{ev} \underline{\in} \mathrm{proj}_{A \pitchfork B}.\mathrm{Im}$, or in display:

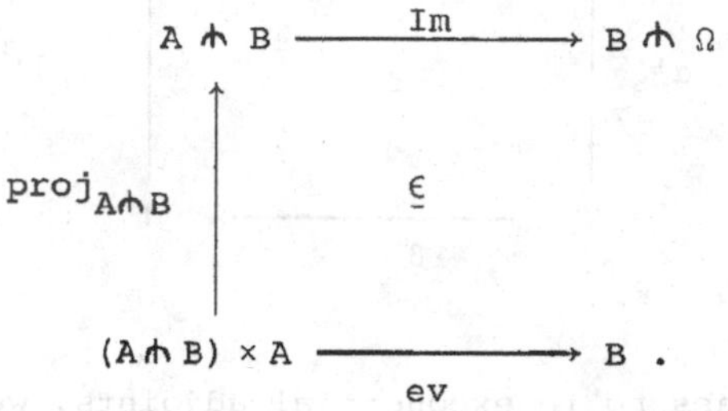

Proof. Set theoretically, this just says that $f(a) \in \mathrm{Im}(f)$ whenever it makes sense. In a general topos, it is a consequence of the Im-principle 1.7. We just have to take (in the notation used in the statement of that principle) $X = Y = (A \pitchfork B) \times A$, $\beta = \mathrm{id}$, $b = \mathrm{ev}_B^A$, $f = \mathrm{proj}_{A \pitchfork B}$, and $a = \mathrm{proj}_A : (A \pitchfork B) \times A \to A$. Then the second condition in 1.7 is satisfied:

$$\langle \mathrm{proj}_{A \pitchfork B}, \mathrm{proj}_A \rangle.\mathrm{ev}_B^A = \mathrm{ev}_B^A,$$

since $\langle \mathrm{proj}_{A \pitchfork B}, \mathrm{proj}_A \rangle$ is an identity map. This proves the lemma, and thereby Proposition 2.11.

We conclude our circle of implications by

2.13 Proposition. Any object A which is D-finite is also K-finite.

<u>Proof</u>. $K(A) \subseteq A \pitchfork \Omega$ is by construction a sub-upper-semilattice of the lattice object $A \pitchfork \Omega$, so in particular, it is a directed ordered object. Since A is D-finite, the diagonal map

$$\Delta\colon K(A) \longrightarrow A \pitchfork K(A)$$

is cofinal. Consider $\{\cdot\}'\colon A \to K(A)$ (the map which composed with $K(A) \subseteq A \pitchfork \Omega$ yields $\{\cdot\}_A$). By cofinality of Δ, there is a diagram of form

$$\begin{array}{ccc} K(A) & \xrightarrow{\ \Delta\ } & A \pitchfork K(A) \\ {\scriptstyle \alpha'}\uparrow & \geq & \uparrow{\scriptstyle \ulcorner\{\cdot\}'\urcorner} \\ Y & \xrightarrow[\ \beta\ ]{} \!\!\!\!\to & 1 \end{array}$$

with β epic. Passing to exponential adjoints, we get an inequality diagram

$$\begin{array}{ccc} K(A) \times A & \xrightarrow{\ \mathrm{proj}\ } & K(A) \\ {\scriptstyle \alpha' \times 1}\uparrow & \geq & \uparrow{\scriptstyle \{\cdot\}'} \\ Y \times A & \xrightarrow[\ \mathrm{proj}_2\ ]{} \!\!\!\!\to & A \end{array}$$

or, equivalently

$$\begin{array}{ccc} Y & \xrightarrow{\ \alpha'\ } & K(A) \\ {\scriptstyle \mathrm{proj}_1}\uparrow & \geq & \uparrow{\scriptstyle \{\cdot\}'} \\ Y \times A & \xrightarrow[\ \mathrm{proj}_2\ ]{} & A \end{array}$$

In particular we have (with α equal to α' followed by the inclusion $K(A) \subseteq A \pitchfork \Omega$)

$$
\text{(2.23)} \qquad
\begin{array}{ccc}
Y & \xrightarrow{\ \alpha\ } & A \pitchfork \Omega \\
{\scriptstyle \text{proj}_1}\uparrow & \geq & \uparrow{\scriptstyle \{\cdot\}} \\
Y \times A & \xrightarrow[\text{proj}_2]{} & A\,.
\end{array}
$$

This, by the "Singleton-domination-Lemma" below, implies that $\alpha =$

$$Y \xrightarrow{\ \beta\ } 1 \xrightarrow{\ulcorner \text{true}_A \urcorner} A \pitchfork \Omega .$$

Since β is epic and α factors through $K(A)$, we conclude that $\ulcorner \text{true}_A \urcorner$ factors through $K(A) \subseteq A \pitchfork \Omega$ so that A is K-finite.

2.14 Singleton-domination Lemma. Suppose we have an inequality diagram of the form (2.23) above. Then

$$\alpha = Y \to 1 \xrightarrow{\ulcorner \text{true}_A \urcorner} A \pitchfork \Omega .$$

Proof. Let $\gamma: Z \to Y$ and $a: Z \to A$ be arbitrary. By extensionality principle, it suffices to prove that $a \underline{\in} \gamma.\alpha$. Consider $<\gamma,a>: Z \to Y \times A$. Clearly

$$a \underline{\in} a.\{\cdot\} = <\gamma,a>.\text{proj}_2.\{\cdot\},$$

and hence by the assumed inequality and the extensionality principle,

$$a \underline{\in} <\gamma,a>.\text{proj}_1.\alpha = \gamma.\alpha,$$

as desired.

Let us sum up the discussion in this section; we proved the implications

K-finite

Proposition 2.13 $\nearrow$ (D-finite $\Rightarrow$ K-finite) $\searrow$ Proposition 2.5 (K-finite $\Rightarrow$ J-finite)

D-finite $\Longleftarrow$ J-finite

Proposition 2.11

so that we have

<u>Theorem</u>. The finiteness notions J,K, and D given in Definitions 2.2, 2.4 and 2.8, respectively, are equivalent.

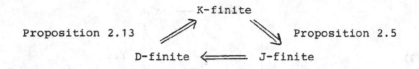

§3. Stability properties and counter examples

In this section, we say that an object A is finite if it verifies one of the three equivalent statements: A is J-finite, A is K-finite, A is D-finite. For any A in the topos $\underline{E}$ under consideration, we denote by k_A the inclusion $K(A) \to A \pitchfork \Omega$. It is not difficult to prove that K is actually a subfunctor of the covariant power "set" functor, by means of k.

Stability properties

3.1 Proposition. The initial object 0 is finite.

Proof. We shall use K-finiteness. We have $0 \pitchfork \Omega = 1$, and thus

$$\ulcorner \mathrm{true}_0 \urcorner = \ulcorner \mathrm{false}_0 \urcorner = \mathrm{id}_1 .$$

Because $\ulcorner \mathrm{false}_0 \urcorner$ factors through $k_0 \colon K(0) \to 0 \pitchfork \Omega$ by construction of K, $\ulcorner \mathrm{true}_0 \urcorner$ does too, and 0 is K-finite.

3.2 Proposition. The terminal object 1 is finite.

Proof. We shall use K-finiteness. We have $1 \pitchfork \Omega \simeq \Omega$. Identifying these objects, we have $\ulcorner \mathrm{true} \urcorner = \mathrm{true}$. Also under the identification $1 = 1 \times 1$, we have $\{\cdot\}_1 = \delta_1$, the characteristic map of $\Delta_1 \colon 1 \to 1 \times 1$; this map is an isomorphism, so $\delta_1 = \{\cdot\}_1 = \mathrm{true}$. Because $\{\cdot\}_1$ factors through k_1 by construction of K, true does too, and 1 is K-finite.

In the proof of the following proposition, round brackets denote maps out of a coproduct (direct sum), pointed brackets denote maps into a product.

3.3 Proposition. We have $K(1) = 1 + 1$, and $k_1 = (\mathrm{true}, \mathrm{false})$.

<u>Proof</u>. We first see that $k_1 \supseteq$ (true,false); for, $k_1 \supseteq \ulcorner \text{false}_1 \urcorner$ = false, by construction of K, and $k_1 \supseteq \{\cdot\}_1$ = true. (It is well-known that (true, false): $1+1 \to \Omega$ is monic.) Conversely, let us prove (true,false) $\supseteq k_1$. Since k_1 is the smallest subobject of $1 \pitchfork \Omega$ containing false_1, containing $\{\cdot\}_1$, and closed under v, it suffices to see that (true,false): $1+1 \to \Omega$ has these three properties. We already know the first two. To see that it is closed under v is essntially just the truth table for disjunction: since X - preserves direct limits, we have

$$(1+1) \times (1+1) \simeq 1+1+1+1 .$$

Denoting this isomorphism $1+1+1+1 \to (1+1) \times (1+1)$ by ρ, we verify that

$$\rho.(\text{true,false}) \times (\text{true,false})$$
$$= (\langle\text{true,true}\rangle, \langle\text{false,true}\rangle, \langle\text{true,false}\rangle, \langle\text{false,false}\rangle).$$

Then

$$(\text{true,false}) \times (\text{true,false}).v = \rho^{-1}.(u_1,u_1,u_1,u_2).(\text{true,false}),$$

where u_1 and u_2 are the structure morphisms of the direct sum 1+1

<u>3.4 Theorem</u>. The product of two finite objects is finite.

We shall use J-finiteness, but there is another proof, using K-finiteness and commutative monads (K being a monoidal submonad of the power "set" monad). Let us first note the following

<u>3.5 Lemma</u>. Let A and B be ordered objects and g: A → B an order-preserving morphism. If F: $X \to A \pitchfork \Omega$ is a 0-directed (respectively 2-directed) family, then $F.\exists_g$ is 0-directed (repsectively 2-directed).

The proof consists in applications of the existence principle, and the fact that the pull-back of two epics gives a commutative square where the composite map is epic. It is quite analogous to the proof of Proposition 2.1, and we omit the details.

Proof of Theorem 3.4. Let A and B be finite objects, and let

$$F\colon X \to ((A\times B)\pitchfork\Omega)\pitchfork\Omega$$

be a directed and covering family. We have to prove that $\omega_X.\ulcorner true_{A\times B}\urcorner \underline{\in} F$. Let us remark that, by virtue of the extensionality principle and union principle, F is covering if and only if for each $\alpha\colon Y \to X$ and each $a\colon Y \to A\times B$, there exists an epic $\beta\colon Z \twoheadrightarrow Y$ and an $h\colon Z \to (A\times B)\pitchfork\Omega$ such that $\beta.a \underline{\in} h$ and $h \underline{\in} \beta.\alpha.F$.

Let us denote by F' the composite morphism

$$X\times A \xrightarrow{F\times\{\cdot\}_A} ((A\times B)\pitchfork\Omega)\pitchfork\Omega\times(A\pitchfork\Omega) \xrightarrow{\psi} (((A\times B)\pitchfork\Omega)\times A)\pitchfork\Omega \xrightarrow{\exists_u} (B\pitchfork\Omega)\pitchfork\Omega,$$

where u corresponds to $ev_{A,B}$ by the isomorphism between $hom((A\times B)\pitchfork\Omega\times A,\ B\pitchfork\Omega)$ and $hom((A\times B)\pitchfork\Omega\times A\times B,\Omega)$. Elementwise, u is characterized by the following property: for

$$g\colon W \to (A\times B)\pitchfork\Omega, \qquad b\colon W \to B, \qquad a\colon W \to A,$$

we have

$$<a,b> \underline{\in}\ g \qquad \text{if} \qquad b \underline{\in} <g,a>.u.$$

We shall prove that F' is a directed covering of A. First, F' is 0-directed. For, F being 0-directed, there exists $\beta\colon Y \twoheadrightarrow X$ and $f\colon Y \to (A\times B)\pitchfork\Omega$ such that $f \underline{\in} \beta.F$; from this and the ψ-principle, we get

$$\begin{aligned} f\times id_A &= <p_Y.f,p_A> \underline{\in} <p_Y.\beta.F,p_A.\{\cdot\}_A>.\psi \\ &= \beta\times id_A.F\times\{\cdot\}_A.\psi, \end{aligned}$$

and by existence principle

$$f\times id_A.u \underline{\in} \beta\times id_A.F\times\{\cdot\}_A.\psi.\exists_u = \beta\times id_A.F'.$$

But $-\times A$ being a left adjoint is epi-preserving, so that $\beta \times id_A$ is epic. Thus, F' is 0-directed.

Next we see that F' is 2-directed. (Set theoretically, if $F(x) = \{A_i\}_{i \in I}$ (with $A_i \subseteq A \times B$), then

$$F'(x,a) = \{A_{i,a}\}_{i \in I} \quad \text{with} \quad A_{i,a} = \{b \in B \mid (a,b) \in A_i\};$$

so since F is directed, we have for each i,j in I a k such that $A_i \cup A_j \subseteq A_k$, and thus such that $A_{i,a} \cup A_{j,a} \subseteq A_{k,a}$.) To prove the general statement, consider α, a_1, and a_2, as displayed in the diagram

$$\begin{array}{ccccc} Y & \xrightarrow{\alpha = \langle x,a \rangle} & X \times A & \xrightarrow{F'} & (B \pitchfork \Omega) \pitchfork \Omega \\ & \searrow^{a_2}_{a_1} \searrow & & & \\ & & B \pitchfork \Omega . & & \end{array}$$

Suppose that $a_i \in \langle x,a \rangle . F$ (for $i = 1,2$). We have to prove that there exists an epic $\delta: W \to Y$ and a $b: W \to B \pitchfork \Omega$ such that $b \geq \delta . a_i$ $(i = 1,2)$, and such that $b \in \delta . \langle x,a \rangle . F'$. By existence principle, there exist epics $\beta_i : Z_i \to Y$ and maps $a_i' : Z_i \to (A \times B) \pitchfork \Omega \times A$ (for $i = 1,2$) such that

$$a_i' \in \beta_i . \langle x,a \rangle . F \times \{\cdot\}_A . \psi$$

and

$$a_i' . u = \beta_i . a_i .$$

Consider the pull-back diagram

$$\begin{array}{ccc} Z & \xrightarrow{\beta_2'} & Z_2 \\ {\scriptstyle \beta_1'} \downarrow & & \downarrow {\scriptstyle \beta_2} \\ Z_1 & \xrightarrow[\beta_1]{} & Y \end{array}$$

and define β and a_i'' by

$$\beta = \beta_1' . \beta_1 = \beta_2' . \beta_2$$

$$a_i'' = \beta_i' . a_i' \qquad (i = 1,2);$$

then we get

$$a_i''.u = \beta.a_i$$

and

$$a_i'' \underline{\in} \beta.\langle x,a\rangle.F \times \{\cdot\}_A.\psi,$$

and then, by principle ψ

$$a_i''.p_1 \underline{\in} \beta.x.F$$

(p_1 being the projection $(((A\times B) ⋔ \Omega) \times A \to (A\times B) ⋔ \Omega)$. But, F being 2-directed, there exists $\gamma: W \twoheadrightarrow Z$ (epic) and $b': W \to (A\times B) ⋔ \Omega$ such that

$$b' \geq \gamma.a_i''.p_1 \qquad (i = 1,2)$$

and

$$b' \underline{\in} \gamma.\beta.x.F.$$

Let δ be the morphism $\gamma.\beta$ and b the morphism $\langle b',\gamma.\beta.a\rangle.u$. Then clearly δ is epic, and further

$$b \geq \delta.a_i \qquad (i = 1,2);$$

for, from $b' \geq \gamma.a_i''.p_1$, we deduce that

$$b = \langle b',\gamma.\beta.a\rangle.u \geq \langle \gamma.a_i''.p_1,\gamma.\beta.a\rangle.u.$$

(In fact, we have more generally from the characteristic property of u that if b' and b'' are maps $W \to (A\times B) ⋔ \Omega$, and $a: W \to A$, then

$$b' \geq b'' \quad \text{implies} \quad \langle b',a\rangle.u \geq \langle b'',a\rangle.u.)$$

Now 2-directedness follows when we prove

(3.1) $$b \underline{\in} \delta.\langle x,a\rangle.F'.$$

From $b' \underline{\in} \gamma.\beta.x.F$ and principle ψ, we get

$$b' \times id_A = \langle p_W.b',p_A\rangle \underline{\in} \langle p_W.\gamma.\beta.x.F,\ p_A.\{\cdot\}\rangle.\psi$$

$$= (\gamma.\beta.x.F \times \{\cdot\}_A).\psi,$$

and thus

$$b = \langle b',\gamma.\beta.a\rangle.u = \langle id_W,\gamma.\beta.a\rangle.b' \times id_A.u$$

$$\underline{\in}\ \langle id_W,\gamma.\beta.a\rangle.(\gamma.\beta.x \times id_A).F \times \{\cdot\}_A.\psi.\exists_u$$

$$= \langle\gamma.\beta.x,\gamma.\beta.a\rangle.F' = \delta.\langle x,a\rangle.F',$$

which proves (3.1).

We now prove F' to be a covering family of B. (Set theoretically, for each $a \in A$ and each $b \in B$, there exists an $i \in I$ such that $(a,b) \in A_i$, and thus also $b \in A_{i,a}$.) Let $\langle x,a\rangle: Y \to X{\times}A$ and $b: Y \to B$ be arbitrary, We have to prove the existence of $\beta: Z \twoheadrightarrow Y$ (epic) and $h: Z \to B \pitchfork \Omega$ such that

$$\beta.b\ \underline{\in}\ b \quad \text{and} \quad h\ \underline{\in}\ \beta.\langle x,a\rangle.F'.$$

Now, F being a covering family, we have a morphism $\beta: Z \twoheadrightarrow Y$ (epic and a morphism $h': Z \to (A{\times}B) \pitchfork \Omega$ such that

$$\beta.\langle a,b\rangle\ \underline{\in}\ h' \quad \text{and} \quad h'\ \underline{\in}\ \beta.x.F.$$

Let h be the morphism $\langle h',\beta.a\rangle.u$. By $\langle\beta.a,\beta.b\rangle\ \underline{\in}\ h'$ and the characteristic property of u, we then have $\beta.b\ \underline{\in}\ h$. To see $h\ \underline{\in}\ \beta.\langle x,a\rangle.F'$, we note that from $h'\ \underline{\in}\ \beta.x.F$ and principle Ψ, we can deduce that

$$h' \times id_A = \langle p_Z.h',p_A\rangle\ \underline{\in}\ \langle p_Z.\beta.x.F,\ p_A.\{\cdot\}_A\rangle.\psi$$

$$= (\beta.x.F) \times \{\cdot\}_A.\psi.$$

Using that, we have

$$\langle h',\beta.a\rangle = \langle id_Z,\beta.a\rangle.h' \times id_A\ \underline{\in}\ \langle id_Z,\beta.a\rangle.(\beta.x.F) \times \{\cdot\}_A.\psi$$

$$= \beta.\langle x,a\rangle.F \times \{\cdot\}_A.\psi.$$

Finally, using existence principle, we get

$$h = \langle h',\beta.a\rangle.u\ \underline{\in}\ \beta.\langle x,a\rangle.F \times \{\cdot\}_A.\psi.\exists_u$$

$$= \beta.\langle x,a\rangle.F'.$$

This proves that F' is a covering family of B. Since we also have seen F' directed, it follows from the assumed J-finiteness of B that

(3.2) $$\omega_{X\times A}.\ulcorner true_B \urcorner \underline{\in} F'.$$

(Set theoretically, for each $a \in A$, there exists an $i_a \in I$ such that $A_{i_a,a} = B$.)

Let us denote by G the morphism

$$X \xrightarrow{F} ((A\times B) \pitchfork \Omega) \pitchfork \Omega \xrightarrow{\exists_{\forall_{p_A}}} (A \pitchfork \Omega) \pitchfork \Omega.$$

(Set theoretically, if $F(x) = \{A_i\}_{i\in I}$, then $G(x)$ is the family $\{A_i'\}_{i\in I}$ with

$$A_i' = \{a \in A \mid \forall b \in B: (a,b) \in A_i\}.)$$

We shall see that G is a directed covering of A. First, G is directed, like F, because of Lemma 3.5 ($\forall_{p_A}$ is order preserving). Also G is covering. (Set theoretically, for each $a \in A$, $A_{i_a a} = B$, and thus

$$a \in A'_{i_a} \subseteq \bigcup_I A_i'.)$$

The formal proof goes as follows.

Let $\alpha: Y \to X$ and $a: Y \to A$ be arbitrary; we have to prove the existence of $\beta: Z \twoheadrightarrow Y$ (epic) and $h: Z \to A \pitchfork \Omega$ so that

$$\beta.a \underline{\in} h \quad \text{and} \quad h \underline{\in} \beta.\alpha.G.$$

From

$$\omega_{X\times A}.\ulcorner true_B \urcorner \underline{\in} F',$$

we deduce that

$$\omega_Y.\ulcorner true_B \urcorner = \langle\alpha,a\rangle.\omega_{X\times A}.\ulcorner true_B \urcorner \underline{\in} \langle\alpha,a\rangle.F'.$$

Since F' was defined by an $\exists$, we get by existence principle the existence of $\beta: Z \twoheadrightarrow Y$ (epic) and $c = \langle c_1,c_2\rangle: Z \to ((A\times B) \pitchfork \Omega) \times A$

such that

(3.3) $c.u = \beta.\omega_Y.\ulcorner true_B \urcorner \quad (= \omega_Z.\ulcorner true_B \urcorner)$

and

$$c \underline{\in} \beta.\langle\alpha,a\rangle.F \times \{\cdot\}_A.\psi.$$

From this, we deduce by principle ψ that

$$c_2 \underline{\in} \beta.a.\{\cdot\}_A$$

(and consequently

$$c_2 = \beta.a),$$

and that

$$c_1 \underline{\in} \beta.\alpha.F.$$

Let us denote by h the morphism

$$Z \xrightarrow{c_2} (A\times B) \pitchfork \Omega \xrightarrow{\forall_{p_A}} A \pitchfork \Omega.$$

It is clear that $h \underline{\in} \beta.\alpha.G$. We want to prove that $\beta.a \underline{\in} h$, i.e. to prove

$$c_2 \underline{\in} c_1.\forall_{p_A}.$$

This is equivalent to

$$c_2.\{\cdot\}_A \leq c_1 . \forall_{p_A},$$

and by the adjunction $p_A \pitchfork id_\Omega \dashv \forall_{p_A}$, this is equivalent to

(3.4) $c_2.\{\cdot\}_A . p_A \pitchfork 1 \leq c_1.$

To prove this inequality, we use extensionality principle. Let $\gamma: Z' \to Z$ and

$$Z' \xrightarrow{\langle a',b\rangle} A\times B$$

be arbitrary and suppose

$$\langle a',b\rangle \underline{\in} \gamma.c_2.\{\cdot\}_A.p_A \pitchfork 1;$$

by pull-back principle

$$a' = \langle a',b\rangle.p_A \underline{\in} \gamma.c_2.\{\cdot\}_A$$

so that

(3.5) $$a' = \gamma . c_2 .$$

We noted above (3.3) that $c.u = \omega_2 . \ulcorner true_B \urcorner$; the characteristic property of u and of $true_B$ implies that

$$<\gamma . c_2 , b> \underline{\in} \gamma . c_1$$

(in fact, for any γ and b where it makes sense). But we have by (3.5) $\gamma . c_2 = a'$; thus

$$<a', b> \in \gamma . c_1$$

which is what is required in order to prove (3.4). So $\beta . a \underline{\in} h$.

Having thus seen that G is a directed covering of A, we deduce from the assumed J-finiteness of A that

(3.6) $$\omega_X . \ulcorner true_A \urcorner \underline{\in} G = F . \exists_{\forall_{p_A}} .$$

(Set theoretically, there exists some $i \in I$ such that $A_i' = A$.) We then finally can prove

$$\omega_X . \ulcorner true_{A \times B} \urcorner \underline{\in} F .$$

(Set theoretically, if $A_i' = A$, then $A_i = A \times B$.)

From (3.6) and existence principle we deduce that there exists an epic $\beta : Y \twoheadrightarrow X$ and a $d : Y \to (A \times B) \pitchfork \Omega$ such that

(3.7) $$d . \forall_{p_A} = \beta . \omega_X . \ulcorner true_A \urcorner = \omega_Y . \ulcorner true_A \urcorner$$

and

$$d \underline{\in} \beta . F .$$

Because $\omega_Y . \ulcorner true_A \urcorner \leq d . \forall_{p_A}$ (by (3.7)), we get by adjointness $p_A \pitchfork id \dashv \forall_{p_A}$ that

$$\omega_Y . \ulcorner true_A \urcorner . p_A \pitchfork id_\Omega \leq d;$$

but $p_A \pitchfork id$ being itself also a right adjoint (to $\exists_{p_A}$) takes maximal elements to maximal elements, so $\ulcorner true_A \urcorner . p_A \pitchfork id = \ulcorner true_{A \times B} \urcorner$.

So $\omega_Y . \ulcorner \text{true}_{A\times B} \urcorner \leq d$, which by maximality of the left hand side implies equality. Thus

$$\beta.\omega_X . \ulcorner \text{true}_{A\times B} \urcorner = \omega_Y . \ulcorner \text{true}_{A\times B} \urcorner = d \mathrel{\underline{\in}} \beta.F,$$

and, cancelling the epic β, we get the desired $\omega_X . \ulcorner \text{true}_{A\times B} \urcorner \mathrel{\underline{\in}} F$. This proves the theorem.

<u>3.6 Theorem</u>. Every quotient object of a finite object is finite.

<u>Proof</u>. We shall use K-finiteness. Let A be a finite object and $p: A \twoheadrightarrow B$ an epic map. Since K is a subfunctor of the power "set" functor, we have

$$k_A . \exists_p = K(p).k_B .$$

Since A is finite, $\ulcorner \text{true}_A \urcorner$ factors through k_A, and thus $\ulcorner \text{true}_A \urcorner . \exists_p$ through k_B. But, for any $f: A \to B$, $\ulcorner \text{true}_A \urcorner . \exists_f$ is the name of the characteristic function of the image of f. In particular, p being epic, $\ulcorner \text{true}_A \urcorner . \exists_p = \ulcorner \text{true}_B \urcorner$. Thus $\ulcorner \text{true}_B \urcorner$ factors through k_B, and B is K-finite.

<u>3.7 Theorem</u>. The direct sum (coproduct) of two objects is finite if and only if both of them is finite.

Proof. We shall use J-finiteness. Consider a direct sum

$$A_1 \xrightarrow{u_1} A_1 + A_2 \xleftarrow{u_2} A_2 .$$

<u>Assume that</u> A_1 <u>and</u> A_2 <u>are J-finite</u>. To prove $A_1 + A_2$ J-finite, consider a directed covering family

$$F: X \to ((A_1 + A_2) \pitchfork \Omega) \pitchfork \Omega .$$

We have to prove that $\omega_X . \ulcorner \text{true}_{A_1+A_2} \urcorner \mathrel{\underline{\in}} F$.

For $j = 1,2$, let F_j' be defined as the composite

$$X \xrightarrow{F} ((A_1 + A_2) \pitchfork \Omega) \pitchfork \Omega \xrightarrow{\exists(u_j \pitchfork id)} (A_j \pitchfork \Omega) \pitchfork \Omega .$$

(Set theoretically, if $F(x) = \{B_i\}_{i\in I}$ with $B_i \subseteq A_1 + A_2$, then $F_1'(x)$ is $\{B_i \cap A_1\}_{i\in I}$, and similarly F_2'.) Then F_j' is directed because F is directed and $u_j \pitchfork id_\Omega$ is order preserving. Also F_j' is covering; to see this, let $\alpha: Y \to X$ and $a_j: Y \to A_j$ be arbitrary. Since F is covering, there exists $\beta: Z \twoheadrightarrow Y$ (epic) and $h: Z \to (A_1 + A_2) \pitchfork \Omega$ such that

$$\beta.a_j.u_j \underline{\in} h \quad \text{and} \quad h \underline{\in} \beta.\alpha.F.$$

From this, using pull-back principle and existence principle, we get

$$\beta.a_j \underline{\in} h.u_j \pitchfork id_\Omega \quad \text{and} \quad h.(u_j \pitchfork id) \underline{\in} \beta.\alpha.F'.$$

So F_j' is covering. Since A_j was assumed J-finite,

$$\omega_X.\ulcorner true_{A_j}\urcorner \in F_j' = F.\exists(u_j \pitchfork id).$$

From this and existence principle, we deduce that there exists an epic $\beta_j: Y_j \twoheadrightarrow X$ and a map $f_j: Y_j \to (A_1 + A_2) \pitchfork \Omega$ such that

$$f_j.u_j \pitchfork id_\Omega = \beta_j.\omega_X.\ulcorner true_{A_j}\urcorner$$

and

$$f_j \underline{\in} \beta_j.F.$$

Considering the pull-back

$$\begin{array}{ccc} Y & \xrightarrow{\beta_j'} & Y_1' \\ {\scriptstyle \beta_2'}\downarrow & & \downarrow{\scriptstyle \beta_1} \\ Y_2' & \xrightarrow{\beta_2} & X \end{array}$$

and denoting the composite $\beta_j'.\beta_j$ by β, we get

$$\beta_j'.f_j \underline{\in} \beta.F.$$

Now, F being 2-directed, there exists $\gamma: Z \twoheadrightarrow Y$ (epic) and $f: Z \to (A_1 + A_2) \pitchfork \Omega$ such that

$$f \geq \gamma.\beta_j'.f_j \qquad (j = 1,2),$$

and

$$f \underline{\in} \gamma.\beta.F.$$

Consequently,

$$f.u_j ⫛ id_\Omega \geq \gamma.\beta_j'.f_j.u_j ⫛ id = \omega_Z.\ulcorner true_A \urcorner ,$$

and thus, by $\exists_{u_j} \dashv u_j ⫛ id$,

$$(3.8) \qquad f \geq \omega_Z.\ulcorner true_{A_j} \urcorner .\exists_{u_j} \qquad (j = 1,2).$$

But

$$(\ulcorner true_{A_1} \urcorner .\exists_{u_1}) \vee (\ulcorner true_{A_2} \urcorner .\exists_{u_2}) = \ulcorner true_{A_1+A_2} \urcorner ,$$

since the two sides in this equation are the names of the characteristic maps of the subobjects $u_1 \cup u_2$ and $A_1 + A_2$ of $A_1 + A_2$, respectively. From the two inequalities $(j = 1,2)$ in (3.8) we therefore get $f \geq \omega_Z.\ulcorner true_{A_1+A_2} \urcorner$. Such an inequality must be an equality, We thus have

$$\gamma.\beta.\omega_X.\ulcorner true_{A_1+A_2} \urcorner = \omega_Z.\ulcorner true_{A_1+A_2} \urcorner = f \underline{\in} \gamma.\beta.F,$$

and cancelling off the epic $\gamma.\beta$, we have

$$\omega_X.\ulcorner true_{A_1+A_2} \urcorner \underline{\in} F.$$

This proves that $A_1 + A_2$ is J-finite.

<u>Assume conversely that</u> $A_1 + A_2$ <u>is J-finite</u>. Suppose we have a directed covering

$$F: X \to (A_1 ⫛ \Omega) ⫛ \Omega.$$

We have to prove that $\omega_X.\ulcorner true_{A_1} \urcorner \underline{\in} F$. We let $G: X \to ((A_1+A_2) ⫛ \Omega) ⫛ \Omega$ be the family which "consists of all subsets $A_i' + A_2$ of $A_1 + A_2$ with $A_i' \subseteq A_1$ being a member of F"; formally, let G be the morphism

$$X \xrightarrow{F} (A_1 ⫛ \Omega) ⫛ \Omega \xrightarrow{\exists_g} ((A_1 + A_2) ⫛ \Omega) ⫛ \Omega,$$

where

$$g\colon A_1 \pitchfork \Omega \to (A_1 + A_2) \pitchfork \Omega$$

is the disjunction (in the lattice $\hom(A_1 \pitchfork \Omega, (A_1 + A_2) \pitchfork \Omega)$) of $\exists u_1$ and $\omega . \ulcorner \mathrm{true}_{A_2} \urcorner . \exists u_2$.

We claim that G is a directed covering of $A_1 + A_2$. Directedness is easy because F is directed and g is order preserving. Indeed, each of the maps in the disjunction defining g is order preserving. To prove that G is covering, let

$$\alpha\colon Y \to X \qquad \text{and} \qquad a\colon Y \to A_1 + A_2$$

be arbitrary. We have to prove the existence of $\beta\colon Z \twoheadrightarrow Y$ (epic) and $h\colon Z \to (A_1 + A_2) \pitchfork \Omega$ such that

$$\beta . a \mathrel{\underline{\in}} h \qquad \text{and} \qquad h \mathrel{\underline{\in}} \beta . \alpha . G.$$

For $i = 1,2$, let us consider the pull-back

(3.9)
$$\begin{array}{ccc} Y_i & \overset{v_i}{\rightarrowtail} & Y \\ {\scriptstyle a_i}\downarrow & & \downarrow{\scriptstyle a} \\ A_i & \underset{u_i}{\rightarrowtail} & A_1 + A_2 \end{array} \quad ;$$

F being a covering of A_1, there exists $\pi_1\colon P_1 \twoheadrightarrow Y_1$ (epic) and $h_1\colon P_1 \to A_1 \pitchfork \Omega$ such that

$$\pi_1 . a_1 \mathrel{\underline{\in}} h_1 \qquad \text{and} \qquad h_1 \in \pi_1 . v_1 . \alpha . F.$$

Now

$$\pi_1 . v_1 . a = \pi_1 . a_1 . u_1 \mathrel{\underline{\in}} h . \exists u_1 ,$$

and, because $h_1 . \exists u_1 \leq h_1 . g$,

(3.10)
$$\pi_1 . v_1 . a \mathrel{\underline{\in}} h_1 . g .$$

On the other hand, we trivially have from $h_1 \mathrel{\underline{\in}} \pi_1 . v_1 . \alpha . F$ that

(3.11)
$$h_1 . g \mathrel{\underline{\in}} \pi_1 . v_1 . \alpha . G .$$

On the other hand, let us consider the pull-back (3.9) for $i=2$. Since F is 0-directed, there exists $\pi_2'\colon P_2' \twoheadrightarrow X$ (epic) and $f\colon P_2' \to A_1 \pitchfork \Omega$ such that $f \underline{\in} \pi_2'.F$. Consider the following diagram, in which the left hand square is constructed as a pull-back

$$\begin{array}{ccccc} P_2 & \xrightarrow{v_2'} & P_2' & \xrightarrow{f} & A_1 \pitchfork \Omega \\ \downarrow{\scriptstyle \pi_2} & & \downarrow{\scriptstyle \pi_2'} & \underline{\in} & \\ Y_2 & \xrightarrow[v_2.\alpha]{} & X & \xrightarrow[F]{} & (A_1 \pitchfork \Omega) \pitchfork \Omega. \end{array}$$

If we denote by h_2 the morphism $v_2'.f$, we get

$$h_2 \underline{\in} \pi_2.v_2.\alpha.F$$

from which we deduce that

$$(3.12) \qquad h_2.g \underline{\in} \pi_2.v.\alpha.G.$$

On the other hand, we know that $\pi_2.a_2 \underline{\in} \omega_{P_2}.\ulcorner \mathrm{true}_{A_2} \urcorner$, and thus

$$\pi_2.v_2.a = \pi_2.a_2.u_2 \underline{\in} \omega_{P_2} \ulcorner \mathrm{true}_{A_2} \urcorner .\exists u_2;$$

but

$$\omega_{P_2}.\ulcorner \mathrm{true}_{A_2} \urcorner .\exists u_2 = h_2.\omega_{A_1 \pitchfork \Omega}.\ulcorner \mathrm{true}_{A_2} \urcorner .\exists u_2 \leq h_2 g,$$

so that

$$(3.13) \qquad \pi_2.v_2.a \underline{\in} h_2.g.$$

By the universality of direct sums, we know that Y is a direct sum of Y_1 and Y_2 by means of v_1, v_2. Let us denote by β the morphism

$$\beta = P_1 + P_2 \xrightarrow{\pi_1 + \pi_2} Y$$

and by h the morphism

$$h = P_1 + P_2 \xrightarrow{(h_1.g,\, h_2.g)} (A_1 + A_2) \pitchfork \Omega.$$

An easy computation shows that if $f_i \underline{\in} F_i$ $(i=1,2)$, where f_1 and f_2 (and thus F_1 and F_2) are coterminal, then $(f_1,f_2) \underline{\in}$

(F_1,F_2). Consequently, we have from the four relations (3.10) - (3.13) that

$$\beta.a = (\pi_1.v_1.a, \pi_2.v_2.a) \underline{\in} h$$

and

$$\underline{h} \underline{\in} \beta.\alpha.G,$$

so that G is covering.

Since now $A_1 + A_2$ is assumed to be J-finite,

$$\omega_X.\ulcorner \mathrm{true}_{A_1+A_2} \urcorner \underline{\in} G.$$

To prove this, that $\omega_X.\ulcorner \mathrm{true}_A \urcorner \underline{\in} F$, we need

3.8 Lemma. The map $g\colon A_1 ⋔ \Omega \to (A_1 + A_2) ⋔ \Omega$ (defined above as disjunction of $\exists u_1$ and $\omega.\ulcorner \mathrm{true}_{A_2} \urcorner.\exists u_2$) is monic.

(Set theoretically, for $A_1' \subseteq A_1$ and $A_1'' \subseteq A_1$, we have that

$$u_1(A_1') \cup u_2(A_2) = u_1(A_1'') \cup u_2(A_2)$$

implies that $A_1' = A_1''$, by disjointness of disjoint sums $A_1 + A_2$.)

The formal proof goes as follows: Let us denote by t the endomorphism of $(A_1 + A_2) ⋔ \Omega$ which has the effect of "intersecting with $A_1 \rightarrowtail A_1 + A_2$":

$$t = \mathrm{id}_{(A_1+A_2) ⋔ \Omega} \wedge \omega.\ulcorner \mathrm{true}_{A_1} \urcorner.\exists u_1.$$

Then

$$\begin{aligned} g.t &= g \wedge (\omega.\ulcorner \mathrm{true}_{A_1} \urcorner.\exists u_1) \\ &= (\exists u_1 \vee \omega.\ulcorner \mathrm{true}_{A_2} \urcorner.\exists u_2) \wedge (\omega.\ulcorner \mathrm{true}_{A_1} \urcorner.\exists u_1) \\ &= (\exists u_1 \wedge (\omega.\ulcorner \mathrm{true}_{A_1} \urcorner.\exists u_1)) \vee ((\omega.\ulcorner \mathrm{true}_{A_2} \urcorner.\exists u_2) \wedge (\omega.\ulcorner \mathrm{true}_{A_1} \urcorner.\exists u_1)) \end{aligned}$$

(since the distributive law of $\wedge$ over $\vee$ holds in any lattice of form $\hom_{\underline{E}}(X, Y ⋔ \Omega)$). The first constituent of this disjunction equals $\exists u_1$ because $\mathrm{id}_{A_1 ⋔ \Omega} \leq \omega.\ulcorner \mathrm{true}_{A_1} \urcorner$ and because $\exists u_1$ is order preserving. The second constituent equals $\omega.\ulcorner \mathrm{false}_{A_1+A_2} \urcorner$, because of $u_1 \wedge u_2 = 0$ (disjointness of coproduct). Thus

$$(3.14) \qquad g.t = \exists u_1 \vee \omega.\ulcorner \text{false}_{A_1+A_2} \urcorner = \exists u_1 .$$

Now $\exists u_1$ is monic since u_1 is monic (this is well known: f monic implies $\exists_f . f \pitchfork 1 = id$, thus $\exists_f$ is split mono). Thus (3.14) implies g monic.

We can now finish the proof that A_1 is J-finite. We have

$$(3.15) \qquad \ulcorner \text{true}_{A_1} \urcorner .g = \ulcorner \text{true}_{A_1+A_2} \urcorner ,$$

because $\ulcorner \text{true}_{A_1} \urcorner .g = (\ulcorner \text{true}_{A_1} \urcorner .\exists u_1) \vee (\ulcorner \text{true}_{A_2} \urcorner .\exists u_2) = \ulcorner \text{true}_{A_1+A_2} \urcorner$. We proved above that $\omega_X . \ulcorner \text{true}_{A_1+A_2} \urcorner \underline{\in} G = F.\exists g$. From the existence principle we get the existence of an epic $\beta : Y \twoheadrightarrow X$ and $c : Y \to A_1 \pitchfork \Omega$ such that

$$c \underline{\in} \beta.F$$

and

$$c.g = \beta.\omega_X . \ulcorner \text{true}_{A_1+A_2} \urcorner = \beta.\omega_X . \ulcorner \text{true}_{A_1} \urcorner .g,$$

the last equality sign by (3.15). Now g being monic by Lemma 3.8, we have

$$\beta.\omega_X . \ulcorner \text{true}_{A_1} \urcorner = c \underline{\in} \beta.F,$$

and finally, β being epic, we get $\omega_X . \ulcorner \text{true}_{A_1} \urcorner \underline{\in} F$.

<u>3.9 Corollary</u>. If a subobject of a finite object has a complement, then it is finite.

<u>Proof</u>. If $A_1 \rightarrowtail A$ is a subobject of A with a complement $A_2 \rightarrowtail A$ (this means $A_1 \cap A_2 = 0$, $A_1 \cup A_2 = A$), then it is well known that $A = A_1 + A_2$. If A is finite, then so is A_1, by the theorem.

<u>3.10 Corollary</u>. In a Boolean topos, every subobject of a finite object is finite.

This is not true in general for non-Boolean toposes. We turn to the counter examples.

Counter examples

3.11 Proposition. In a non-Boolean topos, a subobject of a finite object can be non-finite.

This will be a corollary of Theorem 3.12 below, which is concerned with the category $\mathfrak{S}^{\mathbf{2}}$, i.e., the category whose objects A are maps $A = (A_1 \to A_2)$ in the category of sets, and whose morphisms are commutative squares. If $A = (A_1 \to A_2)$ is an object in $\mathfrak{S}^{\mathbf{2}}$, we shall refer to A_1 as the top of A, and A_2 the bottom set of A.

3.12 Theorem. In the topos $\mathfrak{S}^{\mathbf{2}}$, the finite objects are the surjective maps between finite sets.

The proof we give here will depend on a theorem of Mikkelsen, to be published elsewhere [13]; it says that a left exact left adjoint functor between toposes preserves K-finite objects. Now $\mathfrak{S}^{\mathbf{2}}$ is connected to $\mathfrak{S}$ by five functors

$$\mathfrak{S} \quad \begin{array}{c} \xrightarrow{\quad I \quad} \\ \xleftarrow{\quad D \quad} \\ \xrightarrow{\quad T \quad} \\ \xleftarrow{\quad L \quad} \\ \xrightarrow[\quad R \quad]{} \end{array} \quad \mathfrak{S}^{\mathbf{2}},$$

each left adjoint to the one below. They are defined as follows:

$$I(M) = (0 \to M)$$
$$D(A \to B) = B$$
$$T(M) = (M \xrightarrow{\text{id}} M)$$
$$L(A \to B) = A$$
$$R(M) = (M \to 1).$$

Consequently, D, T, and L preserve finite objects. But we know that the finite objects of $\mathbb{S}$ are exactly the finite sets. Hence, if $f: A \to B$ is finite in $\mathbb{S}^2$, then A and B are finite sets, and if M is a finite set, then id_M is a finite object in $\mathbb{S}^2$. Furthermore, if p is a surjection from the finite set A to B, then p is a finite object by Theorem 3.6 because it is a quotient object of the finite object id_A.

It only remains to prove that:

<u>If</u> $g: A \to B$ <u>is a non-surjective map, then</u> g <u>is non-finite</u>.

Let b_0 be an element of $B - \text{Im}(g)$. Then b_0 determines a subobject of g, namely the one having top set empty and bottom set $\{b_0\}$. It has a complement, namely the object with top set A and bottom set $B - \{b_0\}$. By Corollary 3.9, to see that g is non-finite it suffices to see that the object $h': 0 \to \{b_0\}$ is non-finite. This object is isomorphic to the subobject

$$h: 0 \to 1 \tag{3.16}$$

of 1. So it suffices to prove that h is non-finite. We use J-finiteness. We then have to exhibit a directed covering $X \to (h \pitchfork \Omega) \pitchfork \Omega$ which is not trivially covering. In fact, we can construct one with $X = 1$

$$\overline{F}: 1 \to (h \pitchfork \Omega) \pitchfork \Omega.$$

This will allow a conceptual simplification, since such an $\overline{F}$ is given as the name of the characteristic map of an actual subobject

$F \rightarrowtail h \pitchfork \Omega$ of $h \pitchfork \Omega$. Then it is easy to see that $\overline{F}$ a directed family is equivalent to F being a directed ordered object (with the ordering induced on it from $h \pitchfork \Omega$). Also, $\overline{F}$ factoring through the object of sub-upper-semi-lattice maps is equivalent to F being a sub-upper-semi-lattice of $h \pitchfork \Omega$ (and this implies that $\overline{F}$ is directed). These statements are valid and easy to see in arbitrary topos, for any object h. For the specific object h (see (3.16)) in $\mathcal{S}^2$, we now describe a directed family $\overline{F}\colon 1 \to (h \pitchfork \Omega) \pitchfork \Omega$ by describing a sub-upper-semi-lattice F of $h \pitchfork \Omega$. In general, for $A = (A_1 \xrightarrow{f} A_2)$ an object in $\mathcal{S}^2$, its power object $A \pitchfork \Omega$ has for its top set the set of subobjects of A, and for its bottom set the set of subsets of A_2. The map from top to bottom is given by sending the subobject

$$\begin{array}{ccc} A_1' & \rightarrowtail & A_1 \\ \downarrow & & \downarrow \\ A_2' & \rightarrowtail & A_2 \end{array}$$

to $A_2' \rightarrowtail A_2$. Using this description, it is clear that $h \pitchfork \Omega$ has exactly two elements in its top set, as well as in its bottom set; the map from top to bottom is bijective. The upper semi-lattice structure on $h \pitchfork \Omega$ gives rise to upper semi-lattice structures on the top set as well as on the bottom set of $h \pitchfork \Omega$. Let F consist of the smallest element (with respect to the semi-lattice structure) of the top set, and of both elements of the bottom set. In display (with top set of $h \pitchfork \Omega$ denoted $\{f_1, t_1\}$, bottom set $\{f_2, t_2\}$, with $f_1 \leq t_1$ and $f_2 \leq t_2$):

(3.17)

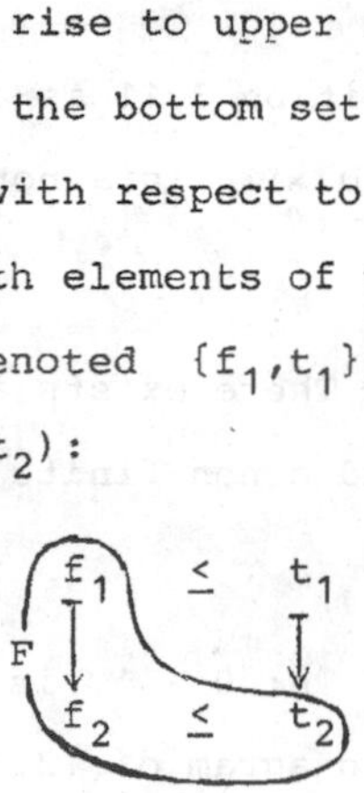

.

Then clearly F is a sub-upper semi-lattice of $h \pitchfork \Omega$, thus the corresponding $\bar{F}: 1 \to (h \pitchfork \Omega) \pitchfork \Omega$ is a directed family. To see that it is covering, we compute

$$\bar{F}.\bigcup_h : 1 \to h \pitchfork \Omega .$$

This can be done directly, but is easier indirectly: there are only two maps from 1 to $h \pitchfork \Omega$ namely $\ulcorner \text{false}_h \urcorner$ and $\ulcorner \text{true}_h \urcorner$ (given, respectively, by the two f's and the two t's in (3.17). We just have to exclude

(3.18) $$\bar{F}.\bigcup_h = \ulcorner \text{false}_h \urcorner .$$

But the functor $D: \mathcal{S}^2 \to \mathcal{S}$ which picks out bottom sets, is logical, and thus commutes with all constructions involved here; applying D to (3.18) yields

$$(1 \xrightarrow{D(\bar{F})} (D(h) \pitchfork 2) \pitchfork 2 \xrightarrow{\bigcup} D(h) \pitchfork 2) = \ulcorner \text{false}_{D(h)} \urcorner$$

in $\mathcal{S}$, which contradicts the fact that $\bigcup F_2 = \{f_2, t_2\} = D(h)$. So $\bar{F}$ is covering. Clearly, it is not trivially covering: we do not have $\ulcorner \text{true}_h \urcorner \in \bar{F}$, since $\ulcorner \text{true}_h \urcorner$ does not factor through F (t_1 being excluded from F). Thus h is not J-finite. This proves the Theorem.

We now get Propposition 3.11 as a corollary of Theorem 3.12; indeed, in $\mathcal{S}^2$, $h: D \to 1$ is a non-finite subobject of the finite object $1 \to 1$.

<u>3.13 Proposition</u>. There exists a finite object A such that $A \pitchfork \Omega$ is non-finite, and a non-finite object A such that $A \pitchfork \Omega$ is finite.

Indeed, in $\mathcal{S}^2$, $h: 0 \to 1$ is non-finite, whereas $h \pitchfork \Omega$, displayed in the total diagram of (3.17), is finite (being isomorphic to $T(2)$). Conversely, to see an example where A is finite and

$A \pitchfork \Omega$ is non-finite, we consider the topos $\mathcal{S}^N$, where N is the natural numbers monoid under addition. Then the objects of $\mathcal{S}^N$ are sets equipped with an endomorphism, and the morphisms from $\alpha: A \to A$ to $\beta: B \to B$ are the maps f from A to B such that $\alpha.f = f.\beta$. The subobject classifier Ω is the set $\bar{N} = -N \cup \{-\infty\}$ equipped with the endomorphism ω defined by the formulas

$$\omega(-\infty) = -\infty$$

$$\omega(0) = 0$$

$$\omega(-(n+1)) = -n \qquad \text{for all } n \in N.$$

In $\mathcal{S}^N$, a necessary (and, in fact, also sufficient) condition for an object $\alpha: A \to A$ to be finite is that A is a finite set and α is a permutation; for the functor

$$\mathcal{S}^N \to \mathcal{S}^{\mathfrak{2}}$$

which sends $\alpha: A \to A$ to $\alpha: A \to A$ has adjoints on both sides since it is induced by a functor between the index categories, $\mathfrak{2} \to N$, in picture

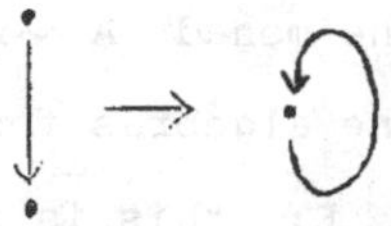

In particular, it is a left exact left adjoint, so preserves finite objects. So if $\alpha: A \to A$ is finite in $\mathcal{S}^N$, it is finite in $\mathcal{S}^{\mathfrak{2}}$, meaning that A is a finite set, and α is a surjective mapping. But a surjective endomorphism of a finite set is bijective.

In particular, in $\mathcal{S}^N$, Ω is not finite; so 1 is finite, whereas $1 \pitchfork \Omega \simeq \Omega$ is not.

Miscellaneous remarks

The morale of Proposition 3.13 is that the power object formation $A \pitchfork \Omega$ is in some respects not the natural thing to consider when dealing with a finite object A. The picture changes if one considers the set of finite subsets of a finite set A, namely K(A). In fact, the following theorem holds.

3.14 Theorem. For any A, K(A) is finite if and only if A is finite.

This was proved by two of the authors (⇒ by P.L., ⇐ by C.J.M.); the proofs will appear elsewhere (hopefully).

On the basis of this, one of the authors (P.L.) has proved that if A and B are finite and B injective, then $A \pitchfork B$ is finite, and that, in the Boolean case, if A and B are finite, then $A \pitchfork B$ is finite. But in a Boolean topos, $\Omega = 1+1$ is finite. Consequently, the finite objects of a Boolean topos define a subtopos.

Using the technique of [8], another of the authors (C.J.M.) has made an analysis of the monad $A \rightsquigarrow K(A)$ on an elementary topos E. He also proved that the algebras for this monad is the category of upper semi-lattices in E; this is of course the category of algebras for a certain (external) finitary algebraic theory $\mathbb{T}$ (generated by a nullary operation 0 and a binary operation $\vee$). In paticular, the category of algebras for this $\mathbb{T}$ is triplable without assuming a natural numbers object in E.

We mentioned briefly in the Introduction the relationship between notions studied in [1], Exposé 6, and our notions. In [1], finiteness (or quasi-compactness, rather) is also discussed for maps in a topos. In this direction, one can prove (A.K.) the result that for a map $f: A \to B$ in an elementary topos, the two conditions

(i) f is a finite object in $\underline{E}/B$

(ii) if $X \to B$ is finite in $\underline{E}/B$, then $f^*(X)$ is finite in $\underline{E}/A$ ("f has finite fibres")

are equivalent.

§ 4. Algebraic lattice objects

In this section we give an example showing how the finiteness- and directedness-notions considered can be used to lift lattice-theoretic theorems into arbitrary elementary toposes. The example is a topos-theoretic version of a generalization of a (not very deep) theorem of Jürgen Schmidt, stating that a closure system is algebraic if and only if it is inductive. Or, alternatively, our example may be seen as liftings of lattice theoretic specializations of recent results concerning categories (Gabriel and Ulmer, [7], § 10).

By an algebraic lattice object in an elementary topos $\underline{E}$, we understand a complete ordered object B, such that the identity map of B can be written as the composite

$$B \xrightarrow{\downarrow \text{seg}} B \pitchfork \Omega \xrightarrow{s \pitchfork 1} S(B) \pitchfork \Omega \xrightarrow{\exists s} B \pitchfork \Omega \xrightarrow{\sup} B$$

(This expresses, set-theoretically, that every $b \in B$ is the supremum of the intranscessible elements $c \leq b$. This is one of the equivalent forms of the classical description of algebraic lattices, see for instance Diener [6]).

If $i: C \rightarrowtail B$ is a subobject of B with order-relation on C induced by that on B, then set-theoretically, i has a left adjoint $\overline{\varphi}: B \to C$ if and only if i preserves all inf's (by adjoint functor theorem), or, equivalently, if and only if C is closed under the formation of inf's in B. So, considering such a situation

(4.1) $\qquad i: C \rightarrowtail B \qquad \overline{\varphi}: B \to C \qquad \overline{\varphi} \dashv i,$

set-theoretically amounts to considering a closure system C on the complete ordered set B, with C the set of closed elements. We consider the situation (4.1) in an arbitrary elementary topos, and denote by φ the composite

$$B \xrightarrow{\overline{\varphi}} C \xrightarrow{i} B.$$

It is a closure operator (monad) on B. Every closure operator arises from such $\overline{\varphi}$, i (by the "Eilenberg-Moore factorization").

Generalizing slightly the terminology of Jürgen Schmidt [17], or Cohn [5], we call the situation (4.1) an <u>inductive</u> closure system if i preserves directed sup's, in the sense that if

$$F: X \longrightarrow C \pitchfork \Omega$$

is directed, then

$$\begin{array}{ccccc} X & \xrightarrow{F} & C\pitchfork\Omega & \xrightarrow{\exists i} & B\pitchfork\Omega \\ & & \downarrow \sup_C & & \downarrow \sup_B \\ & & C & \xrightarrow{i} & B \end{array}$$

commutes. And we call $\varphi: B \to B$ an <u>algebraic closure operator</u> if φ can be written as the composite

$$(4.2) \qquad B \xrightarrow{\downarrow seg} B\pitchfork\Omega \xrightarrow{s\pitchfork 1} S(B)\pitchfork\Omega \xrightarrow{\exists s} B\pitchfork\Omega \xrightarrow{\exists \varphi} B\pitchfork\Omega \xrightarrow{sup} B.$$

(Set-theoretically: for every $b \in B$, $\varphi(b) = \sup\{\varphi(d) \mid d \leq b$ and d intranscessible$\}$).

With notation as in (4.1) and the terminology just introduced, we have the following generalization, and lifting to topos context of Jürgen Schmidt's Theorem

<u>4.1. Theorem</u>. Let B be an algebraic lattice object. Then a closure system on B is inductive if and only if the corresponding closure operator is algebraic.

(These conditions are also equivalent to: $\overline{\varphi}$ preserves intranscessibles - see Theorem 4.6).

Before proving the theorem, we state three lemmas.

<u>4.2. Lemma</u>. Let $\varphi\colon B \to D$ be a monotone map between ordered objects. Then we have the inequality

$$\begin{array}{ccc} B \pitchfork \Omega & \xrightarrow{\ \exists\varphi\ } & D \pitchfork \Omega \\ \downarrow cl \Big| & \leq & \Big| \downarrow cl \\ B \pitchfork \Omega & \xrightarrow[\ \exists\varphi\ ]{} & D \pitchfork \Omega \end{array}$$

The proof is straightforward and omitted.

<u>4.3. Lemma</u>. We have the equality $\downarrow cl \cdot sup_B = sup_B$.

The proof is straightforward and omitted.

<u>4.4. Main Lemma</u>. If $F\colon X \to B \pitchfork \Omega$ is directed, then the total diagram in

$$X \xrightarrow{\ F\ } B \pitchfork \Omega \underset{\downarrow cl}{\overset{\xrightarrow{sup} B \xrightarrow{\downarrow seg}}{\xrightarrow{\qquad\qquad\qquad}}} B \pitchfork \Omega \xrightarrow{\ s \pitchfork 1\ } S(B) \pitchfork \Omega$$

commutes.

<u>Proof</u>. Since $F.{\downarrow}cl.\ sup = F.sup$ (Lemma 4.3), and since clearly F directed implies $F.{\downarrow}cl$ directed, we may as well replace F by $F.{\downarrow}cl$. Thus we just have to prove that, for a directed and $\downarrow$-closed $F\colon X \to B \pitchfork \Omega$, we have

$$F.sup.{\downarrow}seg.\ s \pitchfork 1 \leq F.\ s \pitchfork 1,$$

the other inequality being obvious. We use the extensionality principle 1.2. Let $\alpha\colon Y \to X$ and $b\colon Y \to S(B)$ satisfy

$$b \underset{=}{\in} \alpha.F.sup.\downarrow seg.\ s \pitchfork 1.$$

Then

$$b.s \underset{=}{\in} \alpha.F.sup.\downarrow seg.$$

So

$$b.s \leq \alpha.F.sup.$$

Now $b.s.: Y \to B$ is intranscessible since s is. Since $\alpha.F: Y \to B \pitchfork \Omega$ is directed, we therefore have an epic $\beta: Z \twoheadrightarrow Y$ and $d: Z \to B$ with

$$\beta.b.s \leq d \qquad \text{and} \qquad d \underset{=}{\in} \beta.\alpha.F.$$

Since F is $\downarrow$-closed, $\beta.b.s. \underset{=}{\in} \beta.\alpha.F$, and since β is epic, it cancels off, so that we have $b.s. \underset{=}{\in} \alpha.F$. From this, we conclude

$$b \underset{=}{\in} \alpha.F.s \pitchfork 1$$

as desired.

   <u>Proof of the theorem</u>. Suppose first that the closure system is inductive. We must prove that the following diagram is commutative

(4.3)

$$
\begin{array}{ccccccccc}
B & \xrightarrow{\downarrow seg} & B \pitchfork \Omega & \xrightarrow{s \pitchfork 1} & S(B) \pitchfork \Omega & \xrightarrow{\exists s} & B \pitchfork \Omega & \xrightarrow{\exists \overline{\varphi}} & C \pitchfork \Omega \\
\varphi \downarrow & & & & & \exists \varphi \searrow & & & \downarrow \exists i \\
B & \xleftarrow{\hspace{4cm} sup_B \hspace{4cm}} & & & & & & & B \pitchfork \Omega
\end{array}
$$

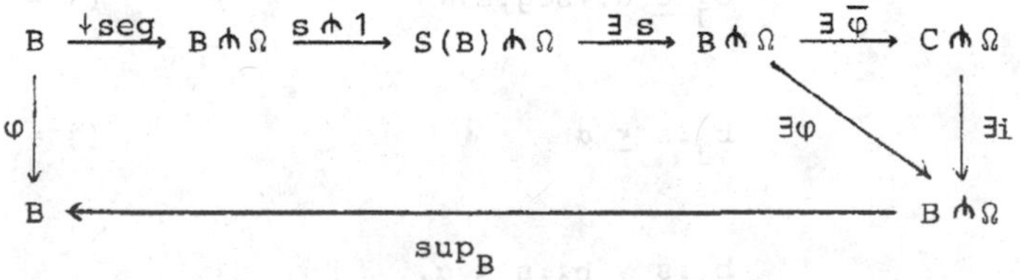

Let γ denote the top row. If we can prove γ directed, then by assumption of inductivity of $i: C \to B$,

$$\gamma.\exists i.sup_B = \gamma.sup_C.i \quad ,$$

and substituting for γ and using that $\bar{\varphi}$ preserves sup's (being a left adjoint), the right hand side of this equation becomes

$$\downarrow\text{seg}.\ s \pitchfork 1.\ \exists s.\ \exists\bar{\varphi}.\ \text{sup}_C.\ i$$

$$= (\downarrow\text{seg}.\ s \pitchfork 1.\ \exists s.\ \text{sup}_B).\ \bar{\varphi}.i$$

$$= \bar{\varphi}.i = \varphi$$

because the bracket part is an identity map by the assumption that B is an algebraic lattice. It thus remains to prove that the top row in (4.3) actually <u>is</u> directed. Existential quantification along an order-preserving map preserves the notion of directedness, so it suffices to prove that

$$\downarrow\text{seg}.\ s \pitchfork 1: B \to S(B) \pitchfork \Omega$$

is directed. This is an easy consequence of the fact that $s: S(B) \to B$ is a "sub-upper-semi-lattice", that is s preserves $\vee$ and 0. For, if now

$$\alpha: X \to B \quad \text{and} \quad b_j: X \to S(B) \qquad (j = 1,2)$$

are given and satisfy

$$b_j \in \alpha.\downarrow\text{seg}.s \pitchfork 1 \qquad (j = 1,2),$$

then

$$b_j.s \leq \alpha \qquad (j = 1,2),$$

thus

$$b_1.s \vee b_2.s \leq \alpha,$$

and since s preserves $\vee$

$$(b_1 \vee b_2).s \leq \alpha,$$

thus

$$b_1 \vee b_2 \underline{\in} \alpha.{\downarrow}\mathrm{seg}.s \pitchfork 1.$$

This witnesses that ${\downarrow}\mathrm{seg}.s \pitchfork 1$ is 2-directed. Similarly, s preserving 0 implies that ${\downarrow}\mathrm{seg}.s \pitchfork 1$ is 0-directed. This completes the proof that "inductive implies algebraic".

Conversely, if φ is an algebraic closure operator $\varphi: B \to B$ on the algebraic lattice object B, then we shall prove that if $F: X \to C \pitchfork \Omega$ is directed, then

$$F.\mathrm{sup}_C.i = F.\exists i.\mathrm{sup}_B.$$

We have

$$F.\mathrm{sup}._C.i = F.\exists i.\exists\bar{\varphi}.\mathrm{sup}_C.i$$

$$= F.\exists i.\mathrm{sup}_B.\bar{\varphi}.i \qquad (\bar{\varphi} \text{ being cocontinuous})$$

$$= F.\exists i.\mathrm{sup}_B.\varphi.$$

so we need only prove that

$$(4.4) \qquad F.\exists i.\mathrm{sup}_B.\varphi = F.\exists i.\mathrm{sup}_B.$$

To prove this, consider the inequalities (and equalities)

$$F.\exists i\, \mathrm{sup}_B.\varphi$$

$$= F.\exists i.\mathrm{sup}_B.{\downarrow}\mathrm{seg}.s \pitchfork 1.\exists s.\exists\varphi.\mathrm{sup}_B$$

(by algebraicity of φ)

$$= F.\exists i.{\downarrow}\mathrm{cl}.s \pitchfork 1.\exists s.\exists\varphi.\mathrm{sup}_B$$

(by Main Lemma 4.4, and directednss of $F.\exists i$)

$$\leq F.\exists i.{\downarrow}\mathrm{cl}.\exists\varphi.\mathrm{sup}_B$$

(by end-adjunction for $\exists s \dashv s \pitchfork 1$)

$$\leq F.\exists i.\exists\varphi.{\downarrow}\mathrm{cl}.\mathrm{sup}$$

(by Lemma 4.2)

$$= F.\exists i.\exists\varphi.\mathrm{sup}$$

(by Lemma 4.3)

$$= F.\exists i.\mathrm{sup},$$

the last equality because i is the equalizer of φ and id_B. Since the other inequality in (4.4) is obvious, equality holds, and the proof that "algebraic implies inductive" is complete.

The theorem may in particular be employed with $B = A \pitchfork \Omega$, giving the form of Jürgen Schmidt's theorem (essentially) which is quoted in Cohn's book [5]. For, it is a consequence of things already proved that $A \pitchfork \Omega$ is an algebraic lattice object: singletons are intranscessible by Lemma 2.7, and every element $X \to A \pitchfork \Omega$ in $A \pitchfork \Omega$ is sup of the intranscessibles below it, because it is already the sup of the singletons below it. Formally, we must prove

$$\downarrow\mathrm{seg}.s \pitchfork 1.\exists s.\bigcup \geq id_{A \pitchfork \Omega},$$

but the left-hand side here is, by end-adjunction, larger than or equal to

$$\downarrow\mathrm{seg}.s \pitchfork 1.\{\cdot\}_A \pitchfork 1.\exists\{\cdot\}_A.\exists s.\bigcup \;.$$

Since $\{\cdot\}.s = \{\cdot\}$, this expression equals

$$\downarrow\mathrm{seg}.\{\cdot\} \pitchfork 1.\exists\{\cdot\}.$$

which is easily seen to be $id_{A \pitchfork \Omega}$.

Before we state the second theorem, we give the topos-theoretic version of the fact that if B is algebraic and $b_i\colon X \to B$ are arbitrary $(i=1,2)$, then the statement $b_1 \leq b_2$ can be tested by intranscessibles:

4.5 Lemma. In an algebraic lattice B, $b_1 \leq b_2$ if for every $\alpha\colon Y \to X$ and every intranscessible $v\colon Y \to B$,

$$v \leq \alpha.b_1 \quad \text{implies} \quad v \leq \alpha.b_2.$$

(The converse is also true).

Proof. By assumption of algebraicity

$$b_1 = b_1.{\downarrow}\mathrm{seg}.s \pitchfork 1.\exists s.\sup.$$

So it suffices to prove

$$b_1.{\downarrow}\mathrm{seg}.s \pitchfork 1.\exists s.\sup \leq b_2$$

which by adjointness $\sup \dashv {\downarrow}\mathrm{seg}$ and $\exists s \dashv s \pitchfork 1$ is equivalent to

(4.5) $$b_1.{\downarrow}\mathrm{seg}.s \pitchfork 1 \leq b_2.{\downarrow}\mathrm{seg}.s \pitchfork 1.$$

To prove this inequality, we use the extensionality-princple: let $\alpha: Y \to X$ and $a: Y \to S(B)$ satisfy

$$a \mathrel{\underline{\in}} \alpha.b_1.{\downarrow}\mathrm{seg}.s \pitchfork 1,$$

that is,

$$a.s \leq \alpha.b_1.$$

Now $a.s$ is intranscessible (s is "the universal intranscessible element"), and the assumption then gives $a.s \leq \alpha.b_2$. From this we conclude $a \mathrel{\underline{\in}} \alpha.b_2.{\downarrow}\mathrm{seg}.s \pitchfork 1$, and (4.5) is proved.

4.6 Theorem. Consider a closure system

$$i: C \rightarrowtail B, \quad \bar{\varphi}: B \to C \qquad\qquad \bar{\varphi} \dashv i$$

as in (4.1), with B an algebraic lattice object. This closure system is inductive if and only if "$\bar{\varphi}$ preserves intranscessibles" in the sense that if $b: X \to B$ is intranscessible, then so is $b.\bar{\varphi}: X \to C$. In this case, C is also an algebraic lattice object.

Proof. Suppose the closure system is inductive. Let $b: X \to B$ be intranscessible in B. We must prove $b.\bar{\varphi}: X \to C$ intranscessible in C. Let $\alpha\colon Y \to X$ and $F\colon Y \to C \pitchfork \Omega$ satisfy: F directed and

$$\alpha.b.\bar{\varphi} \leq F.\sup_C.$$

By adjointness $\bar{\varphi} \dashv i$, we have the inequality in

$$(4.6) \qquad \alpha.b \leq F.\sup_C.i = F.\exists i.\sup_B,$$

the equality sign by the inductiveness-assumption. But $F.\exists i$ is directed, and since $\alpha.b$ is intranscessible, the total inequality in (4.6) implies the existence of an epic $\beta\colon Z \twoheadrightarrow Y$ and $b_o\colon Z \to B$ with

$$b_o \underline{\in} \beta.F.\exists i \quad \text{and} \quad \beta.\alpha.b \leq b_o.$$

Using the existence-principle 1.2, we find yet another epic $\beta'\colon Z' \twoheadrightarrow Z$ and $c\colon Z' \to C$ with

$$c.i = \beta'.b_o \quad \text{and} \quad c \underline{\in} \beta'.\beta.F.$$

We clearly have

$$\beta'.\beta.\alpha.b \leq \beta'.b_o = c.i,$$

hence by adjointness $\bar{\varphi} \dashv i$

$$\beta'.\beta.\alpha.b.\bar{\varphi} \leq c \underline{\in} \beta'.\beta.F.$$

The epic $\beta'.\beta$ and the element $c\colon Z' \to C$ now are witnesses of the intranscessibility of $b.\bar{\varphi}$.

Since now $\bar{\varphi}$ preserves intranscessibles and $\bar{\varphi}$ preserves sup's, and since B is algebraic, we easily conclude that C also is an algebraic lattice object.

Conversely, if $\bar{\varphi}$ preserves intranscessibles, we shall prove that $i\colon C \rightarrowtail B$ preserves directed sup's. Let

$$F: X \to C \pitchfork \Omega$$

be directed. It suffices to prove

$$\text{(4.7)} \qquad F.\sup_C.i \leq F.\exists i.\sup_B,$$

(the other inequality being clear). It suffices to test with intranscessibles, by Lemma 4.5. So let $\alpha: Y \to X$ be arbitrary and $v: Y \to B$ intranscessible, and assume

$$v \leq \alpha.F.\sup_C.i.$$

Then, by adjointness,

$$v.\bar{\varphi} \leq \alpha.F.\sup_C,$$

and since $v.\bar{\varphi}$ by assumption is intranscessible, there is an epic $\beta: Z \twoheadrightarrow Y$ and $c: Z \to C$ with

$$\beta.v.\bar{\varphi} \leq c \underline{\in} \beta.\alpha.F.$$

By existence principle 1.2, and by adjointness, we then have

$$\beta.v \leq c.i \underline{\in} \beta.\alpha.F.\exists i.$$

Then

$$\beta.v \leq \beta.\alpha.F.\exists i.\sup_B,$$

and hence, cancelling the epic β,

$$v \leq \alpha.F.\exists i.\sup_B.$$

This proves the inequality (4.7), and thus the theorem.

BIBLIOGRAPHY

1. M. Artin, A. Grothendieck, and J.L. Verdier, Théorie des topos et cohomologie etale des schemas (SGA 4), Springer Lecture Notes Vol.269 and 270 (1973).

2. M. Barr, Exact categories, in Barr, Grillet, and van Osdol, Exact categories and categories of sheaves, Springer Lecture Notes Vol. 236 (1971).

3. J. Benabou, Categories et logiques faibles, Oberwolfach Tagungsbericht 30/1973.

4. G. Birkhoff and O. Frink, Representation of lattices by sets, Trans.Amer.Math.Soc. 64, 299-316 (1948).

5. P.M. Cohn, Universal algebra, Harper and Row 1965.

6. K.-H. Diener, Über zwei Birkhoff-Frink'sche Struktursätze der allgemeinen Algebra, Archiv der Math. (Basel) 7, 339-345 (1956).

7. P. Gabriel and F. Ulmer, Lokal präsentierbare Kategorien, Springer Lecture Notes Vol. 221 (1971).

8. A. Kock, Strong functors and monoidal monads, Archiv der Math. (Basel) 23, 113-120 (1972).

9. A. Kock and G.C. Wraith, Elementary toposes, Aarhus Lecture Notes Series No. 30 (1971).

10. C. Kuratowski, Sur la notion d'énsemble fini, Fund.Math. 1 129-131 (1920).

11. F.W. Lawvere, Continuously variable sets; Algebraic geometry = Geometric logic, Preprint, Perugia 1973 (to appear in Proc.of the Logic Coll., Bristol 1973).

12. P. Lecouturier, Quantificateurs dans les topos élémentaires, Preprint, Université Nationale du Zaire, Kinshasa 1971-72.

13. C.J. Mikkelsen, Thesis, to appear.

14. C.J. Mikkelsen, On the internal completeness of elementary topoi, Oberwolfach Tagungsbericht 30/1973.

15. W. Mitchell, Boolean topoi and the theory of sets, Journal of pure and appl.algebra 2, 261-274 (1972).

16. G. Osius, The internal and external aspect of logic and set theory in elementary topoi, Oberwolfach Tagungsbericht 30/1973.

17. J. Schmidt, Über die Rolle der transfiniten Schlussweisen in einer allgemeinen Idealtheorie, Math.Nachr. 7, 165-182 (1952).

18. J. Schmidt, Mengenlehre, Bibliographisches Institut Mannheim, 1966.

/ue

Aarhus Universitet, Aarhus, Denmark
and
Université Nationale du Zaïre, Kinshasa, Zaïre.

UNIVERSES IN TOPOI *)

Christian Maurer

Universes lie at the very heart of the foundations of category theory within a set theoretical (Zermelo-Fraenkel) framework (see [1], [5], [11], [12], [17], [18], [26]); topos theory à la Lawvere-Tierney ([2], [7], [14], [16]) seems to be the most effective categorical approach to set theory (apart from the sheaf aspects in geometry). So it is natural to combine these two concepts and consider an object in a topos which plays the role of a universe. This allows for the development of certain categorical notions inside a topos hinging on the distinction between "small" and "large" objects.

We assume the reader to be familiar with the basic methods and results of the theory of elementary topoi (see e.g. [6], [8], [9], [15], [19], [27]). Terminology and notation is the usual one. We shall use the (-)&Co-language (coadjoint means left adjoint etc.). The identity morphism on an object A is also denoted by A ; the exponential transpose (via the evaluation morphism $C^A \times A \xrightarrow{ev_A C} C$) in a cartesian closed category is marked by a bar (in either direction): $B \times A \xrightarrow{f=\overline{g}} C \iff B \xrightarrow{g=\overline{f}} C^A$. $!_A$ or just $! : A \longrightarrow 1$ denotes the unique morphism into the terminal object.

By a topos we mean a category $\underline{E}$ with a terminal object 1 , pullbacks for any two morphisms with common codomain, a subobject classifier $1 \xrightarrow{true} \Omega$ and power objects $PA \times A \xrightarrow{ev_A} \Omega$. All these data have got to be chosen once for all. (By [8], [21] and [27], this is equivalent to the "old" definition.)

---

*) This paper is a concise version of the author's dissertation [19] at the University of Bremen, Germany. Some of the results were announced at the Berlin Topos Seminar, 1973.

Some topos theoretic notations: for any $A \in \mathrm{Ob}\underline{E}$, let 0_A, 1_A: $1 \longrightarrow PA$, $\neg_A$: $PA \longrightarrow PA$ and $\Rightarrow_A$, $\wedge_A$, $\vee_A$: $PA\times PA \longrightarrow PA$ be the Heyting algebra operations and $PA\times PA \xrightarrow{\supseteq_A} \Omega$ the order (inverse inclusion) relation with its transpose $PA \xrightarrow{P_A} P^2A$ (internal power formation; dually $\subset_A$ and $ꟼ_A$). With respect to this ordering, let $P^2A \xrightarrow{\cup_A} PA$ denote the coadjoint of P_A (internal union operator; dually $\cap_A$ the internal intersection operator, adjoint on the right to $ꟼ_A$). For any two morphisms f, g: $X \longrightarrow PA$ such that $\subset_A(f,g) = \mathrm{true}_X = \mathrm{true}!_X$ we write $f \leq g$. $\in_A \rightarrowtail PA\times A$ is the subobject classified by $PA\times A \xrightarrow{ev_A} \Omega$. For any morphism $A \xrightarrow{f} B$, let $Pf = \Omega^f$ and $\exists f$, $\forall f$: $PA \longrightarrow PB$ its coadjoint and adjoint resp. The characteristic morphism $A \longrightarrow \Omega$ of a mono $B \overset{f}{\rightarrowtail} A$ is denoted by $ch(f)$. For any A, $A \xrightarrow{\{\}_A} PA$ is the transpose of $ch(\Delta_A)$, where Δ_A: $A \longrightarrow A\times A$ is the diagonal.

Lemma 1: If $A \xrightarrow{f} B$ is monic, then

$$\begin{array}{ccc}
\in_A & \longrightarrow & 1 \\
\downarrow & & \\
PA\times A & & \Big\downarrow \mathrm{true} \\
{\scriptstyle PA\times f}\downarrow & & \\
PA\times B & \xrightarrow{\overline{\exists f}} & \Omega
\end{array}$$

is a pullback.

Definition 1: A relation $A \xrightarrow{r} PA$ (or $A\times A \xrightarrow{\bar{r}} \Omega$) is called extensional, if r is monic. An extensional relation is called power closed, if there is a factorization

$$\begin{array}{ccc}
A & \overset{p}{\dashrightarrow} & A \\
{\scriptstyle r}\downarrow & & \downarrow{\scriptstyle (\exists r)r} \\
PA & \xrightarrow{P_A} & P^2A
\end{array}$$

(p then is uniquely determined and monic).

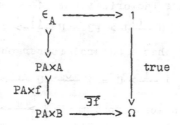

From now on, let $U \xrightarrow{r} PU$ be an extensional power closed relation in the topos $\underline{E}$. Elementary spoken, add to the language of $\underline{E}$ (in the sense of [13]) a constant $r \in \underline{E}$ with $cod(r) = P(dom(r))$ satisfying the axioms

(EXT) r is monic

and (POW) there is a factorization $U \xrightarrow{p} U$ as in def. 1.

As a further axiom we postulate the existence of enough global sections of U :

(GS) there is some $1 \longrightarrow U$.

Lemma 2: If $X \xrightarrow{a} PU$ is a morphism such that $a \leq rb$ for some $X \xrightarrow{b} U$, then there is a (unique) factorization $X \xrightarrow{a} PU = X \dashrightarrow U \overset{r}{\rightarrowtail} PU$.

Proof. Consequence of lemma 1 and the fact that the diagram defining p is a pullback.

Corollary: There are factorizations $1 \xrightarrow{0_U} PU = 1 \overset{0}{\dashrightarrow} U \overset{r}{\rightarrowtail} PU$ and $U \xrightarrow{\{\}_U} PU = U \overset{\{\}}{\dashrightarrow} U \overset{r}{\rightarrowtail} PU$. Hence, by [6], theorem 5.44, $\underline{E}$ has a natural number object $N \rightarrowtail U$.

Definition 2: Let $1 \xrightarrow{1} U = 1 \xrightarrow{0} U \xrightarrow{p} U$.

Lemma 3: There is a monomorphism $U+U \overset{m}{\rightarrowtail} U \times U$.

Proof. $U \xrightarrow{(U,1!)} U \times U$ and $U \xrightarrow{(U,0!)} U \times U$ are disjoint.

Let $PU \times PU \overset{l}{\rightarrowtail} P(U \times U)$ be the composition of the canonical isomorphism $PU \times PU \xrightarrow{j} P(U+U)$ with $\exists m$. As a further axiom we now postulate the existence of a (recursive) pairing morphism

(PA) there is a monomorphism $U \times U \overset{s}{\rightarrowtail} U$ such that

$$\begin{array}{ccc} U \times U & \overset{s}{\rightarrowtail} & U \\ {\scriptstyle l(r \times r)} \downarrow & & \downarrow {\scriptstyle r} \\ P(U \times U) & \overset{\exists s}{\rightarrowtail} & PU \end{array}$$

commutes.

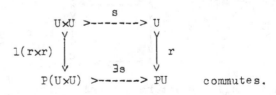

(The set theoretic construction in ZF of this map hinges on the fact that the relation R given by

$$(x',y')\,R\,(x,y) \;\Leftrightarrow\; (x'\in x \wedge y'=1) \vee (x'\in y \wedge y'=0)$$

is extensional and well founded (see [20]). Thus, using the theorem of Mostowski (see e.g. [10], ch. III), there is an ordered pair $\langle .,. \rangle$ such that $\langle a,b\rangle = \{\langle x,1\rangle \mid x\in a\} \cup \{\langle y,0\rangle \mid y\in b\}$ for $0 = \emptyset$ and $1 = \{0\}$ and one has $\langle a,b\rangle = \langle a',b'\rangle \Leftrightarrow a = a'$ and $b = b'$.)

By means of the internal product operator $k_U = \wedge_{U\times U}(Ppr_0, Ppr_1)$: $PU\times PU \longrightarrow P(U\times U)$ composed with $\exists s$ there is the <u>product morphism</u> q_1: $PU\times PU \longrightarrow PU$. As next axiom we have

(PROD) there exists a factorization (automatically unique)

$$\begin{array}{ccc} U\times U & \overset{q}{\dashrightarrow} & U \\ {\scriptstyle r\times r}\downarrow & & \downarrow{\scriptstyle r} \\ PU\times PU & \xrightarrow{q_1} & PU \end{array} .$$

<u>Definition 3</u>: An object U with an extensional power closed relation r satisfying moreover the axioms (GS), (PROD) and (PA) is called a <u>weak universe</u>. - For the transposes of the relations r and rp we write $U\times U \xrightarrow{\ni} \Omega$ and $U\times U \xrightarrow{\supset} \Omega$, resp. (called (inverse) <u>element</u> and <u>inclusion relation</u>).

<u>Definition 4</u>: For any two objects A , B in <u>E</u> , let f_{AB}: $P(A\times B) \longrightarrow PA$ be the transpose of the composition $P(A\times B)\times A \xrightarrow{\bar{e}_{AB}} PB \xrightarrow{ch(\{\}_B)} \Omega$, where $\bar{e}_{AB}$ is the transpose of $ev_{A\times B}$. Then it is known (see [8]) that there is the pullback

$$\begin{array}{ccc} B^A & \dashrightarrow & 1 \\ \downarrow & & \downarrow{\scriptstyle 1_A} \\ P(A\times B) & \xrightarrow{f_{AB}} & PA \end{array} .$$

Definition 5: For a weak universe, let $PU\times PU \xrightarrow{e_1} P^2U$ be the intersection of the two morphisms $PU\times PU \xrightarrow{q_1} PU \xrightarrow{P_U} P^2U$ and $PU\times PU \xrightarrow{pr_o} PU \xrightarrow{\{\}_U} P^2U \xrightarrow{Pf_{UU}} P^2(U\times U) \xrightarrow{\exists^2 s} P^2U$.

Lemma 4: There is a factorization

$$\begin{array}{ccc} U\times U & \xrightarrow{e} & U \\ {\scriptstyle r\times r}\downarrow & & \downarrow{\scriptstyle (\exists r)r} \\ PU\times PU & \xrightarrow{e_1} & P^2U \end{array} \quad .$$

Intuitively, this lemma expresses that a weak universe is exponentially closed.

Proof. By looking at (POW) and (PROD) and applying lemma 2 twice.

Now, one of the main aspects of this note is that with the help of this operation, U can be given the structure of an internal category in the topos $\underline{E}$. (For the definition of internal categories see [4], [9], [19] or [27].)

Definition 6: We consider the transpose of $U\times U \xrightarrow{e_o} PU =$
$= U\times U \xrightarrow{e} U \xrightarrow{r} PU$, and define the morphism object U' and the morphisms "domain", "codomain" and "graph" by the pullback

$$\begin{array}{ccc} U' & \dashrightarrow & 1 \\ {\scriptstyle (dom,cod,gra)}\downarrow & & \downarrow{\scriptstyle true} \\ U\times U\times U & \xrightarrow{\overline{e}_o} & \Omega \end{array} \quad .$$

Obviously, one has for the internal product the inequality $\exists\Delta_U \leq k_U\Delta_{PU}$, from which we get by the definition of q_1
$\exists s\exists\Delta_U \leq q_1\Delta_{PU}$, therefore $\exists(s\Delta_U)r \leq rq\Delta_U$. Thus, by lemma 2 there is a factorization

$$\begin{array}{ccc} U & \xrightarrow{s'} & U \\ {\scriptstyle r}\downarrow & & \downarrow{\scriptstyle r} \\ PU & \xrightarrow{\exists(s\Delta_U)} & PU \end{array} \quad .$$

For this operation one has $U \xrightarrow{(U,U,s')} U\times U\times U \xrightarrow{\bar{e}_o} \Omega = true_U$ as an obvious consequence of lemma 4.

Definition 7: Let $id\colon U \longrightarrow U'$ be the morphism which arises out of the definition 6 by the above equality.

Theorem 1: If $U \xrightarrow{r} PU$ is a weak universe, then $U \xleftarrow{dom} U' \xrightarrow{cod} U$ is an internal category in $\underline{E}$.

Proof. The identity morphism (unit in the corresponding monad) is given in def. 7; the equations dom id = U = cod id are immediate. The object "of composable pairs of morphisms" is given by the pull-back

$$\begin{array}{ccc} U'' & \xrightarrow{cod'} & U' \\ {\scriptstyle dom'}\downarrow & & \downarrow{\scriptstyle dom} \\ U' & \xrightarrow{cod} & U \end{array} .$$

Then the composition morphism $comp\colon U'' \longrightarrow U'$ is defined using the fact that the composition of morphisms in a topos is but the relational composition $P(U\times U)\times P(U\times U) \xrightarrow{Ppr_{01}\times Ppr_{12}} P(U\times U\times U)\times P(U\times U\times U) \longrightarrow$ $\xrightarrow{\wedge_{U\times U\times U}} P(U\times U\times U) \xrightarrow{\exists pr_{02}} P(U\times U)$ where pr_{ik} denotes projection into the (i+1)-th and (k+1)-th factor (see e.g. [8]). For the details of this construction and the proof that the data given do indeed yield a category object (both involving somewhat lengthy and boring computations) the reader is referred to [19].

It is pretty obvious that such a weak universe has much more structure than just that of an internal category. Thinking intuitively of "elements" of U as of U-sets and of "elements" of U' as of morphisms between U-sets, we can describe internal versions of all topos axioms by constructing appropriate operators between (finite) limits of diagrams, starting with the basic data U , U' , U" , dom , cod , id and comp , then using the derived data dom' , cod' etc. and the ones given by the axioms (r , p , 0 , 1 , s and q).

However, this procedure (similar to the techniques used in [4]) will be rather long and computational. So, avoiding chasing through large diagrams, one should employ the more suggestive and easier to handle methods which are due to Mitchell, Bénabou [3] and Osius [24], [25] to get the following result:

<u>Theorem 2</u>: U is an internal topos in $\underline{E}$.

Of course a weak universe can be made more set like. For this purpose we first recall a definition from [23], ch. 6:

<u>Definition 8</u>: A relation $A \xrightarrow{r} PA$ is called <u>recursive</u> if for any $PB \xrightarrow{g} B$ there exists a unique morphism $A \xrightarrow{f} B$ such that

$$\begin{array}{ccc} A & \xrightarrow{f} & B \\ {\scriptstyle r}\downarrow & & \uparrow{\scriptstyle g} \\ PA & \xrightarrow{\exists f} & PB \end{array}$$

commutes. (In the set case, this property expresses that r is well founded.)

For any relation $A \xrightarrow{r} PA$ there is the object "of (r-)transitive subobjects of A ", given by $PA \xrightarrow{(\exists r, P_A)} P^2A \times P^2A \xrightarrow{\subseteq_{PA}} \Omega$ as a subobject of PA . Let $PA \xrightarrow{T_{A,r}} P^2A$ be the transpose of $PA \times PA \xrightarrow{(\subseteq_A, (\subseteq_{PA}(\exists r, P_A))pr_1)} \Omega \times \Omega \xrightarrow{\wedge} \Omega$ and $PA \xrightarrow{t_{A,r}} PA =$
$= PA \xrightarrow{T_{A,r}} P^2A \xrightarrow{\cap_A} PA$, intuitively the transitive hull operator.

<u>Definition 9</u>: An extensional relation $A \rightarrowtail^{r} PA$ is called <u>transitive</u> <u>closed</u> if there is a factorization

$$\begin{array}{ccc} A & \xrightarrow{t} & A \\ {\scriptstyle r}\downarrow & & \downarrow{\scriptstyle r} \\ PA & \xrightarrow{t_{A,r}} & PA \end{array} .$$

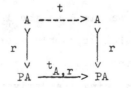

Using the (r-)union operator $PA \xrightarrow{\exists r} P^2A \xrightarrow{\cup_A} PA$ we define an extensional relation r to be <u>union closed</u> analogously.

<u>Lemma 5</u>: A transitive closed relation is union closed.

<u>Proof</u>. The adjunctions between $\cup_A$, P_A and $\cap_A$, $¶_A$ resp., yield $\cup_A \exists r \leq t_{A,r}$ from which the assertion follows by lemma 2.

Finally, for a relation $A \rightarrowtail^{r} PA$ there is an appropriate formulation of the replacement axiom. For that, we consider the subobject $FA \rightarrowtail P(A \times A)$ defined by the pullback

$$\begin{array}{ccc} FA & \dashrightarrow & A \\ \downarrow & & \downarrow (r,r) \\ P(A\times A) & \xrightarrow{(\exists pr_0, f_{AA})} & PA\times PA \end{array}$$

<u>Definition 10</u>: An extensional relation (A,r) is called <u>replacement closed</u> if there is a factorization

$$\begin{array}{ccc} FA & \overset{ran}{\dashrightarrow} & A \\ \downarrow & & \downarrow r \\ P(A\times A) & \xrightarrow{\exists pr_1} & PA \end{array}$$

(intuitively, if the range of a functional relation defined on an element of A is again an element of A).

<u>Definition 11</u>: A weak universe (U,r) is called a <u>universe</u> if the following axioms hold:

(UN) r is union closed

and (REP) r is replacement closed.

It is called a <u>set theory object</u> if, moreover, the following axioms are valid:

(REC) r is recursive

and (TRH) r is transitive closed.

In the set case, elements of a universe are sets themselves, therefore a universe is a subworld. In the absence of actual elements in a topos, as a first approximation to that notion we have the one of global sections. To imitate the above situation, let us consider the subsystem $\underline{E}_0 = \underline{E}(1,U) \subset \underline{E}$ of such global elements of U .

<u>Theorem 3</u>: If U is a weak universe then $\underline{E}_0$ is a category.

<u>Proof</u>. Objects are given by $\underline{E}_0$, the morphisms by $\underline{E}(1,U')$ etc., i.e. for any two a , $b \in \underline{E}_0$, $\underline{E}_0(a,b) = \underline{E}(1,E_0(a,b))$ is given by the pullback

$$\begin{array}{ccc} E_0(a,b) & \dashrightarrow & 1 \\ \downarrow & & \downarrow{\scriptstyle (a,b)} \\ U' & \xrightarrow{(dom,cod)} & U\times U \end{array} .$$

The statement then is easily deduced from theorem 1.

<u>Theorem 4</u>: $\underline{E}_0$ is a topos.

<u>Proof</u>. Corollary to theorem 2.

Let $\Phi: \underline{E}_0 \longrightarrow \underline{E}$ be the function (in the language of the category $\underline{E}$) which sends a global section $1 \xrightarrow{a} U$ to the (canonically chosen) subobject of U classified by $U \xrightarrow{\overline{ra}} \Omega$, followed by the forgetful $\underline{E}/U \longrightarrow \underline{E}$.

<u>Theorem 5</u>: $\Phi: \underline{E}_0 \longrightarrow \underline{E}$ is a logical embedding, i.e. a fully faithful functor preserving all the topos structure.

<u>Proof</u>. Out of the definition of the operators p , 0 , s and q , it is straightforward to see that for any a , $b \in \underline{E}_0$ there are isomorphisms $\Phi(pa) \cong P(\Phi(a))$, $\Phi(0) \cong 0$, $\Phi(s(a,b)) \cong \Phi(a)+\Phi(b)$ and $\Phi(q(a,b)) \cong \Phi(a)\times\Phi(b)$. Furthermore, by looking into the construction of 1 and e we get $\Phi(1) \cong 1$, $\Phi(p1) \cong \Omega$ and $\Phi(e(a,b)) \cong \Phi(b)^{\Phi(a)}$. As an example, the last isomorphism is proved by showing that there is a pullback

$$\begin{array}{ccc} \Phi(b)^{\Phi(a)} & \longrightarrow & 1 \\ \downarrow & & \\ P(\Phi(a)\times\Phi(b)) & & \downarrow \text{true} \\ \downarrow & & \\ U & \xrightarrow{\overline{e}_0(a,b)} & \Omega \end{array} .$$

Since $\overline{e}_0(a,b,\mathrm{graf}) = \mathrm{true}$ for any $f \in \underline{E}_0(a,b)$, this yields some $1 \longrightarrow \Phi(b)^{\Phi(a)}$, the transpose of which is defined to be $\Phi(f)$. It is not hard to see that this function $\underline{E}_0(a,b) \longrightarrow \underline{E}(\Phi(a),\Phi(b))$ is a bijection in $\underline{E}$ and that Φ preserves identities and equalizers. The proof that the composition is preserved is a bit more nasty.

As a corollary, we get:

Theorem 6: If U is a set theory object then $\underline{E}_0$ is a topos of Z-sets in the sense of [23], 8.21 f.

Proof. Φ is logical, hence it preserves the notion of transitivity. Thus, the axiom 8.20 (APT) of [23] follows by (REC) from theorem 6.3 loc.cit. and from the next result which we mention without proof:

Lemma 6: $PA \xrightarrow{t_{A,r}} PA \xrightarrow{\subseteq_{PA}(\exists r, P_A)} \Omega = \mathrm{true}_{PA}$ for any relation (A,r).

With the same methods as in the proofs of the theorems 1 and 2 we get the following generalization:

Theorem 7: If U is a weak universe then PU is an internal category and $U \xrightarrow{r} PU$ an internal functor. Furthermore, PU is internally finitely complete and cocomplete and has an internal subobject classifier, and r preserves all these things.

Corollary: $\underline{E}_1 = \underline{E}(1,PU)$ is a finitely complete and cocomplete category with subobject classifier and there are logical embeddings $\underline{E}(1,r)\colon \underline{E}_0 \longrightarrow \underline{E}_1$ and $\Phi_1\colon \underline{E}_1 \longrightarrow \underline{E}$ such that $\Phi_1\underline{E}(1,r)$ and Φ_0 are naturally isomorphic.

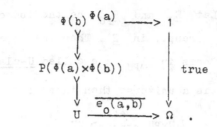

<u>Definition 12</u>: Let $\underline{E}_U$ and $\underline{E}_{PU}$ be the isomorphism closures of $\Phi\, \underline{E}_o$ and $\Phi_1\, \underline{E}_1$ resp., in $\underline{E}$. Then the objects in $\underline{E}_U$ are called <u>U-sets</u> and the ones in $\underline{E}_{PU}$ are called <u>U-classes</u>.

<u>Lemma 7</u>: If U is a universe then there is a factorization

$$\begin{array}{ccc} U\times PU & \overset{e'}{\dashrightarrow} & PU \\ {\scriptstyle r\times PU}\big\downarrow & & \big\downarrow{\scriptstyle \exists r} \\ PU\times PU & \xrightarrow{e_1} & P^2U \end{array} \quad .$$

In particular, B^A is a U-class for B a U-class and A a U-set.

<u>Proof</u>. From the construction of $FU \rightarrowtail P(U\times U)$ we can derive the inequality $e_1(r\times PU) \leq \overline{FU}!$ where $\overline{FU}\colon 1 \longrightarrow P^2U$ is the transpose of the characteristic morphism of $FU \rightarrowtail P(U\times U) \overset{\exists s}{\rightarrowtail} PU$.

By (REP), we get a factorization $FU \rightarrowtail P(U\times U) \overset{\exists s}{\rightarrowtail} PU =$ $= FU \dashrightarrow U \overset{r}{\rightarrowtail} PU$ and, therefore, the assertion as an application of lemma 2.

With this lemma at hand, one should expect $\underline{E}_1$ to be some sort of model for Neumann-Bernays-Gödel set theory. Though we have not checked the details we are convinced that in case U is a set theory object, $\underline{E}_1$ (or $\underline{E}_{PU}$, resp.) is a <u>category of classes and maps</u> in the sense of [22].

Last not least, there is the obvious result:

<u>Theorem 8</u>: Let $\underline{S}$ be the topos of ZF-sets. Then our notion of set theory objects coincides up to relational isomorphism with the (classical) one of universes.

<u>Proof</u>. A set with an extensional and recursive relation is isomorphic to a transitive set with the $\in$-relation. The rest of the proof is a straightforward exercise in passing from topos theoretic statements to set theoretic ones and vice versa.

References

[1] Artin, M. - Grothendieck, A. - Verdier, J. L.: Théorie des Topos et Cohomologie Étale des Schémas (SGA 4, 1963/64). Springer Lecture Notes 269, 270 (1972).

[2] Bénabou, J. - Celeyrette, J.: Généralités sur les Topos des Lawvere et Tierney. Preprint, Séminaire Bénabou (1970).

[3] Bénabou, J.: Catégories et logiques faibles. Talk at the Oberwolfach Conference (1973).

[4] Diaconescu, R.: Change of Base for some Toposes. Ph.D. thesis, Dalhousie University (1973).

[5] Feferman, S.: Set-Theoretical Foundations of Category Theory. Reports of the Midwest Category Seminar III, Springer Lecture Notes 106, 201 - 247 (1969).

[6] Freyd, P.: Aspects of topoi. Bull. Austral. Math. Soc. 7, 1 - 76 (1972).

[7] Gray, J.: The meeting of the Midwest Category Seminar in Zürich, August 24-30, 1970. Reports of the Midwest Category Seminar V, Springer Lecture Notes 195, 248 - 255 (1971).

[8] Kock, A. - Mikkelsen, Ch. J.: Topos theoretic factorizationsof nonstandard extensions. Victoria Symposium on Nonstandard Analysis, Springer Lecture Notes 369, 122 - 143 (1974).

[9] Kock, A. - Wraith, G. C.: Elementary Toposes. Lecture Notes Series No. 30, Aarhus University (1971/72).

[10] Krivine, J.-L.: Introduction to Axiomatic Set Theory. Dordrecht, Holland (1971).

[11] Kruse, A. H.: Grothendieck universes and the super-complete models of Shepherdson. Compositio Math. 17, 96 - 101 (1965).

[12] Kühnrich, M.: Über den Begriff des Universums. Z. math. Logik Grundlagen Math. 12, 37 - 50 (1966).

[13] Lawvere, F.W.: The Category of Categories as a Foundation for Mathematics. Proc. Conf. Categorical Algebra (La Jolla 1965), 1 - 20, Berlin-Heidelberg-New York (1966).

[14] Lawvere, F.W.: Quantifiers and sheaves. Actes du Congrès international des Mathématiciens (Nice 1970) 1, 329 - 334. Paris (1971).

[15] Lawvere, F.W.: Toposes, Algebraic Geometry and Logic. Introduction. Springer Lecture Notes 274, 1 - 12 (1972).

[16] Lawvere, F.W. - Tierney, M.: Talks at the Midwest Category Seminar, Zürich 1970. (see [7])

[17] MacLane, S.: Foundations for categories and sets. Category Theory, Homology Theory and their Applications II, Springer Lecture Notes 92, 146 - 164 (1969).

[18] MacLane, S.: One Universe as a Foundation for Category Theory. Reports of the Midwest Category Seminar III, Springer Lecture Notes 106, 192 - 200 (1969).

[19] Maurer, Ch.: Universen als interne Topoi. Dissertation, University of Bremen (1974).

[20] Maurer, Ch.: Ein rekursiv definiertes geordnetes Paar. In preparation.

[21] Mikkelsen, Ch. J.: Thesis. To appear. - Talks at the Oberwolfach Conferences (1972, 1973).

[22] Osius, G.: Eine Charakterisierung der Kategorie der Klassen und Abbildungen. Preprint, University of Bielefeld (1972).

[23] Osius, G.: A characterization of the category of sets. To appear in J. Pure Applied Algebra.

[24] Osius, G.: The internal and external aspect of logic and set theory in elementary topoi. To appear in Cahiers Top. Géom. Diff.

[25] Osius, G.: Logical and set theoretical tools in elementary topoi. This volume.

[26] Tarski, A.: Über unerreichbare Kardinalzahlen. Fund. Math. 30, 68 - 89 (1938).

[27] Wraith, G. C.: Lectures on Elementary Topoi. This volume.

Institut für Mathematik II
Freie Universität Berlin
West Germany

LOGICAL AND SET THEORETICAL TOOLS IN ELEMENTARY TOPOI

GERHARD OSIUS

Contents

0. Introduction

It has often been pointed out that the elmentary topoi introduced by Lawvere and Tierney [11,12,14] serve as the right generalization of "the" category of sets. Consequently many successful attempts have been made to lift results well understood for the category of sets (or set theory) to arbitrary topoi, using various more or less general techniques to establish such liftings (see bibliography). The purpose of this paper is to present a detailed exposition (and some applications) of logical and set theoretical tools which turn out to be extremely useful for establishing results in arbitrary topoi. The method originates from W. Mitchell [20] but has underwent changes, precisions and further development, some of them due to the author's discussions with J. Bénabou, A. Kock, F. W. Lawvere, Ch. Maurer and Ch. J. Mikkelsen.

The basic idea of this set theoretical method is that we imagine the objects of an (arbitrary) topos to have unspecified "elements" which behave in much the same way as the elements of sets in the category of sets. Formally the introduction of these "elements" amounts to the construction of a many-sorted set theoretical language L(SET) over the language L(ET) of the theory ET of elementary topoi (which corresponds to the language L($\underline{E}$) in [20,23] defined over a model $\underline{E}$ of ET).

The language L(SET) admits a natural "internal" interpretation in topos theory ET which gives rise to a notion of truth, called internal validity, and hence to a "set theory" SET defined over topos theory ET (for a natural "external" interpretation of L(SET) the reader is referred to Osius [23]). In fact SET is an Ω-valued set theory, Ω being the Heyting-algebra of subobjects of 1 in ET. The first important result is that the axioms and deductive rules of many-sorted intuitionistic (and even classical for the boolean case) logic and the axioms of many-sorted set theory hold in SET. Furthermore, the complete topos structure can be characterized in the set theory SET, so that any "property" in ET (e.g. equality or existence of maps, diagrams being commutative, squares being pullbacks) holds if and only if a corresponding set theoretical property in SET is internally valid. Hence results in topos theory can be established by showing that their "translation" in SET holds. This can be phrased by the slogan: "Topos theory is contained in intuitionistic many-sorted Heyting-valued set theory".

This set theoretical method of investigations in topos theory has the advantage, that - once the set theory SET has been developed to a certain extent - it allows to immediatly proceed from a heuristical set theoretical idea or construction to the corresponding result in the topos without having to wrestle with lots of diagrams (getting bigger and bigger). To illustrate the method thoroughly we prove a few results for recursive relations (due to Mikkelsen [19]) and natural number objects using our set theoretical arguments.

Particular care has been taken in order to present a detailed and sound approach to the set theory SET, which may even appear pedantic at some places. Some material on intuitionistic logic has been included to facilitate further applications and to keep the paper as self-contained as possible.

Independently of our investigations J. Bénabou [1] has constructed a formal language over more general types of categories (rather than topoi) and has achieved some of our results in section 1-3 by specializing his formalism to topoi.

1. The theory ET of elementary topoi

An elementary topos is - in the original definition given by Lawvere and Tierney [14] - a finitely bicomplete cartesian closed category with a subobject-classifier. Mikkelsen [17] has shown that finite bicompleteness can be reduced to finite completeness (later Paré [24] has given a different proof), and Kock [6] has proved that cartesian closedness can be weakend to existence of power-objects. Hence an elementary topos is a finitely complete category with a subobject-classifier and power-object formation (an equivalent definition is given by Wraith [29]). To be definite, we give the full (elementary) definition.

1.1 Definition An elementary topos $\underline{E}$ consists of a collection $Obj(\underline{E})$ of objects and a collection $Map(\underline{E})$ of maps together with

(1) unary operators "dom" (domain), "cod" (codomain), "id" (identity map), and a partial binary operator "·" (composition) such that $\langle Obj(\underline{E}), Map(\underline{E}), dom, cod, id, \cdot \rangle$ is an elementary category.

(2) a terminal object "1" and a unary operator "ter" assigning to any object A the unique map $A \longrightarrow 1$.

(3) partial binary projection-operators pr_1, pr_2 such that for all pairs of maps $\langle A \xrightarrow{f} C, B \xrightarrow{g} C \rangle$ the diagram

$$\begin{array}{ccc} pb(f,g) & \xrightarrow{pr_2(f,g)} & B \\ {\scriptstyle pr_1(f,g)}\downarrow & & \downarrow{\scriptstyle g} \\ A & \xrightarrow{f} & C \end{array}$$

is a pullback, and a partial operator pb* assigning to any commutative square

$$\begin{array}{ccc} D & \xrightarrow{h} & B \\ {\scriptstyle k}\downarrow & & \downarrow{\scriptstyle g} \\ A & \xrightarrow{f} & C \end{array}$$

the unique map $D \longrightarrow pb(f,g)$ induced by $\langle k,h \rangle$.

(4) a *subobject-classifier* $1 \xrightarrow{\text{true}} \Omega$ and a unary operator χ assigning to any monomorphism $B \overset{m}{\rightarrowtail} A$ its unique characteristic map, i.e.

$$\begin{array}{ccc} B & \longrightarrow & 1 \\ {\scriptstyle m}\downarrow & & \downarrow{\scriptstyle \text{true}} \\ A & \xrightarrow{\chi m} & \Omega \end{array}$$

is a pullback.

(5) two unary operators P, ev assigning to an object A its *power-object* PA and the *evaluation* $PA\times A \xrightarrow{ev_A} \Omega$, and one further unary operation p* which assigns to any (relation) $C\times A \xrightarrow{R} \Omega$ the unique map $C \longrightarrow PA$ induced by R, i.e. $R = ev_A\,(p^*R\times A)$. (The *product functor* $\times$ is defined as usual in terms of 1, ter, pr_1, pr_2, pb*.) For convenience: $P1=\Omega$ and $P1\times 1 \xrightarrow{ev_1} \Omega$ is the first projection (this is not essential since it always holds "up to isomorphisms").

It is obvious that elementary topoi are precisely the models of an appropriate first-order theory, the *theory* ET *of elementary topoi*. We only give a brief description of ET (the exact definition can easily be worked out by the reader familiar with formal theories): ET is two-sorted (i.e. the terms are devided into *objects* and *maps*) and has as primitive notions the operators dom, cod, id, $\cdot$, 1, ter, pr_1, pr_2, pb*, true, χ, P, ev, p* and two equality predicates (one for objects, one for maps). The nonlogical axioms of ET are the formal translations of 1.1.1-5. Freyd [4] points out, that ET is an essentially algebraic theory (in fact, the operators ter, pb*, χ, p* were only introduced to avoid existential quantification in the axioms for topoi).

Unless otherwise mentioned all our considerations take place in the elementary theory ET and can be formalized there. However for intuitive reasons we sometimes pretend to work in a fixed topos $\underline{E}$ (i.e. a model of ET) rather than in the theory ET itself.

The basic development of the theory ET of elementary topoi will be presupposed (see e.g. Lawvere-Tierney [14], Freyd [4], Kock-Wraith [9]) but to explain some notations let us briefly mention some results which turn out to be important for our considerations.

Kock [6] (p.5) has constructed exponentials B^A (for arbitrary objects A, B) and evaluation maps $B^A \times A \xrightarrow[ev_{AB}]{} B$, and Mikkelsen [17] has constructed an initial object 0 and pushouts, so that all finite colimits exist. However we will not need coproducts and coequalizers until their construction will be given (section 5) but assume only the existence of unions of monomorphisms and images of maps (see [17]).

Passage from a map $C \times A \xrightarrow{f} B$ to its exponential adjoint $C \xrightarrow{g} B^A$ and conversely will be denoted by $g := \overline{f}$ resp. $f := \overline{g}$.

The subobject - classifier Ω is an internal Heyting-algebra (i.e. pseudo boolean algebra) with respect to the maps $1 \xrightarrow{true} \Omega$, $1 \xrightarrow{false} \Omega$, $\Omega \xrightarrow{\neg} \Omega$ (negation), $\Omega \times \Omega \xrightarrow{\wedge} \Omega$ (conjunction), $\Omega \times \Omega \xrightarrow{\vee} \Omega$ (disjunction) and $\Omega \times \Omega \xrightarrow{\Rightarrow} \Omega$ (implication). By a subobject of a given object A we understand a map $A \longrightarrow \Omega$ rather than the corresponding equivalence class of monos into A, however sometimes a mono $B \rightarrowtail A$ will also be called a subobject.

The structure of Ω induces a Heyting-algebra structure on the subobjects of A having the operations $true_A$, $false_A$, $\neg_A$ (complement), $\cap_A$ (intersection), $\cup_A$ (union) and $\Rightarrow_A$ (implication).

A map $A \xrightarrow{f} B$ induces an operation of inverse image under f, denoted $f^{-1}(-)$, from subobjects of B to those of A , and three operations of direct image under f from subobjects of A to those of B:

1. direct existential image under f, denoted $\exists f(-)$
2. direct universal image under f, denoted $\forall f(-)$
3. direct unique-existential image under f, denoted $\exists ! f(-)$.

Indeed, for monic maps $C \overset{m}{\rightarrowtail} A$, $D \overset{n}{\rightarrowtail} B$ with characters $A \xrightarrow{M} \Omega$, $B \xrightarrow{N} \Omega$ we have:

1.2 $f^{-1}(N)$ is the character of pulling n along f.

1.3 $\exists f(M)$ is the character of the image of fm.

1.4 $\forall f(M)$ is the character of $\Pi_f(m)$ (Π_f is the right adjoint of pulling-back-along f).

1.5 $\exists! f(M)$ is the character of the unique-existentiation part of fm, i.e. the pullback of $C \xrightarrow{\{-\}} PC$ along $B \xrightarrow{\{-\}} PB \xrightarrow{\Omega^{fm}} PC$ (see Freyd [4], Prop.2.21).

In some places we will also consider the stronger theory EBT of <u>elementary</u> <u>boolean</u> <u>topoi</u> which we get from ET by adding the following

<u>1.6 Axiom of booleaness</u> $\Omega \xrightarrow{\neg} \Omega \xrightarrow{\neg} \Omega = id_\Omega$

In EBT Ω is an internal boolean algebra and the algebra of subobjects of a given object A is boolean.

Finally a convention concerning the notation. Although we frequently introduce subscripts (or indices) for a better understanding, we will omit these subscripts whenever no confusion seems to be possible.

2. The language L(SET) and its internal interpretation

Let us proceed to the construction of the set theory SET defined over ET which will serve as a powerful tool to translate set theoretical arguments and constructions into topos theory. First we describe the language L(SET) of SET which is essentially due to W. Mitchell (who denoted it L($\underline{E}$) in [20] for a given topos $\underline{E}$). The idea behind this language is that we imagine the objects in ET to have unspecified "elements" (as if we were working in the topos of sets) having the following important properties:

a) 1 has a (unique) element.

b) any map $A \xrightarrow{f} B$ induces the operation "value under f" from elements of A to those of B.

c) the elements of $A \times B$ are "ordered pairs" of elements of A, B. Using the predicate of equality and first-order logic we can formulate enough "properties" of elements.

Formally the language L(SET) will be constructed over the language L(ET) of the theory ET of elementary topoi as follows.

L(SET) is a many-sorted first-order language having the objects of ET as <u>types</u>, i.e. there is a type-operator τ assigning to each term x of L(SET) an object (term) τx of ET. The <u>terms</u> of L(SET) and the type-operator are given recursively in the usual way by the following rules 2.1-4.

2.1 0° is a constant with $\tau 0^{\circ} = 1$.

2.2 For any object A there is a countable number of variables of type A.

2.3 For any map $A \xrightarrow{f} B$ there is an operator f(-) "value under f" from terms x of type A to those of type B : $\tau f(x) = B$.

2.4 For any pair $\langle A,B\rangle$ of objects there is an "ordered-pair-operator" $\langle -,-\rangle$ assigning to terms x, y with ${}_\tau x=A$, ${}_\tau y=B$ a term $\langle x,y\rangle$ with ${}_\tau\langle x,y\rangle = A\times B$.

The only primitive notions of L(SET) are the constant predicate "False" and the predicate of equality "=" (which may hold only between terms of equal types), i.e.

2.5 The atomic formulas of L(SET) are:

(1) False , (2) $x=y$, provided ${}_\tau x={}^\tau y$

The (well-formed) formulas of L(SET) are generated from the atomic ones in the standard way allowing the connectives $\wedge$ (conjunction), $\vee$ (disjunction), $\Longrightarrow$ (implication) and the quantifiers $\forall x$ (for all x), $\exists x$ (there exists x) , provided the variable x occours free in the formula following the quantifier. Negation $\neg$, True, equivalence $\Longleftrightarrow$ and unique-existentiation $\exists!$ are defined as usual:

$\neg\varphi$	means	$\varphi \Longrightarrow$ False
True	means	$\neg$False
$(\varphi \Longleftrightarrow \psi)$	means	$(\varphi \Longrightarrow \psi) \wedge (\psi \Longrightarrow \varphi)$
$\exists! x\, \varphi(x)$	means	$\exists x(\varphi(x) \wedge \forall y(\varphi(y) \Longleftrightarrow x=y))$

2.6 Remark It should be pointed out that the types (being the terms of ET) are countable, and that the operators generating the terms of L(SET) are countable. Hence the language L(SET) is countable and can in fact (in various ways) be explicitly constructed over the same alphabet of L(ET). In the semantical approach where L(E) is constructed over a topos E, the language L(E) will not be countable (unless E is). The latter approach is adopted in [20] and [23].

For intuitive reasons we call the terms resp. variables of L(SET) from now (except in a formal context) simply elements resp. element-variables (defined over ET), and for objects A and elements x let us us write "$x\varepsilon A$" (read: x is an A-element) instead of "${}_\tau x=A$".

Note that $x \varepsilon A$ is a metastatement and not a formula of L(SET). The Ω-elements will also be called truth-values. Furthermore, if $x \varepsilon A$ we frequently write $\forall x \varepsilon A$ resp. $\exists x \varepsilon A$ instead of $\forall x$ resp. $\exists x$ to emphasize that the quantifiers are actually restricted.

Let us now give a few definitions in L(SET) which show that this language is fairly "rich" and has a set theoretical character.

2.7 To any global section $1 \xrightarrow{a} A$ corresponds an A-element $a^\circ := a(0^\circ) \varepsilon A$. In particular we have the truth-values true°, false°.

2.8 For $x \varepsilon A$, $Y \varepsilon PA$ the membership relation is defined

$$x \in Y \quad :\Longleftrightarrow \quad (PA \times A \xrightarrow{ev} \Omega)\langle Y, x\rangle = \text{true}^\circ \quad .$$

2.9 For $x \varepsilon A$, $F \varepsilon B^A$ the value $Fx \varepsilon B$ is defined as

$$Fx := (B^A \times A \xrightarrow{ev} B)\langle F, x\rangle \; .$$

2.10 For any map $A \xrightarrow{f} B$ the exponential adjoint $1 \xrightarrow{\overline{f}} B^A$ gives an B^A-element $f^\circ := \overline{f}^\circ \varepsilon B^A$, and in particular we have for any subobject $A \xrightarrow{M} \Omega$ a PA-element $M^\circ = \overline{M}^\circ \varepsilon PA$.

2.11 Remark By 2.8, 2.10 any subobject $A \xrightarrow{M} \Omega$ induces a unary predicate $(-) \in M^\circ$ for A-elements. These predicates are taken as primitive notions in [20] and [23].

Furthermore, the notion of ordered pairs extends in a standard way to ordered n-tuples, whose definition is given by:

2.12 $\langle x\rangle := x$ and $\langle x_1, \ldots x_{n+1}\rangle := \langle\langle x_1, \ldots x_n\rangle, x_{n+1}\rangle$.

The most remarkable feature of the language L(SET) is that it admits a natural "internal" interpretation in the language L(ET) of ET, which in fact goes back to W. Mitchell [20] and runs as follows (for another interesting "external" interpretation see Osius [23]).

First, let $t \varepsilon A$ be a term of L(SET) such that all variables occurring in t are among the variables $x_1 \varepsilon A_1, .. x_n \varepsilon A_n$ ($n \geq 0$). By induction on the length of the term t we define a map

$$\{\langle x_1,..x_n\rangle \mapsto t\} : A_1 \times .. \times A_n \longrightarrow A \quad ,$$

which represents the term t with respect to $x_1,..x_n$, by 2.13-16 .

2.13 $\{\langle x_1,..x_n\rangle \mapsto 0^\circ\}$ is the unique map $A_1 \times .. \times A_n \longrightarrow 1$.

2.14 $\{\langle x_1,..x_n\rangle \mapsto x_i\}$ is the projection $A_1 \times .. \times A_n \longrightarrow A_i$.

2.15 For $A \xrightarrow{f} B$: $\{\langle x_1,..x_n\rangle \mapsto f(t)\} := f \cdot \{\langle x_1,..x_n\rangle \mapsto t\}$

2.16 For $r \varepsilon B$, $s \varepsilon C$ the map $\{\langle x_1,..x_n\rangle \mapsto \langle r,s\rangle\}$ into $B \times C$ is the unique map induced by $\{\langle x_1,..x_n\rangle \mapsto r\}$ and $\{\langle x_1,..x_n\rangle \mapsto s\}$.

Second, let φ be a formula of L(SET) such that all free variables occurring in φ are among the variables $x_1 \varepsilon A_1, .. x_n \varepsilon A_n$ ($n \geq 0$). By induction on the length of the formula φ we define a subobject

$$\{\langle x_1,..x_n\rangle | \varphi\} : A_1 \times .. \times A_n \longrightarrow \Omega \quad ,$$

which represents the "property" φ with resp. to $x_1,..x_n$, by 2.17-20.

2.17 $\{\langle x_1,..x_n\rangle | \text{False}\} := \text{false}_{A_1 \times .. \times A_n}$

2.18 $\{\langle x_1,..x_n\rangle | r = s\} := \{\langle x_1,..x_n\rangle \mapsto \langle r,s\rangle\}^{-1}(A \times A \xrightarrow{\Delta} \Omega)$, provided $r,s \varepsilon A$ and Δ is the diagonal of A.

2.19 $\{\langle x_1,..x_n\rangle | \varphi \wedge \psi\} := \{\langle x_1,..x_n\rangle | \varphi\} \cap \{\langle x_1,..x_n\rangle | \psi\}$ and similiar for $\vee$, $\Longrightarrow$ (replace $\cap$ by $\cup$, $\Rightarrow$).

2.20 $\{\langle x_1,..x_n\rangle | \forall x \varepsilon A\ \varphi(x)\} := \forall p \ \{\langle x_1,..x_n,y\rangle | \varphi(y)\}$

$\{\langle x_1,..x_n\rangle | \exists x \varepsilon A\ \varphi(x)\} := \exists p \ \{\langle x_1,..x_n,y\rangle | \varphi(y)\}$

where $A_1 \times .. \times A_n \times A \xrightarrow{p} A_1 \times .. \times A_n$ is the projection and $y \varepsilon A$ is distinct from $x_1,..x_n$ (see also 2.25).

For the defined notions we get immediatly (for $\exists!$ see 4.23.1) :

2.21 (1) $\{\langle x_1,..x_n\rangle | \neg \varphi\} = \neg \{\langle x_1,..x_n\rangle | \varphi\}$

(2) For $t \in A$, $A \xrightarrow{M} \Omega$: $\{\langle x_1,..x_n\rangle \mid t \in M^\circ\} = \{\langle x_1,..x_n\rangle \mapsto t\}^{-1}(M)$.

(3) $\{\langle x_1,..x_n\rangle \mid \varphi \Longleftrightarrow \psi\} = \{\langle x_1,..x_n\rangle \mid \varphi\} \Leftrightarrow \{\langle x_1,..x_n\rangle \mid \psi\}$. □

To facilitate the computation of the operators $\{...\}$ we note some technical points.

2.22 (Superfluous variables) If $x_{n+1} \in A_{n+1},..\ x_{n+k} \in A_{n+k}$ do not occur (free) in t resp. φ , and $A_1 \times .. \times A_{n+k} \xrightarrow{p} A_1 \times .. \times A_n$ is the canonical projection, then

(1) $\{\langle x_1,..x_{n+k}\rangle \mapsto t\} = \{\langle x_1,..x_n\rangle \mapsto t\} \cdot p$

(2) $\{\langle x_1,..x_{n+k}\rangle \mid \varphi\} = p^{-1}\ \{\langle x_1,..x_n\rangle \mid \varphi\}$

2.23 (Permuting the variables) If σ is a permutation of $1,..n$ and $f_\sigma = \{\langle x_1,..x_n\rangle \mapsto \langle x_{\sigma 1},..x_{\sigma n}\rangle\} : \prod_i A_i \longrightarrow \prod_i A_{\sigma i}$ is the corresponding isomorphism, then

(1) $\{\langle x_1,..x_n\rangle \mapsto t\} = \{\langle x_{\sigma 1},..x_{\sigma n}\rangle \mapsto t\} \cdot f_\sigma$

(2) $\{\langle x_1,..x_n\rangle \mid \varphi\} = f_\sigma^{-1}\ \{\langle x_{\sigma 1},..x_{\sigma n}\rangle \mid \varphi\}$.

2.24 (Substitution) If the variable $x \in B$ occurs in $t(x)$ resp. free in $\varphi(x)$, and if $s \in B$ is a term whose variables are not bounded in $\varphi(x)$, then

$$\{\langle x_1,..x_n\rangle \mapsto t(s)\} = \{\langle x_1,..x_n,x\rangle \mapsto t(x)\} \cdot \{\langle x_1,..x_n\rangle \mapsto \langle x_1,..x_n,s\rangle\}$$

$$\{\langle x_1,..x_n\rangle \mid \varphi(s)\} = \{\langle x_1,..x_n\rangle \mapsto \langle x_1,..x_n,s\rangle\}^{-1}\ \{\langle x_1,..x_n,x\rangle \mid \varphi(x)\}\ .$$

2.25 If $x \in A$ and $A_1 \times .. \times A_n \times A \xrightarrow{p} A_1 \times .. \times A_n$ is the projection, then

(1) $\{\langle x_1,..x_n,x\rangle \mid \exists x\ \varphi(x)\} = p^{-1}(\ \exists p\ \{\langle x_1,..x_n,x\rangle \mid \varphi(x)\}\ )$

(2) $\{\langle x_1,..x_n,x\rangle \mid \forall x\ \varphi(x)\} = p^{-1}(\ \forall p\ \{\langle x_1,..x_n,x\rangle \mid \varphi(x)\}\ )$.

Proofs: 2.22-24 are straight-forward by induction on the length of t resp. φ , using the so called Beck-condition for quantification (i.e. 1.36 of [9] or 5.3 of [22]). 2.25 follows from 2.20, 2.24 . □

Furthermore, we immediatly conclude from the corresponding definitions:

2.26 For $x \varepsilon A$, $Y \varepsilon PA$: $\{\langle Y,x\rangle \mid x \in Y\} = PA \times A \xrightarrow{ev} \Omega$. □

2.27 For $x \varepsilon A$, $F \varepsilon B^A$: $\{\langle F,x\rangle \mapsto Fx\} = B^A \times A \xrightarrow{ev} B$. □

2.28 For $A \xrightarrow{f} B$, $x \varepsilon A$: $f = \{x \mapsto f^{\circ}x\} = \{x \mapsto f(x)\}$. □

2.29 For $A \xrightarrow{M} \Omega$, $x \varepsilon A$: $M = \{x \mid x \in M^{\circ}\}$. □

2.30 For $x,y \varepsilon A$: $\{\langle x,y\rangle \mid x = y\} = A \times A \xrightarrow{\Delta} \Omega$. □

2.31 For any global section $1 \xrightarrow{a} A$ we have

(1) $\{y \varepsilon B \mapsto a^{\circ}\} = B \longrightarrow 1 \xrightarrow{a} A$,

and adopting the usual notation $\{a^{\circ}\} := \{x \varepsilon A \mid x = a^{\circ}\}$ we get

(2) $\{a^{\circ}\} = \ulcorner\{a\}\urcorner = \chi(a)$,

where $\ulcorner\{a\}\urcorner$ is the exponential adjoint of $\{a\}: 1 \xrightarrow{a} A \xrightarrow{\{-\}} PA$. □

In view of the above results the superscript "$^{\circ}$" (read: internal) will be omitted from now on if no confusion is possible.

3. Internal validity and intuitionistic logic

In this section we consider a notion of truth (called internal validity) for formulas of the language L(SET) which naturally arises from the internal interpretation given in the last section. We start off with the following definition (going back to W. Mitchell [20]).

3.1 Definition For any formula φ resp. term $t \varepsilon A$ of L(SET) let $x_1 \varepsilon A_1, \ldots x_n \varepsilon A_n$ ($n \geq 0$) be exactly all free variables of φ resp. t in their natural order of their first ocurrance in φ resp. t .

(1) $\tau\varphi := A_1 \times \ldots \times A_n$ is called the type of φ , resp.

$\tau t := A_1 \times \ldots \times A_n$ is called the type of t .

(2) $\|\varphi\| : \tau\varphi \longrightarrow \Omega := \{\langle x_1, \ldots x_n \rangle \mid \varphi\}$ is called the internal interpretation of φ , resp.

$\|t\| : \tau t \longrightarrow A := \{\langle x_1, \ldots x_n \rangle \mapsto t\}$ is called the internal interpretation of t .

(3) φ is called internally valid (or: internally true) , noted $\models \varphi$, iff $\|\varphi\| = \mathrm{true}_{\tau\varphi}$ (i.e. $\|\varphi\|$ factors through $1 \xrightarrow{\mathrm{true}} \Omega$) . Note that the order of the variables is not important here, indeed for any permutation σ of 1,..n the formula φ is internally valid iff $\{\langle x_{\sigma 1}, \ldots x_{\sigma n} \rangle \mid \varphi\} = \mathrm{true}$ (cf. 2.23).

3.2 Criterion The formula φ is internally valid iff for all sequences of variables $y_1, \ldots y_m$ containing all free variables of φ $\{\langle y_1, \ldots y_m \rangle \mid \varphi\} = \mathrm{true}$ holds. Note that it is not sufficient, if the condition holds only for some sequence, indeed for $y \varepsilon 0$ one always has $\{\langle y_1, \ldots y_n, y \rangle \mid \varphi\} = \mathrm{true}_0$.

The criterion follows from 2.22.□ Some immediate properties of internal validity are the following

3.3 If $x_1,..x_n$ are <u>exactly</u> the variables occuring in the term $\langle r,s\rangle$, then (1) $\models r=s$ iff

$$\{\langle x_1,..x_n\rangle \mapsto r\} = \{\langle x_1,..x_n\rangle \mapsto s\} \quad .$$

In particular (2) $1 \xrightarrow{a} A = 1 \xrightarrow{b} A$ iff $\models a=b$,

(3) $1 \xrightarrow{a} A \xrightarrow{M} \Omega = \text{true}$ iff $\models a \in M$.

By slight abuse of notation we sometimes write simply $a \in M$ instead of $\models a \in M$.

3.4 $\models \neg\varphi$ iff $\|\varphi\| = \text{false}_{\tau\varphi}$.

3.5 $\models \varphi\wedge\psi$ iff $\models \varphi$ and $\models \psi$,

$\models \forall x\, \varphi(x)$ iff $\models \varphi(x)$.

3.6 If $x_1,..x_n$ are <u>exactly</u> the free variables of $\varphi\vee\psi$, then

$\models \varphi\vee\psi$ iff $\{\langle x_1,..x_n\rangle | \varphi\} \cup \{\langle x_1,..x_n\rangle | \psi\} = \text{true}$

$\models \varphi \Longrightarrow \psi$ iff $\{\langle x_1,..x_n\rangle | \varphi\} \subset \{\langle x_1,..x_n\rangle | \psi\}$

$\models \varphi \Longleftrightarrow \psi$ iff $\{\langle x_1,..x_n\rangle | \varphi\} = \{\langle x_1,..x_n\rangle | \psi\}$.

The straight-forward proofs are omitted.□ The following results are concerned with the relationship between internal validity and intuitionistic logic.

<u>3.7 Proposition</u> The formulas of L(SET) which are intuitionistical propositional tautologies (i.e. are valid in any Heyting-algebra, see Rasiowa-Sikorski [25] Chap.IX) are internally valid.

<u>Proof</u>: We illustrate the general method by a particular example, namely we prove that the intuitionistical tautology $(\varphi\wedge\neg\varphi)\Longrightarrow\psi$ is internally valid. Using 3.2 let $x_1 \varepsilon A_1,..x_n \varepsilon A_n$ contain all free variables of φ and ψ, and let $A = A_1\times..\times A_n$. Then $M := \{\langle x_1,..x_n\rangle | \varphi\}$ and $N := \{\langle x_1,..x_n\rangle | \psi\}$ are subobjects of A and we have to establish $(M \cap \neg M) \Rightarrow N = \text{true}_A$, which holds since it is an interpretation of the given tautology in the Heyting-algebra of subobjects of A. □

<u>3.8 Proposition</u> The axiom of booleaness (1.6) holds if and only if all classical propositional tautologies in L(SET) are internally valid.

<u>Proof:</u> If 1.6 holds then the proof of 3.7 in fact proves that all propositional tautologies are internally valid. Conversly, for $p\varepsilon\Omega$ the formula $(p=\mathrm{true} \vee \neg(p=\mathrm{true}))$ is a (classical) tautology and hence internally valid, i.e. $\{\mathrm{true}\} \cup \neg\{\mathrm{true}\} = \mathrm{true}_\Omega$. This implies 1.6 since $\{\mathrm{true}\} = \chi(\mathrm{true})$, $\neg\{\mathrm{true}\} = \chi(\mathrm{false})$ by 2.31.2 . □

We now turn to the axioms and rules for quantification.

<u>3.9 Lemma</u> The following <u>axioms of quantification</u> are internally valid:

$$(a\exists)\quad \varphi(x) \Longrightarrow \exists x\, \varphi(x)$$
$$(a\forall)\quad \forall x\, \varphi(x) \Longrightarrow \varphi(x)$$

<u>Proof:</u> Let $x\varepsilon A, x_1\varepsilon A_1, .. x_n\varepsilon A_n$ be all free variables of $\varphi(x)$, and let $A_1\times..\times A_n\times A \xrightarrow{p} A_1\times..\times A_n$ be the projection. Using 3.6 and 2.25, we define $M := \{\langle x_1, .. x_n, x\rangle \mid \varphi(x)\}$ and have to show $M \subset p^{-1}(\exists p(M))$ and $p^{-1}(\forall p(M)) \subset M$, which are well known to hold. □

<u>3.10 Lemma</u> The following <u>rules of quantification</u> (and the converse rules) are internally valid:

<u>If</u> x <u>is not free in</u> ψ, then

$$(r\exists)\quad \frac{\varphi(x) \Longrightarrow \psi}{\exists x\, \varphi(x) \Longrightarrow \psi} \qquad\qquad (r\forall)\quad \frac{\psi \Longrightarrow \varphi(x)}{\psi \Longrightarrow \forall x\, \varphi(x)}$$

<u>Proof:</u> Let $x\varepsilon A, x_1\varepsilon A_1, .. x_n\varepsilon A_n$ be all free variables of $\varphi(x) \Longrightarrow \psi$, and put $M:=\{\langle x_1, .. x_n, x\rangle \mid \varphi(x)\}$, $N:=\{\langle x_1, .. x_n\rangle \mid \psi\}$. For the projection $A_1\times..\times A_n\times A \xrightarrow{p} A_1\times..\times A_n$ we have

$$M \subset p^{-1}(N) \text{ iff } \exists p(M) \subset N \quad , \quad p^{-1}(N) \subset M \text{ iff } N \subset \forall p(M)$$

which in view of 3.6 and 2.22.2 proves the rules and their converse. □

Concerning substitution, 2.24 immediatly gives

<u>3.11 Corollary</u> For any formula $\varphi(x)$ with a free variable $x \epsilon A$ and for any term $t \epsilon A$ the following <u>substitution rule</u> is internally valid:

$$(\text{Subst}) \quad \frac{\varphi(x)}{\varphi(t)}$$ □

So far we have proved, that internal validity satisfies all axioms and deductive rules of intuitionistic logic (see e.g. Rasiowa-Sikorski [25]) except for the rule of "<u>modus ponens</u>"

$$(\text{Mp}) \quad \frac{\varphi \quad , \quad \varphi \Longrightarrow \psi}{\psi}$$

which is <u>not</u> internally true. Indeed, for $x \epsilon 0$ the formulas $x = x$ and $x = x \Longrightarrow \exists x \epsilon 0 \; x = x$ are internally valid, but $\exists x \epsilon 0 \; x = x$ is not (provided of course $0 \neq 1$). More generally, for any formula $\varphi(x)$ with free x we have $\| \exists x \epsilon 0 \; \varphi(x) \| = \text{false}$ and $\| \forall x \epsilon 0 \; \varphi(x) \| = \text{true}$.

For a better understanding of this situation let us split up the modus ponens (Mp) into two rules, the

<u>3.12 (Restricted modus ponens)</u> <u>If all free variables of</u> φ <u>are among those of</u> ψ , then

$$\frac{\varphi \quad , \quad \varphi \Longrightarrow \psi}{\psi} \quad (\text{Rmp})$$

and the following rule for existentiation: $\quad (r^*\exists) \quad \dfrac{\varphi(x)}{\exists x \; \varphi(x)}$.

Clearly (Mp) and (a∃) imply (Rmp) and (r*∃). Conversly (Rmp), (r*∃), (r∃) will now be shown to imply (Mp): By hypothesis of (Mp) φ and $\varphi \Longrightarrow \psi$ hold. Hence, if $x_1,..x_k$ are all <u>free</u> variables of φ which are <u>not free</u> in ψ, then $\exists x_1..\exists x_k \varphi$ and $\exists x_1..\exists x_k \varphi \Longrightarrow \psi$ hold by (r*∃), (r∃), and thus ψ holds by (Rmp).□

The example just given actually shows that (r*∃) is not internally valid. More generally, for $x \epsilon A$ the formula $x = x$ is internally valid and $\| \exists x \; x = x \|$ is the <u>support</u> of A (i.e. the image of $A \longrightarrow 1$, see e.g. [23]). Hence, $\exists x \epsilon A \; x = x$ is internally valid iff $A \longrightarrow 1$ is epic (which is certainly not true for all A). However, the important part of (Mp), namely (Rmp) is internally valid.

3.13 Lemma The restricted modus ponens (3.12) is internally valid.

Proof: Let $x_1,..x_n$ be all free varables of ψ, and hence of $\varphi \Longrightarrow \psi$. Since φ and $\varphi \Longrightarrow \psi$ are internally valid, we have
$\{\langle x_1,..x_n\rangle | \varphi\} = \text{true}$, $\{\langle x_1,..x_n\rangle | \varphi\} \subset \{\langle x_1,..x_n\rangle | \psi\}$,
which implies that ψ is internally valid. □

Concerning the axioms of equality, we observe

3.14 Lemma The following axioms of equality are internally valid : (Eq1) $\forall x\varepsilon A \;\; x = x$

(Eq2) $\forall x,y\varepsilon A \; (x = y \Longrightarrow y = x)$

(Eq3) $\forall x,y,z\varepsilon A \; (x = y \wedge y = z \Longrightarrow x = z)$

(Eq4) $\forall x,u\varepsilon A \; \forall y,v\varepsilon B \; (x = u \wedge y = v \Longrightarrow \langle x,y\rangle = \langle u,v\rangle)$

(Eq5) For $A \xrightarrow{f} B$: $\forall x,y\varepsilon A \; (x = y \Longrightarrow f(x) = f(y))$.

The straight-forward proof is omitted. □

For convenience let us now introduce a weaker notion of truth for formulas of L(SET).

3.15 Definition A formula φ is said to be intuitionistically valid (or true) resp. classically valid (or true), denoted $\vdash \varphi$ resp. $\vdash_c \varphi$, iff it is among

(i) the intuitionistically resp. classically propositional tautologies ,

(ii) the axioms (a∃) and (a∀) of quantification (see 3.9) ,

(iii) the axioms (Eq1-5) of equality (see 3.14) ,

or can be deduced from the formulas in (i-iii) using the rules (r∃), (r∀) of quantification (see 3.10) , the substitution rule (Subst) (see 3.11), and the restricted modus ponens (Rmp) (see 3.12).

Notice, that the rule (r*∃) of existentiation is not allowed to deduce intuitionistically resp. classically valid formulas and hence the full modus ponens (Mp) is not allowed. However the rule (r*∃), which is in fact equivalent to the single axiom $\forall x\, \varphi(x) \Longrightarrow \exists x\, \varphi(x)$, does not seem very intuitive to us anyway and its absense does not inflict most of the deductions in usual intuitionistic logic (see e.g. 3.18-22).

From our preceeding considerations (3.7-13) we conclude

3.16 Theorem

(1) Intuitionistically valid formulas of L(SET) are internally valid.

(2) The internal valid formulas of L(SET) are closed under the intuitionistically valid rules of deduction.

(3) If the axiom of booleaness (1.6) holds, then we can replace "intutionistically valid" in (1-2) by "classically valid". □

In order to apply this theorem (i.e. to show that some formula is internally valid) we need some standard knowledge of intuitionistic logic (in the restricted sense employed here). It is without the scope of this paper to develop the relevant material (including proofs) on intuitionistical validity. Let us however state some basic facts (without proofs) to which we can refer when we apply later theorem 3.16 in order to prove results in topos theory ET.

First , we slightly strengthen the restricted modus ponens. Let us call the types of the free variables of a formula φ briefly the free types of φ .

3.17 If all free types of φ are among those of ψ, then the rule (Rmp') $\dfrac{\varphi \;,\; \varphi \Longrightarrow \psi}{\psi}$ is intuitionistically (and internally) valid.

Proof: Replace all free variables of φ which are not free in ψ by a free variable of ψ with same type. Then φ becomes φ', and by assumption and (Subst) (3.11) φ' and $\varphi' \Rightarrow \psi$ are valid. Hence ψ is valid by (Rmp). □

Second, we state without proof some standard results of logic.

3.18 Substitution of equivalent formulas

If α, β, $\varphi(\alpha)$, $\varphi(\beta)$ are formulas such that

(1) α is a subformula of $\varphi(\alpha)$,

(2) β is a subformula of $\varphi(\beta)$,

(3) $\varphi(\alpha)$ and $\varphi(\beta)$ are alike, except that $\varphi(\alpha)$ contains α wherever $\varphi(\beta)$ contains β ,

then the following rule is intuitionistically (and internally) valid :

$$\frac{\alpha \Leftrightarrow \beta}{\varphi(\alpha) \Leftrightarrow \varphi(\beta)} .$$

Furthermore, if the free types of α are among those of $\varphi(\beta)$, then

$$\frac{\varphi(\alpha) \quad , \quad \alpha \Leftrightarrow \beta}{\varphi(\beta)}$$

is intuitionistically (and internally) valid. □

Concerning the propositional calculus we note

3.19 Proposition

The following rules are intuitionistically (and internally) valid :

(1) $$\frac{\varphi \Rightarrow \psi}{(\psi \Rightarrow \theta) \Rightarrow (\varphi \Rightarrow \theta) \quad , \quad (\theta \Rightarrow \varphi) \Rightarrow (\theta \Rightarrow \psi)}$$

(2) If all free types of ψ are among those of $\varphi \Rightarrow \theta$, then

$$\frac{\varphi \Rightarrow \psi \quad , \quad \psi \Rightarrow \theta}{\varphi \Rightarrow \theta}$$

(3) $$\frac{\varphi}{\varphi \vee \psi}$$

(4) $$\frac{\varphi \quad , \quad \psi}{\varphi \wedge \psi}$$

(5) If all free types of ψ are among those of φ, then

$$\frac{\varphi \wedge \psi}{\varphi}$$

(6) $$\frac{\varphi \Rightarrow \psi \quad , \quad \varphi' \Rightarrow \psi'}{\varphi \wedge \varphi' \Rightarrow \psi \wedge \psi' \quad , \quad \varphi \vee \varphi' \Rightarrow \psi \vee \psi'}$$

(7) $$\frac{\varphi \Rightarrow \psi}{\varphi \wedge \theta \Rightarrow \psi}$$

(8) $$\frac{\varphi \Rightarrow (\psi \Rightarrow \theta)}{\varphi \wedge \psi \Rightarrow \theta}$$ and conversly

(9) $$\frac{\varphi \Rightarrow \neg \varphi}{\neg \varphi}$$

(10) $$\frac{\varphi \Rightarrow \psi}{\neg \psi \Rightarrow \neg \varphi}$$

(11) $$\frac{\varphi \Rightarrow \neg \psi}{\psi \Rightarrow \neg \varphi}$$

(12) $$\frac{\varphi \Leftrightarrow \psi}{\varphi \Rightarrow \psi \quad , \quad \psi \Rightarrow \varphi}$$ and conversly.

Proof: Apply the modus ponens (version 3.17) to the corresponding propositional tautology. □

3.20 Proposition The following rules for quantification are intuitionistically (and internally) valid :

(1) $$\frac{\forall x\, \varphi(x)}{\varphi(x)}$$ and conversly

(2) If the type of x is a free type of $\exists x\, \varphi(x)$, then

$$\frac{\varphi(x)}{\exists x\, \varphi(x)}$$

(3) $$\frac{\varphi(x) \Rightarrow \psi(x)}{\forall x\, \varphi(x) \Rightarrow \forall x\, \psi(x) \quad , \quad \exists x\, \varphi(x) \Rightarrow \exists x\, \psi(x)}$$

(4) If the type of x is a free type of $\psi \Rightarrow \exists x\, \varphi(x)$, then

$$\frac{\psi \Rightarrow \varphi(x)}{\psi \Rightarrow \exists x\, \varphi(x)}$$ □

3.21 Proposition The following formulas concerning quantification are intuitionistically (and internally) valid :

(1) $\exists x \exists y\, \varphi(x,y) \Leftrightarrow \exists y \exists x\, \varphi(x,y)$

(2) $\forall x \forall y\, \varphi(x,y) \Leftrightarrow \forall y \forall x\, \varphi(x,y)$

(3) $\exists x \forall y\, \varphi(x,y) \Leftrightarrow \forall y \exists x\, \varphi(x,y)$

(4) $\forall x(\varphi(x) \wedge \psi(x)) \Leftrightarrow \forall x\, \varphi(x) \wedge \forall x\, \psi(x)$

(5) $\exists x(\varphi(x) \vee \psi(x)) \Leftrightarrow \exists x\, \varphi(x) \vee \exists x\, \psi(x)$

(6) $\forall x\, \varphi(x) \vee \forall x\, \psi(x) \Rightarrow \forall x(\varphi(x) \vee \psi(x))$

(7) $\exists x(\varphi(x) \wedge \psi(x)) \Rightarrow \exists x\, \varphi(x) \wedge \exists x\, \psi(x)$

(8) $\forall x\ (\ \varphi(x) \Rightarrow \psi(x)\ ) \Rightarrow (\ \forall x\, \varphi(x) \Rightarrow \forall x\, \psi(x)\ )$

(9) $\forall x\ (\ \varphi(x) \Rightarrow \psi(x)\ ) \Rightarrow (\ \exists x\, \varphi(x) \Rightarrow \exists x\, \psi(x)\ )$

(10) $\exists x\, \neg \varphi(x) \Rightarrow \neg \forall x\ \varphi(x)$

(11) $\exists x\ \varphi(x) \Rightarrow \neg \forall x\, \neg\, \varphi(x)$

(12) $\neg \exists x\ \varphi(x) \Rightarrow \forall x\, \neg\, \varphi(x)$

Furthermore, if x is not free in θ then :

(13) $\theta \wedge \forall x\ \varphi(x) \Rightarrow \forall x(\theta \wedge \varphi(x))$

(14) $\theta \vee \forall x\ \varphi(x) \Rightarrow \forall x(\theta \vee \varphi(x))$

(15) $\exists x(\theta \vee \varphi(x)) \Rightarrow \theta \vee \exists x\ \varphi(x)$

(16) $\exists x(\theta \wedge \varphi(x)) \Leftrightarrow \theta \wedge \exists x\ \varphi(x)$

(17) $\exists x\ (\ \varphi(x) \Rightarrow \theta\ ) \Rightarrow (\ \forall x\, \varphi(x) \Rightarrow \theta\ )$

(18) $\exists x\ (\ \theta \Rightarrow \varphi(x)) \Rightarrow (\ \theta \Rightarrow \exists x\, \varphi(x)\ )$

(19) $\forall x\ (\ \theta \Rightarrow \varphi(x)) \Leftrightarrow (\ \theta \Rightarrow \forall x\, \varphi(x)\ )$

(20) $\forall x\ (\ \varphi(x) \Rightarrow \theta\ ) \Leftrightarrow (\ \exists x\, \varphi(x) \Rightarrow \theta\ )$

The converse of (13)(15) hold under the additional assumption that the <u>type</u> of x is a <u>free type</u> of θ. The converse of (10)(14)(17)(18) hold if we assume the axiom of booleaness (1.6). □

Finally, concerning equality and unique existentiation, we note

<u>3.22 Proposition</u> Intuitionistically (and internally) valid are

(1) $x = y \Rightarrow (\ \varphi(x) \Leftrightarrow \varphi(y)\ )$

(2) $\exists x\ (\ x = y \wedge \varphi(x)\ ) \Leftrightarrow \varphi(y)$

(3) $\forall x\ (\ x = y \Rightarrow \varphi(x)\ ) \Leftrightarrow \varphi(y)$

(4) $\exists! x\, \varphi(x) \Leftrightarrow \exists x\ (\ \varphi(x) \wedge \forall y(\varphi(y) \Rightarrow x=y\ )\ )$

(5) $\exists! x\ \varphi(x) \Leftrightarrow \exists x\, \varphi(x) \wedge \forall y,z\ (\ \varphi(y) \wedge \varphi(z) \Rightarrow y=z\ )$

(6) $\exists! x\ (\ x = y \wedge \varphi(x)\ ) \Leftrightarrow \varphi(y)$

(7) $\exists! x\ x = y$ (x distinct from y) . □

4. The set theory SET

In this section we establish the basic properties of internal validity which do not hold for purely logical reasons but involve the topos structure. Our aim is to characterize some basic notion of topos theory ET internally and to derive set theoretical properties of the language L(SET). First we observe that ordered pairs of elements behave as they should and that the maps act on elements as expected.

4.1 Lemma For $x \varepsilon A$, $y \varepsilon B$, $u \varepsilon A\times B$ we have :

(1) $\models \quad pr_1\langle x,y\rangle = x \;\wedge\; pr_2\langle x,y\rangle = y$

(2) $\models \quad u = \langle pr_1 u, pr_2 u\rangle$

4.2 Lemma For $x \varepsilon A$ and $1 \xrightarrow{a} A \xrightarrow{f} B \xrightarrow{g} C$ we have :

(1) $\models \quad id_A x = x \;\wedge\; g(fx) = (gf)x$

(2) $\models \quad f^\circ x = f(x) \;\wedge\; f(a^\circ) = (fa)^\circ$

4.3 Lemma Let $A \xrightarrow{f} B$, $A \xrightarrow{g} C$, $C \xrightarrow{h} D$ and $x \varepsilon A$, $y \varepsilon C$.

(1) For the induced map $A \xrightarrow{(f,g)} B\times C$: $\models \quad (f,g)(x) = \langle fx, gx\rangle$.

(2) For the induced map $A\times C \xrightarrow{f\times h} B\times D$: $\models \quad (f\times h)\langle x,y\rangle = \langle fx, hy\rangle$.

The proofs are straight-forward (using 3.3). □

A standard consequence of the existence of ordered pairs is that successive quantifiers (of same sort) can be reduced to one quantifier.

4.4 Reduction of quantifiers If $x \varepsilon A$, $y \varepsilon B$ are free in $\varphi(x,y)$, then :

(1) $\models \quad \exists x \varepsilon A\; \exists y \varepsilon B\; \varphi(x,y) \iff \exists u \varepsilon A\times B\; \varphi(pr_1 u, pr_2 u)$

(2) $\models \quad \forall x \varepsilon A\; \forall y \varepsilon B\; \varphi(x,y) \iff \forall u \varepsilon A\times B\; \varphi(pr_1 u, pr_2 u)$ □

4.5 Quantifiers over products If $u \varepsilon A\times B$ is free in $\psi(u)$, then :

(1) $\models \quad \exists u \varepsilon A\times B\; \psi(u) \iff \exists x \varepsilon A\; \exists y \varepsilon B\; \psi(\langle x,y\rangle)$

$$(2) \quad \models \quad \forall u \varepsilon A \times B \ \psi(u) \quad \Longleftrightarrow \quad \forall x \varepsilon A \ \forall y \varepsilon B \ \psi(\langle x,y \rangle) \qquad . \square$$

An important point is that maps are determined by their values and subobjects by their elements:

<u>4.6 Principle of extensionality</u>

(1) $A \xrightarrow{f} B = A \xrightarrow{g} B$ iff $\models \forall x \varepsilon A \ \ fx = gx$

(2) $A \xrightarrow{M} \Omega = A \xrightarrow{N} \Omega$ iff $\models \forall x \varepsilon A \ (x \in M \Longleftrightarrow x \in N)$

<u>Proof:</u> By 2.28 and 3.3.1 we have $f=g$ iff $fx = gx$ is internally valid, which proves (1). The proof of (2) is similiar. □

The internal interpretation of terms in formulas (given in section 2) behaves as expected:

<u>4.7 Lemma</u> If the free variables of a formula φ resp. a term t are among $x_1 \varepsilon A_1, \ldots x_n \varepsilon A_n$, then

(1) $\models \quad t = \{ \langle x_1, \ldots x_n \rangle \mapsto t \} \langle x_1, \ldots x_n \rangle$

(2) $\models \quad \varphi \quad \Longleftrightarrow \quad \langle x_1, \ldots x_n \rangle \in \{ \langle x_1, \ldots x_n \rangle | \ \varphi \}$,

and for $x \varepsilon A_1 \times \ldots \times A_n$:

(3) $\{ \langle x_1, \ldots x_n \rangle \mapsto t(x_1, \ldots x_n) \} = \{ x \mapsto t(pr_1 x, \ldots pr_n x) \}$

(4) $\{ \langle x_1, \ldots x_n \rangle | \ \varphi(x_1, \ldots x_n) \} = \{ x \mid \varphi(pr_1 x, \ldots pr_n x) \}$.

Note, that the equivalent formulas in (2) may have different free variables.

<u>Proof:</u> (1) follows from 3.3.1, (2) from 3.6, 2.21.2, and (1)(2) imply (3)(4) in view of 4.5-6. □

Moreover, 4.7 tells us that that every term of L(SET) has a representation of the form $f\langle x_1, \ldots x_n \rangle$ (where $x_1, \ldots x_n$ are <u>variables</u>) and every formula is equivalent to an atomic one of the form $\langle x_1, . x_n \rangle \in M$.

Using the principle of extensionality we proceed to characterize the operations on subobjects.

4.8 Lemma For subobjects M and N of A we have:

(1) $\text{false}_A = \{x \varepsilon A \mid \text{False}\}$, $\text{true}_A = \{x \varepsilon A \mid \text{True}\}$

(2) $\neg M = \{x \varepsilon A \mid \neg x \in M\}$

(3) $M \cap N = \{x \varepsilon A \mid x \in M \wedge x \in N\}$

(4) $M \cup N = \{x \varepsilon A \mid x \in M \vee x \in N\}$

(5) $M \Rightarrow N = \{x \varepsilon A \mid x \in M \Longrightarrow x \in N\}$

(6) $M \Leftrightarrow N = \{x \varepsilon A \mid x \in M \Longleftrightarrow x \in N\}$

(7) $M \subset N$ iff $\models \forall x \varepsilon A\, (x \in M \Longrightarrow x \in N)$

Proof: (1-6) follow from 2.29, and (7) follows from (4) since $M \subset N$ is equivalent to $M \cap N = N$. □

4.9 Proposition

For a map $A \xrightarrow{f} B$ and subobjects $A \xrightarrow{M} \Omega$, $B \xrightarrow{N} \Omega$ we have :

(1) $f^{-1}(N) = \{x \varepsilon A \mid fx \in N\}$

(2) $\exists f(M) = \{y \varepsilon B \mid \exists x \varepsilon A\, (fx = y \wedge x \in M)\}$

(3) $\forall f(M) = \{y \varepsilon B \mid \forall x \varepsilon A\, (fx = y \Longrightarrow x \in M)\}$

In particular, for the image of f

(4) $\text{im}(f) := \exists f(\text{true}_A) = \{y \varepsilon B \mid \exists x \varepsilon A\ fx = y\}$.

Proof: (1) follows from 2.22.2. To prove (2,3) we establish the universal properties, namely

(2') $\{y \mid \exists x\, (fx = y \wedge x \in M)\} \subset L$ iff $M \subset f^{-1}(L)$

(3') $L \subset \{y \mid \forall x\, (fx = y \Longrightarrow x \in M)\}$ iff $f^{-1}(L) \subset M$.

Using 4.8.7 and (1) it is sufficient to show

(2") $\models \exists x\, (fx = y \wedge x \in M) \Longrightarrow y \in L$ iff

$\models x \in M \Longrightarrow fx \in L$

(3") $\models y \in L \Longrightarrow \forall x\, (fx = y \Longrightarrow x \in M)$ iff

$\models fx \in L \Longrightarrow x \in M$,

which are easily seen to hold (use the logical calculus of section 3, in particular 3.21.19-20, 3.22.3). □

We are now in the position to describe monic, epic and iso maps internally.

4.10 $A \xrightarrow{f} B$ is monic iff $\models \forall x,u\varepsilon A\ (\ fx = fu \Rightarrow x = u\ )$

4.11 $A \xrightarrow{f} B$ is epic iff $\models \forall y\varepsilon B\ \exists x\varepsilon A\ fx = y$

4.12 $A \xrightarrow{f} B$ is iso iff $\models \forall y\varepsilon B\ \exists ! x\varepsilon A\ fx = y$

<u>Proofs:</u> f is monic iff $(f\times f)^{-1}(\Delta_B) \subset \Delta_A$, and hence 4.10 follows from 4.8.7, 4.9.1. f is epic iff $\mathrm{im}(f) \supset \mathrm{true}_B$, which gives 4.11 by 4.8.7, 4.9.4. Finally, 4.12 is a consequence of 4.10-11. □

<u>4.13 Quantification along maps</u> If $A \xrightarrow{f} B$ is a map and $\varphi(y)$ a formula with free $y\varepsilon B$, then

(1) $\models \quad \forall x\varepsilon A\ \varphi(fx) \Leftrightarrow \forall y\varepsilon B\ (\ y \in \mathrm{im}(f) \Rightarrow \varphi(y)\ )$

(2) $\models \quad \exists x\varepsilon A\ \varphi(fx) \Leftrightarrow \exists y\varepsilon B\ (\ y \in \mathrm{im}(f) \wedge \varphi(y)\ )$.

Furthermore, if f <u>is monic</u>, then

(3) $\models \quad \exists ! x\varepsilon A\ \varphi(fx) \Leftrightarrow \exists ! y\varepsilon B\ (\ y \in \mathrm{im}(f) \wedge \varphi(y)\ )$.

<u>Proof:</u> Apply 4.9.4, 3.21-22 and 4.10. □

Our next step is an internal description of maps into power-objects and arbitrary exponentials.

<u>4.14 Characterization of exponential adjoints</u>

(1) The following diagram commutes

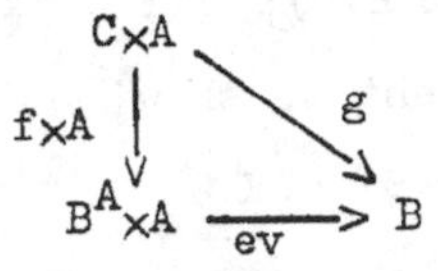

iff $\models \forall x\varepsilon C\ \forall y\varepsilon A\ (fx)y = g\langle x,y\rangle$.

(2) The following diagram commutes

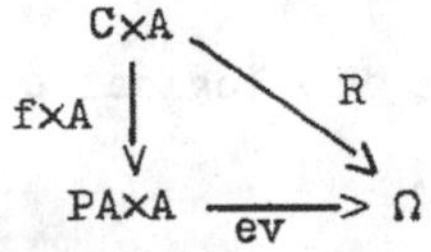

iff $\models \forall x\varepsilon C\ \forall y\varepsilon A\ (\ y \in fx \Leftrightarrow \langle x,y\rangle \in R\ )$.

The proof follows from the principle of extensionality (4.6). □
An important consequence is the internal extensionality principle (which generalizes 4.6).

4.15 Strong extensionality principle

(1) $\models \quad \forall F,G\varepsilon B^A \; (F=G \iff \forall x\varepsilon A \; Fx=Gx)$

(2) $\models \quad \forall Y,Z\varepsilon PA \; (Y=Z \iff \forall x\varepsilon A \, (x \in Y \iff x \in Z))$

Proof: To prove (1) we wish to show

$$L := \{\langle F,G\rangle \mid \forall x \; Fx=Gx \} \subset \{\langle F,G\rangle \mid F=G\} =: \Delta \quad .$$

Take a monic map $C \xrightarrow{(m,n)} B^A \times B^A$ with character $L = \mathrm{im}(m,n)$, then by 4.9.4 and 4.3.1 for $y\varepsilon C$

$$\models \quad \langle my,ny\rangle = (m,n)(y) \in L \quad .$$

Hence $\models \quad \forall x\varepsilon A \; (my)(x) = (ny)(x) \quad .$

Thus m and n have by 4.14.1, the same exponential adjoint, which gives $m=n$, resp. $L \subset \Delta$. The proof of (2) is similiar. □

Another immediate consequence of the characterization of exponential adjoints (4.14) is the following useful principle for defining maps into exponentials.

4.16 Principle for defining maps into exponentials

If φ is a formula resp. t a term of L(SET) with free variables among $x_1\varepsilon C_1$, .. $x_n\varepsilon C_n$, $y\varepsilon A$, then

(1) There exists a unique map $f_t : C_1\times..\times C_n \longrightarrow B^A$ (namely the exponential adjoint of $\{\langle x_1,..x_n,y\rangle \mapsto t\}$) such that

$$\models \quad (f_t\langle x_1,..x_n\rangle)(y) = t \quad .$$

(2) There exists a unique map $f_\varphi : C_1\times..\times C_n \longrightarrow PA$ (namely the exponential adjoint of $\{\langle x_1,..x_n,y\rangle \mid \varphi\}$) such that

$$\models \quad y \in f_\varphi\langle x_1,..x_n\rangle \iff \varphi \quad .$$ □

To illustrate this principle let us characterize (resp. define) some important maps into powerobjects.

4.17 Singleton The singleton map $A \xrightarrow{\{-\}} PA$ is characterized by : $\models \; x \in \{y\} \iff x = y \qquad (x,y \varepsilon A)$. □

4.18 Implication, binary union and intersection

(1) The internal implication $PA \times PA \xrightarrow{\Rightarrow} PA$ is characterized by

$$\models \quad x \in Y \Rightarrow Z \iff (x \in Y \Longrightarrow x \in Z)$$

(2) The internal union $PA \times PA \xrightarrow{\cup} PA$ is characterized by

$$\models \quad x \in Y \cup Z \iff (x \in Y \;\vee\; x \in Z)$$

(3) The internal intersection $PA \times PA \xrightarrow{\cap} PA$ is characterized by

$$\models \quad x \in Y \cap Z \iff (x \in Y \;\wedge\; x \in Z) \quad , (x \varepsilon A,\; Y,Z \varepsilon PA).$$ □

4.19 Arbitrary union and intersection

(1) The union map $PPA \xrightarrow{\bigcup} PA$ is characterized by

$$\models \quad x \in \bigcup Z \iff \exists Y \varepsilon PA \,(Y \in Z \;\wedge\; x \in Y)$$

(2) The intersection map $PPA \xrightarrow{\bigcap} PA$ is characterized by

$$\models \quad x \in \bigcap Z \iff \forall Y \varepsilon PA \,(Y \in Z \Longrightarrow x \in Y) \quad , (x \varepsilon A, Y \varepsilon PA).$$ □

4.20 Inclusion and powersets

The relation $PA \times PA \xrightarrow{\subset} \Omega$ of inclusion on PA is defined

$$\subset_A \; := \{\langle Y,Z\rangle \mid \forall x \varepsilon A \,(x \in Y \Longrightarrow x \in Z)\} = \{\langle Y,Z\rangle \mid Y \cap Z = Y\} \quad .$$

The internal power operator $PA \xrightarrow{\mathring{P}} PPA$ (which is the downward segment of $\subset_A$) is characterized by

$$\models \quad Y \in \mathring{P}Z \iff Y \subset Z \qquad , \qquad (Y,Z \varepsilon PA).$$ □

4.21 Images For any map $A \xrightarrow{f} B$, $x \varepsilon A$, $Y \varepsilon PA$, $Z \varepsilon PB$:

(1) The internal inverse image map $PB \xrightarrow{f^{-1}} PA$ (also denoted Ω^f) is is characterized by $\models \; x \in f^{-1}Z \iff fx \in Z$.

(2) The internal existential image map $PA \xrightarrow{\exists f} PB$ is characterized by $\models \; x \in \exists f Y \iff \exists y \varepsilon A \,(fx = y \;\wedge\; x \in Y)$.

(3) The internal universal image map $PA \xrightarrow{\forall f} PB$ is characterized by $\models \; x \in \forall f Y \iff \forall y \varepsilon A \,(fx = y \Longrightarrow y \in Y)$. □

Let us stop for a moment to realize, that we have already established the internal validity of the following axioms of many-sorted set theory:

- Axiom of extensionality (4.15.2)
- Axiom of empty sets (4.8.1)
- Axiom of singletons (4.17)
- Axiom of binary unions (4.18.2)
- Axiom of arbitrary unions (4.19.1)
- Axiom of powersets (4.20)
- Axiomscheme of separation (4.16.2) .

Since the usual axioms of set theory are internally valid, we will refer to the language L(SET), together with the "internal validity" as a notion of truth, as the natural set theory SET defined over topos theory ET.

Actually SET is not just an ordinary (many-sorted) set theory, but a Heyting-valued set theory:

For the predicates "=" of equality and "∈" of membership we have a "realization"

$$A\times A \xrightarrow{\Delta} \Omega \quad \text{resp.} \quad PA\times A \xrightarrow{ev} \Omega \quad ,$$

assigning to all pairs $\langle x,y\rangle \in A\times A$ resp. $\langle Y,x\rangle \in PA\times A$ a truth value in the (internal) Heyting-algebra Ω , such that

$$\models \quad x = y \quad \Longleftrightarrow \quad \Delta\langle x,y\rangle = \text{true}$$

$$\models \quad x \in Y \quad \Longleftrightarrow \quad ev\langle Y,x\rangle = \text{true} \quad .$$

It should be clear by now, that most (if not all) investigations of ordinary set theory which involve only intuitionistic logic (and make sense in SET) can already be carried out in the set theory SET. We will from now on presuppose some basic results of set theory in SET (in particular the "algebra of classes and relations") which the reader may easily establish by stepwise translating any introductionary text for ordinary set theory into SET. In particular some concepts of set theory will be used without giving their evident definitions in SET (controversial concepts however will be explicitly defined).

We conclude with a few results on unique existentiation.

4.22 Proposition The unique-existential image of $A \xrightarrow{M} \Omega$ under a map $A \xrightarrow{f} B$ is given by

$$(\exists!f)M = \{ y\varepsilon B \mid \exists!x\varepsilon A (fx=y \wedge x\in M) \} \quad .$$

Proof: For a monic map $C \rightarrowtail^{m} A$ with character $M = \chi m = \mathrm{im}(m)$ we have by definition 1.5

$$(\exists!f)M = (B \xrightarrow{\{-\}} PB \xrightarrow{f^{-1}} PA \xrightarrow{m^{-1}} PC)^{-1} \mathrm{im} (C \xrightarrow{\{-\}} PC) \quad .$$

Now we get the following internally valid formulas

$$\begin{aligned} y \in (\exists!f)M &\Leftrightarrow \exists u\varepsilon C\ f^{-1}m^{-1}\{y\} = \{u\} &&, \text{ by } 4.9.4 \\ &\Leftrightarrow \exists u\varepsilon C\ \forall v\varepsilon C\ (fmv = y \Leftrightarrow v = u) &&, \text{ by } 4.15.2,\ 4.17 \\ &\Leftrightarrow \exists!u\varepsilon C\ fmu = y \\ &\Leftrightarrow \exists!x\varepsilon A\ (x \in \mathrm{im}(m) \wedge fx = y) &&, \text{ by } 4.13.3 \quad . \quad \square \end{aligned}$$

Finally, we can give a description of unique existentiation in the internal interpretation, which is similiar to the definition 2.20.

4.23 Theorem Let $\varphi(x)$ be a formula of L(SET) with free $x\varepsilon A$ and other free variables among $x_1\varepsilon A_1, \ldots x_n\varepsilon A_n$.

(1) If $A_1\times\ldots\times A_n\times A \xrightarrow{p} A_1\times\ldots\times A_n$ is the projection, then
$\{\langle x_1,\ldots x_n\rangle \mid \exists!x\varepsilon A\ \varphi(x)\} = \exists!p\ \{\langle x_1,\ldots x_n,y\rangle \mid \varphi(y)\}$,
where $y\varepsilon A$ is distinct from $x_1,\ldots x_n$.

(2) If $\models \exists!x\varepsilon A\ \varphi(x)$,
then there exists a unique map $A_1\times\ldots\times A_n \xrightarrow{h} A$ such that $\models \varphi(h\langle x_1,\ldots x_n\rangle)$.

Proof: (1) Applying 4.6.2 we have to show (using 4.5 and 4.7.2)

$$\models \exists!x\varepsilon A\ \varphi(x) \Leftrightarrow \langle x_1,\ldots x_n\rangle \in \exists!p\ \{\langle x_1,\ldots x_n,y\rangle \mid \varphi(y)\} \quad ,$$

which follows easily from 4.22, using 3.20.4 and 4.5 .

(2) From (1) we conclude $(\exists ! p)\{\langle x_1,..x_n,y\rangle | \varphi(y)\} = \text{true}$, and by the characteristic property of unique existentiation (see Freyd [4] Prop. 2.21) we get a map g such that

$$A_1\times..\times A_n \xrightarrow{g} C \xrightarrow{m} A_1\times..\times A_n\times A \xrightarrow{p} A_1\times..\times A_n = \text{id} \quad ,$$

where m has the character $\chi m = \{\langle x_1,..x_n,y\rangle | \varphi(y)\}$. Then

$$A_1\times..\times A_n \xrightarrow{g} C \xrightarrow{m} A_1\times..\times A_n\times A \xrightarrow{pr} A$$

is the desired map h. The uniqueness of h follows from 4.6.1 . □

In some sense 4.23.2 states, that <u>internal</u> <u>unique</u> <u>existence</u> <u>implies</u> <u>actual</u> (<u>unique</u>) <u>existence</u> in ET. In particular we have

<u>4.24 Corollary</u> If $x \in A$ is the <u>only</u> free variable of $\varphi(x)$, then $\models \exists ! x \in A\ \varphi(x)$ implies the unique existence of a global section $1 \xrightarrow{a} A$ such that $\models \varphi(a^\circ)$. □

5. An internal characterization of the topos structure

In this section we are going to characterize the topos structure internally (i.e. using the set theory SET) in a way one would expect. The first basic observation in this direction (going back to W. Mitchell [20]) is a 1-1 correspondence between maps in the topos and "functional relations".

5.1 Proposition

(1) For any map $A \xrightarrow{f} B$, the graph of f , defined as

$$\mathrm{graph}(f) := \{\langle x,y\rangle \mid fx = y\}$$

(or, as the character of $A \rightarrowtail^{(A,f)} A\times B$) , satisfies

$$\models \forall x\varepsilon A\ \exists! y\varepsilon B\ \ \langle x,y\rangle \in \mathrm{graph}(f) \qquad .$$

(2) For any relation $A\times B \xrightarrow{R} \Omega$ such that

$$\models \forall x\varepsilon A\ \exists! y\varepsilon B\ \ \langle x,y\rangle \in R$$

holds, there exists a unique map $A \xrightarrow{\mathrm{map}(R)} B$ such that

$$\mathrm{graph}(\mathrm{map}(R)) = R \qquad .$$

Proof: (1) is evident. (2) By 4.23.2 there exists a unique map $A \xrightarrow{h} B$ such that $\langle x,hx\rangle \in R$ is internally valid, i.e. $\mathrm{graph}(h) \subset R$. But the latter condition is equivalent to $\mathrm{graph}(h) = R$. □

Moreover, obviously the graphs of identity maps are equality relations, and composing maps corresponds to relational composition of the graphs.

5.2 $\quad \mathrm{graph}(A \xrightarrow{\mathrm{id}} A) = \Delta_A$ □

5.3 $\quad$ For $A \xrightarrow{f} B \xrightarrow{g} C$: $\quad \mathrm{graph}(gf) = \mathrm{graph}(g) \circ \mathrm{graph}(f)$

Proof:
$$\begin{aligned}\mathrm{graph}(g)\circ\mathrm{graph}(f) &= \{\langle x,u\rangle \mid \exists y\ (\ fx = y \wedge gy = u\ )\}\\ &= \{\langle x,u\rangle \mid g(fx) = u\} \qquad \text{by 3.22.2}\\ &= \mathrm{graph}(gf)\end{aligned}$$
□

In a certain sense 5.1-3 describe the category structure of the topos internally (in terms of SET) and we proceed to give an internal characterization of the remaining topos structure, starting with finite limits.

The terminal object 1 has a unique element, namely 0° :

5.4 $\models \quad \forall x \varepsilon 1 \; x = 0^\circ$ □

5.5 For a formula $\varphi(x)$ with free $x \varepsilon 1$ we have

$$\models \quad \forall x \varepsilon 1 \; \varphi(x) \quad \Longleftrightarrow \quad \varphi(0^\circ) \quad ,$$

$$\models \quad \varphi(x) \quad \Longleftrightarrow \quad \varphi(0^\circ) \quad .$$ □

5.6 Characterization of terminal objects

(1) $\text{graph}(A \longrightarrow 1) = \{\langle x,y\rangle \mid y = 0^\circ\}$

(2) $\text{graph}(1 \xrightarrow{a} A) \quad \{\langle 0^\circ, a^\circ\rangle\}$

(3) A is a terminal object iff $\models \; \exists! x \varepsilon A \; x = x$.

Proof: (1,2) follow from 5.5 , to show (3) we only note, that by 4.12 the map $A \longrightarrow 1$ is iso iff $\models \exists! x \; x = x$. □

5.7 Characterization of products

(1) $\text{graph}(A\times B \xrightarrow{pr_1} A) = \{\langle\langle x,y\rangle,u\rangle \mid u = x \}$

(2) $\text{graph}(A\times B \xrightarrow{pr_2} B) = \{\langle\langle x,y\rangle,u\rangle \mid u = y \}$

For maps $C \xrightarrow{f} A$, $C \xrightarrow{g} B$ we have :

(3) $\text{graph}(C \xrightarrow{(f,g)} A\times B) = \{\langle w,\langle x,y\rangle\rangle \mid fw = x \wedge gw = y \}$

(4) $\langle C \xrightarrow{f} A , C \xrightarrow{g} B \rangle$ is a product of $\langle A,B\rangle$ iff

$$\models \quad \forall x \varepsilon A \; \forall y \varepsilon B \; \exists! w \varepsilon C \; (fw = x \wedge gw = y)$$

Proof: (1,2) follow from 4.1-3, and the condition $\models \ldots$ in (3) holds iff $(f,g): C \longrightarrow A\times B$ is iso (by 4.12). □

5.8 Characterization of equalizers

For maps $A \xrightarrow{f} B$, $A \xrightarrow{g} B$ we have :

(1) A map $D \xrightarrow{h} A$ is an equalizer of $\langle f,g\rangle$ iff

$\models\quad \forall u \varepsilon D\ fhu = ghu\ \wedge\ \forall x \varepsilon A\ (\ fx = gx\ \Rightarrow\ \exists ! v \varepsilon D\ hv = x\ )$

(2) A monic map $C \overset{m}{\rightarrowtail} A$ is an equalizer of $\langle f,g\rangle$, iff

$$im(m) = \{\ x \varepsilon A \mid fx = gx\ \} \quad .$$

<u>Proof:</u> (2) The mono m is an equalizer of $\langle f,g\rangle$, iff m is the inverse image of $B \xrightarrow{(B,B)} B\times B$ under $A \xrightarrow{(f,g)} B\times B$, i.e. iff $im(m) = (f,g)^{-1}(\Delta_B) = \{x \varepsilon A \mid fx = gx\}$. (1) follows from (2) since the condition $\models \ldots$ is easily seen to be equivalent to the following two conditions $\models\ \forall u,v \varepsilon D\ (\ hu = hv\ \Rightarrow\ u = v\ )$ (h is monic, 4.10)

$\models\ \forall x \varepsilon A\ (\ fx = gx\ \Leftrightarrow\ x \in im(h)\ )$. □

5.9 Characterization of pullbacks

A commutative diagram
$$\begin{array}{ccc} D & \xrightarrow{k} & B \\ {\scriptstyle h}\downarrow & & \downarrow{\scriptstyle g} \\ A & \xrightarrow{f} & C \end{array}$$
is a pullback, iff

$\models\quad \forall x \varepsilon A\ \forall y \varepsilon B\ (\ fx = gy\ \Rightarrow\ \exists ! u \varepsilon D\ (\ hu = x\ \wedge\ ku = y\ )\ )$.

<u>Proof:</u> We construct a pullback P as the equalizer of $A\times B \xrightarrow{pr_1} A \xrightarrow{f} C$ and $A\times B \xrightarrow{pr_2} B \xrightarrow{g} C$. Then the condition $\models \ldots$ holds by 4.12 iff the obvious map $D \longrightarrow P$ is iso. □

Having internally described all finite limits, let us consider the subobject classifier. Viewing Ω as the power $P1$, we get from 2.8

5.10 $\models\ \forall p \varepsilon \Omega\ (\ 0 \in p\ \Leftrightarrow\ p = true\ )$

resp. $\{true\} = \{\ p \varepsilon \Omega \mid 0 \in p\ \}$. In particular: $true = \{0\}$. □

5.11 Characterization of subobject classifiers

(1) For a monic map $B \overset{m}{\rightarrowtail} A$ with character $A \xrightarrow{M} \Omega$ we have

$graph(M) = \{\langle x,p\rangle \mid p = true \Leftrightarrow \exists y \varepsilon B\ my = x\ \}$.

In particular: $\models\quad M(x) = true\ \Leftrightarrow\ x \in M$.

(2) $1 \xrightarrow{a} A$ is a subobject classifier, iff

$\models\quad \forall p \varepsilon \Omega\ \exists ! x \varepsilon A\ (\ p = true\ \Leftrightarrow\ x = a\ )$.

Proof

(1) We get $\models \quad M(x) = p \iff (0 \in M(x) \iff 0 \in p)$, 4.15, 5.5

$\iff (M(x) = \text{true} \iff p = \text{true})$, 5.10

and $\models \quad M(x) = \text{true} \iff x \in M = \text{im}(m)$, by def. 2.8

$\iff \exists y \in B \; my = x$, by 4.9.4 .

(2) The condition $\models \ldots$ holds in view of (1) and 4.12 iff $A \xrightarrow{\chi_a} \Omega$ is iso. □

Viewing Ω again as the power P1 of 1, it is an easy exercise to compute the set theoretical operations $\{-\}$, $\cap$, $\cup$ $\bigcap$, $\bigcup$ on Ω .

5.12 $\quad 1 \xrightarrow{\{-\}} P1 \; = \; 1 \xrightarrow{\text{true}} \Omega$ □

5.13 $\quad P1 \times P1 \xrightarrow{\cap} P1 \; = \; \Omega \times \Omega \xrightarrow{\wedge} \Omega$ □

5.14 $\quad P1 \times P1 \xrightarrow{\cup} P1 \; = \; \Omega \times \Omega \xrightarrow{\vee} \Omega$ □

5.15 $\quad P1 \times P1 \xrightarrow{\subset} \Omega \; = \; \Omega \times \Omega \xrightarrow{\Rightarrow} \Omega$ □

5.16 $\quad P\Omega \xrightarrow{\bigcup} \Omega \; = \; \{ X \in P\Omega \mid \{\text{true}\} \subset X \}$ □

5.17 $\quad P\Omega \xrightarrow{\bigcap} \Omega = \{ X \in P\Omega \mid X \subset \{\text{true}\} \}$ □

In a certain sense, Ω is internally two-valued :

5.18 (1) $\models \quad \forall p \in \Omega \quad p \neq \text{true} \iff p = \text{false}$

i.e. $\quad \neg\{\text{true}\} = \{\text{false}\}$

(2) $\models \quad \text{true} = \text{false} \iff \forall p \in \Omega \; p = \text{true}$

i.e. $\quad \text{false} = \forall(\Omega \longrightarrow 1) \{\text{true}\}$ (This serves Mikkelsen [17] as a definition of $0 \longrightarrow 1$.) □

Returning for a moment to the fundamentals of the Ω-valued set theory SET, we observe that the object Ω of truth-values is in fact a complete (internal) Heyting-algebra with respect to $\bigcap$ and $\bigcup$. Having a complete Heyting-algebra of truth-values and "realizations" for atomic formulas (cf. section 4, right after 4.21), there is a a standard way (described e.g. in Rasiowa-Sikorski [25], X.2) for assigning to any formula φ <u>with free variables</u> $x_1 \varepsilon A_1, \ldots x_n \varepsilon A_n$ (by induction on the length of φ) a <u>realization</u>

$$[x_1 \ldots x_n | \varphi] : A_1 \times \ldots \times A_n \longrightarrow \Omega$$

using only the operations of the Heyting-algebra structure of Ω .
As expected, one can establish (again by induction on the length)

$$[x_1 \ldots x_n | \varphi] = \{\langle x_1, \ldots x_n \rangle | \varphi\}$$

by applying the following description (5.19) of existential and universal images to the definition 2.20 . Consequently, the natural notion of truth (or "satisfaction") arising from this realization coincides with internal validity.

<u>5.19 Description of existential and universal images</u>

For a map $A \xrightarrow{f} B$ and a subobject $A \xrightarrow{M} \Omega$ let $B \xrightarrow{h} P\Omega$ be the exponential adjoint of the image of $A \xrightarrow{(f,M)} B \times \Omega$, i.e.

$$\models \quad p \in hy \iff \exists x \varepsilon A \; (fx = y \wedge Mx = p) \quad , \text{ for } p \varepsilon \Omega, \; y \varepsilon B \; .$$

Then (1) $\exists f(M) = B \xrightarrow{h} P\Omega \xrightarrow{\bigcup} \Omega$

(2) $\forall f(M) = B \xrightarrow{h} P\Omega \xrightarrow{\bigcap} \Omega$.

The proof, using 5.16-17 and simple arguments in SET is omitted. □

Returning to the description of the fundamental notion of topoi, we consider power-objects.

<u>5.20 Characterization of power-objects</u>

(1) $PA \times A \xrightarrow{ev} \Omega = \{\langle Y, x \rangle | x \in Y\}$

For any relation $C \times A \xrightarrow{R} \Omega$ we have

(2) The graph of the exponential adjoint $C \xrightarrow{\overline{R}} PA$ of R is

$$\text{graph}(\overline{R}) = \{\langle u, Y \rangle \mid \forall x \varepsilon A \, (x \in Y \iff \langle u, x \rangle \in R)\} \quad .$$

(3) $C \times A \xrightarrow{R} \Omega$ is a power-formation of A , iff

$$\models \quad \forall Y \varepsilon PA \; \exists ! u \varepsilon C \; \forall x \varepsilon A \; (x \in Y \iff \langle u, x \rangle \in R) \quad .$$

<u>Proof</u>: (1) is trivial, (2) follows from 4.14.2, 4.15.2 , and to prove (3) observe that the condition $\models \ldots$ holds (by 4.12) iff

the exponential adjoint $\overline{R}$ of R is iso. □

We have now characterized all fundamental notions of topos theory ET internally (i.e. in set theory SET) and of course it is then possible to give such characterizations for all defined notions of ET (like arbitrary exponentiation, finite colimits etc.) by simply following each step of the definition within SET. We illustrate the method by some important examples: exponentials and colimits.

As to exponentials a complete analogue of 5.20 holds, but we rather give another description of exponentials which internalizes the 1-1 correspondence between maps and functional relations (see 5.1) and is essentially Kock's construction of exponentials in [6].

<u>5.21 Description of exponentials (Kock)</u>

For objects A and B the <u>internal graph map</u> $B^A \xrightarrow{\Gamma} P(A\times B)$ is defined through (cf.4.16) : $\models \forall F\varepsilon B^A\ \forall x\varepsilon A\ \forall y\varepsilon B\ \ \langle x,y\rangle \in \Gamma F \iff Fx = y$.

(1) The following diagram commutes

$$\begin{array}{ccc} B^A\times A & \xrightarrow{ev} & B \\ {\scriptstyle \Gamma\times A}\downarrow & & \downarrow{\scriptstyle \{-\}} \\ P(A\times B)\times A & \xrightarrow{\overline{ev}} & PB \end{array}$$

where $\overline{ev}$ is the adjoint of $P(A\times B)\times(A\times B) \xrightarrow{ev} A\times B$, i.e.

$$\models \forall R\varepsilon P(A\times B)\ \forall x\varepsilon A\ \forall y\varepsilon B\ \ y \in \overline{ev}\langle R,x\rangle \iff \langle x,y\rangle \in R .$$

(2) Γ is a monomorphism with character

$$\chi(\Gamma) = \{ R\varepsilon P(A\times B) \mid \forall x\varepsilon A\ \exists! y\varepsilon B\ \ \langle x,y\rangle \in R \} .$$

<u>Proof:</u> (1) is straight-forward. (2) Γ is by 4.15 monic. Let $C \overset{m}{>\!\!\longrightarrow} P(A\times B)$ be a monic map with character $\{ R \mid \forall x\ \exists! y\ \langle x,y\rangle \in R \}$. For the map $C \xrightarrow{h} B^A$ defined through

$$\models (hu)(x) = y \iff \langle x,y\rangle \in mu ,$$

we have $\Gamma h = m$ and hence $\chi m \subset \Gamma$. The converse $\Gamma \subset \chi m$ is evident. □

Turning towards a description of finite colimits we start with the initial object.

5.22 Characterization of initial objects

A is an initial object iff $\models \neg \exists x \varepsilon A \; x = x$.

The proof is straight-forward.□

Our following characterizations of coproducts and coequalizers differ from our previous characterizations since we will not assume that these colimits exist. In fact, we will actually construct coproducts and coequalizers, following the ideas of Mikkelsen [17].

5.23 Characterization of coproducts

$\langle A \xrightarrow{f} C \, , \, B \xrightarrow{g} C \rangle$ is a coproduct of $\langle A,B \rangle$ iff the two conditions hold :

$$\models \forall u \varepsilon C \; (\exists ! x \varepsilon A \; fx = u \;\vee\; \exists ! y \varepsilon B \; gy = u)$$

$$\models \forall x \varepsilon A \; \forall y \varepsilon B \quad fx \neq gy \qquad .$$

Proof: The two conditions say that f and g are monic and $true_C$ is the disjoint union of im(f) and im(g). The coproduct is known to have this property. Conversly, if the conditions hold we wish to show for given maps $A \xrightarrow{h} D$, $B \xrightarrow{k} D$ the existence of a unique map $C \longrightarrow D$ such that

$$\begin{array}{ccccc} A & \xrightarrow{f} & C & \xleftarrow{g} & B \\ & {\scriptstyle h} \searrow & \downarrow & \swarrow {\scriptstyle k} & \\ & & D & & \end{array}$$

commutes. Now

$\{ \langle u,v \rangle \mid \exists x \; fx = u \wedge hx = v \} \cup \{ \langle u,v \rangle \mid \exists y \; gy = u \wedge ky = v \}$ is easily seen to be a graph of the desired map. □

5.24 Construction of coproducts (Mikkelsen)

For objects A and B we consider the monic maps

$$m_1 = \{ x \varepsilon A \mapsto \langle \{x\}, false_B \rangle \} : A \rightarrowtail PA \times PB$$

$$m_2 = \{ y \varepsilon B \mapsto \langle false_A, \{y\} \rangle \} : B \rightarrowtail PA \times PB$$

and let $A+B \overset{m}{\rightarrowtail} PA \times PB$ be their "union" (for definitness : $m := pr_2(true, \chi m_1 \cup \chi m_2)$, see 1.1.3).

Then the unique maps in_1 and in_2 such that

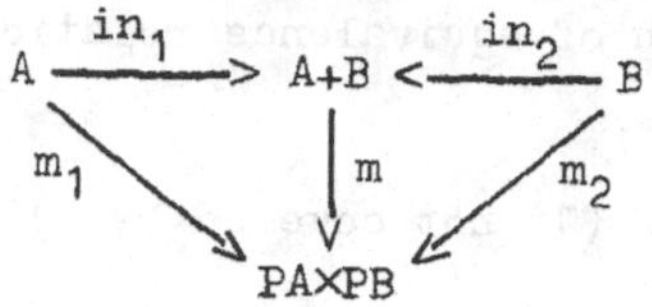

commutes is a coproduct of $\langle A,B\rangle$ by 5.23 . □

For the internal description and construction of coequalizers we need some basic knowledge of equivalence relation in our set theory SET which we state without proof.

<u>5.24 Equivalence relations</u>

For any object A let $\mathrm{Eqrel}_A\colon P(A\times A) \longrightarrow \Omega$ be the <u>subobject of all internal equivalence relations on</u> A (defined as usual).

(1) Any map $A \xrightarrow{f} B$ induces an equivalence relation on A :

$$\sim_f \;:=\; (f\times f)^{-1}(\Delta_B) \qquad , \qquad \models\ \sim_f \in \mathrm{Eqrel}_A$$

(2) Let $A\times A \xrightarrow{\pi} \Omega$ be an (external) equivalence relation on A , i.e. $\models\ \pi\in\mathrm{Eqrel}_A$, and let $A \xrightarrow{\pi^*} A/\pi$ be the epic part of the exponential adjoint $A \xrightarrow{\bar{\pi}} PA$ of π . Then $\sim_{\pi^*} = \pi$. □

<u>5.25 Coequalizers (Mikkelsen)</u>

For a pair $A \xrightarrow{f} B$, $A \xrightarrow{g} B$ of maps we consider the relation

$$\pi := \bigcap \{\, R\varepsilon P(A\times B) \mid R\in \mathrm{Eqrel}_B \wedge \exists(f\times g)(\Delta_A)\subset R \,\} \quad .$$

(1) <u>Characterization of coequalizers:</u>

$B \xrightarrow{h} C$ is a coequalizer of $\langle f,g\rangle$ iff h is epic and $\sim_h = \pi$.

(2) <u>Construction of coequalizers:</u>

π is an equivalence relation, and $B \xrightarrow{\pi^*} B/\pi$ is a coequalizer of $\langle f,g\rangle$.

<u>Proof:</u> We only scetch the proof.

a) Suppose first h in (1) is epic and $\sim_h = \pi$. Then for any map $B \xrightarrow{k} D$ with $kf = kg$ a unique factorization of k through h is given by the graph $\{\langle u,v\rangle \,\varepsilon\, C\times D \mid \exists y\varepsilon B\,(\, hy = u \wedge ky = v\,)\,\}$.

b) Let us now establish (2) using a). π is clearly an equivalence relation (any intersection of equivalence relations is one), π^* is epic and $\sim_{\pi^*} = \pi$, by 5.24.

c) The remaining part of (1) not covered by a) follows from (2). □

Having now established the preceding characterization of the basic notion of topos theory ET, it should be clear that all considerations within ET can be translated into the set theory SET. Hence, topos theory may be viewed as a part of intuitionistic many-sorted Heyting-valued set theory. However one should bear in mind that the notion of truth (internal validity) in SET refers to the notion of truth in ET, so that all arguments in SET have corresponding arguments in ET and hence working in set theory SET means actually working in topos theory ET from a different point of view : the set theoretical (or internal) one.

Unfortunatly, set theorists have neglected to study intuitionistic set theories in detail (at least to our knowledge), otherwise their results could now be applied to get new insights into topos theory. But for the time being it seems that one has to put the wagon before the horses and develop parts of intuitionistic set theory mainly for applications in topos theory. However at least in the boolean case (i.e. 1.6 holds) all classical results of (many-sorted) set theory which can be formulated in SET can be transferred into boolean topos theory EBT.

6. Applications to recursive relations and natural number objects

In the preceding section we have to some extent outlined the set theory SET defined over topos theory ET and proved some fundamental properties. Now it becomes necessary to convince the reader that the set theoretical machinery works as it should when it comes down to simplify heavy proofs and constructions in elementary topoi and to provide "new" results in topos theory by translating set theoretical results into topos theory. This of course presupposes some familiarity with the set theory SET which will be assumed here.

One application in this direction has already been given when we reformulated Mikkelsen's construction of coproducts and coequalizers. It is in fact our conviction that the set theory SET (or rather a fragment of it) should be introduced at the very beginning of the investigations in topos theory ET in order to already simplify its basic development (which was presupposed here). For example, a fragment of SET (not containing "False", $\vee$, $\exists$) can be useful to establish the existence of the initial object 0, unions of subobjects and images of maps following Mikkelsen's construction [17].

In order to get further applications for the set theoretical method let us first prove some results on inductive resp. recursive relations which are essentially due to Mikkelsen [19] (namely our following 6.1, 6.3-5).

6.1 Proposition For a relation $A \xrightarrow{r} PA$ the following conditions are equivalent :

(1) For all $A \xrightarrow{N} \Omega$: $\quad r^{-1}(\mathring{P}N) \subset N \implies N = true_A$

(2) For all $A \xrightarrow{L} \Omega$: $\quad r^{-1}(\mathring{P}L) = L \implies L = true_A$

(3) $\models \quad \forall X \varepsilon PA \; (\; r^{-1}\mathring{P}X \subset X \implies X = true_A \;)$

(4) $\models \quad \forall Y \varepsilon PA \; (\; r^{-1}\mathring{P}Y = Y \implies Y = true_A \;)$.

The relation $A \xrightarrow{r} PA$ is called <u>inductive</u> iff any of (1)-(4) holds.

<u>Proof</u>: Clearly $(3) \Rightarrow (1) \Rightarrow (2)$, $(3) \Rightarrow (4) \Rightarrow (2)$, so that it remains to show $(2) \Rightarrow (3)$. Consider the subobject of PA $M := \{ X \varepsilon PA \mid r^{-1}\mathring{P}X \subset X \}$ and the subobject $L := \bigcap M$ of A. For $x \varepsilon A$ we get

$$\begin{aligned} \models \quad x \in r^{-1}\mathring{P}L \quad & \Rightarrow \quad rx \subset L \\ & \Rightarrow \quad \forall X (X \in M \Rightarrow rx \subset X) \\ & \Rightarrow \quad \forall X (X \in M \Rightarrow x \in r^{-1}\mathring{P}X) \\ & \Rightarrow \quad \forall X (X \in M \Rightarrow x \in X) \\ & \Rightarrow \quad x \in L \end{aligned}$$

i.e $r^{-1}PL \subset L$ resp. $L \in M$. From this we conclude

$$r^{-1}\mathring{P}(r^{-1}\mathring{P}L) \subset r^{-1}\mathring{P}L \quad , \text{ i.e. } \quad r^{-1}\mathring{P}L \in M \quad .$$

Hence $L \subset r^{-1}\mathring{P}L$, and thus $L = r^{-1}\mathring{P}L$. By (2) we get $L = \bigcap M = true_A$ which implies (3). □

Our aim is to show that the inductive relations coincide with the recursive relations which were introduced in Osius [22] in order to build a model of set theory in well-pointed topoi. Recalling the

<u>6.2 Definition</u> A relation $A \xrightarrow{r} PA$ is called <u>recursive</u> iff for any map $PB \xrightarrow{g} B$ there exist a unique (<u>by</u> g r-<u>recursively defined</u>) map $A \xrightarrow{f} B$ such that $f = g(\exists f)r$, i.e the diagram

$$\begin{array}{ccc} A & \xrightarrow{f} & B \\ r \big\downarrow & & \big\uparrow g \\ PA & \xrightarrow{\exists f} & PB \end{array}$$

commutes.

Let us now prove

<u>6.3 Proposition</u> A relation $A \xrightarrow{r} PA$ is inductive if and only if for any map $PB \xrightarrow{g} B$ <u>there exists at most one</u> map $A \xrightarrow{f} B$ such that $f = g(\exists f)r$.

<u>Proof</u>: Suppose r is inductive and let f' and f" be such maps. For $N := \{ x \varepsilon A \mid f'x = f''x \}$ one clearly has : $\models \; rx \subset N \Rightarrow x \in N$.

Hence $r^{-1}\mathring{P}N \subset N$ and by 6.1.1 we get $N = true_A$, which implies $f' = f''$. Conversly, if the condition is satisfied we wish to prove 6.1.2 . Now for a subobject L of A the condition $r^{-1}\mathring{P}L = L$ holds iff the diagram

$$\begin{array}{ccc} A & \xrightarrow{\ L\ } & \Omega \\ {\scriptstyle r}\downarrow & \overset{\mathring{P}L}{\nearrow} & \uparrow{\scriptstyle \cap} \\ PA & \xrightarrow{\ \exists L\ } & P\Omega \end{array} \qquad \text{commutes ,}$$

since $(\exists L)^{-1}(\cap) = \mathring{P}L$ (which can easily be established). Hence by assumption there is at most one L such that $r^{-1}\mathring{P}L = L$. On the other hand $r^{-1}\mathring{P}(true_A) = true_A$ which proves 6.1.2 . □

<u>6.4 Corollary</u> Recursive relations are inductive. □

As in ordinary set theory, maps can be defined recursivly along an inductive relation.

<u>6.5 Recursion theorem (Mikkelsen)</u>

If $A \xrightarrow{r} PA$ is an inductive relation, then for every map $P(A\times B)\times B \xrightarrow{h} B$ there exists a unique (<u>by</u> h <u>r-recursivly defined</u>) map f such that

$$\begin{array}{ccc} A & \xrightarrow{\qquad f \qquad} & B \\ {\scriptstyle (r,A)}\downarrow & & \uparrow{\scriptstyle h} \\ PA\times A & \xrightarrow{\ \exists(A,f)\times A\ } & P(A\times B)\times A \end{array}$$

commutes, i.e. internally $\models \quad \forall x \varepsilon A \quad fx = h\langle graph(f) | rx, x\rangle$.

<u>Proof</u>: We only give the important steps of the proof, leaving some elementary details to the reader. First, the graph of f will be constructed. Let $R, F \varepsilon P(A\times B)$, $x \varepsilon A$, $y \varepsilon B$, and define a subobject of $P(A\times B)$ by

$M := \{ R \mid \forall F \ \forall x \quad F \text{ function} \wedge dom(F) = rx \wedge F \subset R \Rightarrow \langle x, h\langle F,x\rangle\rangle \in R \}$

and the subobject $G := \bigcap M$ of $A\times B$. G will be shown to be the graph of the desired map f. It is easily seen that (0) $G \in M$. Defining

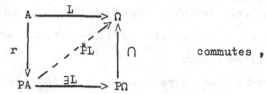

$G^* := \{\langle x,y\rangle \mid \exists F \quad F \text{ function} \wedge dom(F) = rx \wedge F \subset G \wedge y = h\langle F,x\rangle \}$, we conclude from (0) $G^* \subset G$, which in turn gives $G^* \in M$ and hence $G \subset G^*$, so that (1) $G = G^*$.

Now we establish by induction on r, that G is a graph of a map $A \longrightarrow B$, i.e. defining $N := \{ x \mid \exists ! y \ \langle x,y \rangle \in G \}$ we prove $r^{-1}PN \subset N$, resp. internally (2) $\models \quad rx \subset N \Rightarrow x \in N$.

To prove (2) we observe :

$$\models \ rx \subset N \Rightarrow G|rx \text{ function} \wedge \mathrm{dom}(G|rx) = rx$$
$$\Rightarrow \langle x, h\langle G|rx, x\rangle\rangle \in G \qquad (3)$$

and $\models \ rx \subset N \Rightarrow (F \text{ function} \wedge \mathrm{dom}(F) = rx \wedge F \subset G \Rightarrow F = G|rx)$.

From both we conclude :

$$\models \ rx \subset N \Rightarrow (\langle x,y \rangle \in G^* = G \Rightarrow y = h\langle G|rx , x\rangle)$$

which gives (2). Now by 6.1.1 we get $N = true_A$ which makes G the graph of a map $A \xrightarrow{f} B$, for which we conclude from (3)

$$\models \ fx = h\langle graph(f)|rx , x\rangle \quad .$$

This proves that f is the desired map. The uniqueness of f follows similiar as the proof of 6.3 . □

6.6 Corollary inductive relations = recursive relations

Proof: By 6.4 we have to show that inductive relations are recursive. Let $A \xrightarrow{r} PA$ be inductive and let $PB \xrightarrow{g} B$ be any map. Apply the recursion theorem to the map

$$h := P(A\times B)\times B \xrightarrow{pr_1} P(A\times B) \xrightarrow{pr_2^{-1}} PB \xrightarrow{g} B$$

in order to get the r-recursivly defined map $A \xrightarrow{f} B$ by g . □

It should be pointed out, that 6.6 and 6.1.3-4 provide an internal characterization of recursive resp. inductive relations.

As a nice application of the recursion theorem (and the set theoretical method) let us now turn to an internal characterization of natural number objects, namely through the internal "Peano axioms".

6.7 Definition

A sequence $1 \xrightarrow{o} N \xrightarrow{s} N$ is called a Peano object iff the following conditions hold :

(P1)
$$\begin{array}{ccc} 0 & \longrightarrow & N \\ \downarrow & & \downarrow s \\ 1 & \xrightarrow{o} & N \end{array} \quad \text{is a pullback.}$$

(P2) s is monic

(P3) Principle of induction. For all $N \xrightarrow{M} \Omega$:

$$o \in M \;\wedge\; \exists s(M) \subset M \quad \Longrightarrow \quad M = \mathrm{true}_N$$

or, equivalently (but internally) :

(P1') $\models \; o \notin \mathrm{im}(s)$

(P2') $\models \; \forall m,n \varepsilon N \;\; (\; sm = sn \; \Longrightarrow m = n \;)$

(P3') $\models \; \forall X \varepsilon PN \;\; (\; o \in X \wedge (\exists s)X \subset X \;\; \Longrightarrow X = \mathrm{true}_N \;)$.

The equivalences (P1) $\Longleftrightarrow$ (P1') , (P2) $\Longleftrightarrow$ (P2') are obvious, and (P3) $\Longleftrightarrow$ (P3') follows similiar to 6.1.□

Concerning the existence of Peano objects we recall the classical criterion.

6.8 Proposition If there exists maps $1 \xrightarrow{a} A \xrightarrow{f} A$ such that f is monic and

$$\begin{array}{ccc} 0 & \longrightarrow & A \\ \downarrow & & \downarrow f \\ 1 & \xrightarrow{a} & A \end{array} \quad \text{is a pullback,}$$

then there exists a Peano object $1 \xrightarrow{o} N \xrightarrow{s} N$ and a monic $N \rightarrowtail A$ such that

$$\begin{array}{ccccc} 1 & \xrightarrow{o} & N & \xrightarrow{s} & N \\ \downarrow & & \downarrow & & \downarrow \\ 1 & \xrightarrow{a} & A & \xrightarrow{f} & A \end{array} \quad \text{commutes.}$$

Proof: For the subobject $N := \bigcap \{ X \varepsilon PA \mid a \in X \wedge (\exists f) X \subset X \}$ we clearly have $a \in N$ and $(\exists f) N \subset N$. Hence, for a monic $N \overset{h}{\rightarrowtail} A$ with character $\chi h = N$ there exist a sequence $1 \xrightarrow{o} N \xrightarrow{s} N$ such that the above diagram commutes. The properties P1, P2 follow from the assumptions on the maps a, f and P3 follows from the construction of N . □

Our aim is to show that Peano objects and natural number objects (see e.g. Freyd [4]) coincide.

6.9 Successor relation

For a Peano object $1 \xrightarrow{o} N \xrightarrow{s} N$ the successor relation $N \xrightarrow{r} PN$ is defined as the exponential adjoint of $\mathrm{graph}(s): N \times N \longrightarrow \Omega$.

Then (1) $\models \; ro = \mathrm{false}_N$

(2) $\models \; \forall n \in N \;\; r(sn) = \{n\}$

(3) r is recursive .

Proof: (1,2) follow from P1,2 and to prove (3) from P3 it suffices to show for a subobject M of N (cf. 6.1.1) :

(0) $\models \quad r^{-1} \circ PM \subset M \iff o \in M \wedge (\exists s) M \subset M$.

Now by P3 and (1,2) we have

$\models \; \forall n \, (rn \subset M \Rightarrow n \in M)$

$\iff (ro \subset M \Rightarrow o \in M) \wedge \forall n \, (r(sn) \subset M \Rightarrow sn \in M)$

$\iff o \in M \wedge \forall n \, (n \in M \Rightarrow sn \in M)$,

which proves (0). □

Since the successor relations of Peano objects are recursive (inductive) we can directly apply the recursion theorem to get the usual recursion property for a natural number object (see the proof of 6.10.) .

6.10 Theorem Peano objects are natural number objects.

Proof: Given a Peano object $1 \xrightarrow{o} N \xrightarrow{s} N$ and maps $1 \xrightarrow{a} A \xrightarrow{h} A$, we wish to show the unique existence of a "sequence" f' such that

$$\begin{array}{ccccc} 1 & \xrightarrow{o} & N & \xrightarrow{s} & N \\ \big\downarrow & & \big\downarrow{\scriptstyle f'} & & \big\downarrow{\scriptstyle f'} \\ 1 & \xrightarrow{a} & A & \xrightarrow{h} & A \end{array}$$

commutes.

i.e. $\models \quad f'o = a \wedge \forall n\ f'sn = hf'n \qquad (0)$.

Using the partial map classifier (see e.g. [4,9]) $A \xrightarrow{\eta} \tilde{A}$, let g be the unique map such that

$$\begin{array}{ccccc} 1+A & \xrightarrow{(\mathrm{false},\{-\})} & PA & \xrightarrow{\exists\eta} & P\tilde{A} \\ \big\downarrow{\scriptstyle (a,h)} & & & & \big\downarrow{\scriptstyle g} \\ A & & \xrightarrow{\qquad\eta\qquad} & & \tilde{A} \end{array}$$

is a pullback. In particular for $x \varepsilon A$

(1) $\models \quad g(\mathrm{false}) = \eta a$

(2) $\models \quad g\{\eta x\} = g(\exists\eta)\{x\} = \eta hx$.

Now by the recursion property (6.2) of the successor relation r of the Peano object, there exists a unique map $\mathbb{N} \xrightarrow{f} \tilde{A}$ such that

(3) $\models \quad \forall n \varepsilon \mathbb{N} \quad fn = g(\exists f)rn$.

From (1-3) and 6.9.1-2 we conclude

(4) $\models \quad fo = \eta a \wedge \forall n \varepsilon \mathbb{N}\ (\ fn = \eta x \Rightarrow f(sn) = \eta hx\ )$

which implies $\{\ n \mid fn \in \mathrm{im}(\eta)\ \} = \mathrm{true}_{\mathbb{N}}$ by P3. Hence $\mathrm{im}(f) \subset \mathrm{im}(\eta)$ and there exists a unique factorization $\mathbb{N} \xrightarrow{f} \tilde{A} = \mathbb{N} \xrightarrow{f'} A \xrightarrow{\eta} \tilde{A}$. Since η is monic, (4) implies (0). The uniqueness of f' follows from P3 similiar to the proof of 6.3 . □

Now we establish the converse of 6.10 .

<u>6.11 Theorem</u> Natural number objects are Peano objects.

<u>Proof</u>

P1: There exists a subobject M of $\mathbb{N}$ such that

$$\begin{array}{ccccc} 1 & \xrightarrow{o} & \mathbb{N} & \xrightarrow{s} & \mathbb{N} \\ \big\downarrow & & \big\downarrow{\scriptstyle M} & & \big\downarrow{\scriptstyle M} \\ 1 & \xrightarrow[\mathrm{false}]{} & \Omega & \xrightarrow[\mathrm{true}_\Omega]{} & \Omega \end{array}$$ commutes.

Hence for $n \varepsilon \mathbb{N}$: $\models \quad o = sn \Rightarrow \mathrm{true} = \mathrm{false}$,

which proves $\models \quad o \notin \mathrm{im}(s)$, by 5.18.1 .

P2: We define the "<u>predecessor</u>" $\mathbb{N} \xrightarrow{p} \mathbb{N}$ and prove $ps = id_{\mathbb{N}}$, which makes s monic. Now there exist maps p and f such that

$$\begin{array}{ccccc}
1 & \xrightarrow{o} & N & \xrightarrow{s} & N \\
\downarrow & & \downarrow{\scriptstyle (f,p)} & & \downarrow{\scriptstyle (f,p)} \\
1 & \xrightarrow[(o,o)]{} & N\times N & \xrightarrow[(spr_1,pr_1)]{} & N\times N
\end{array}$$ commutes.

From $fo = o$ and $fs = sf$ we conclude $f = id_N$ (by uniqueness property) and hence $ps = f = id$.

P3: Let M be a subobject of $\mathbb{N}$ with $o \in M$ and $(\exists s)M \subset M$. Then M is the character of a monic map $A \overset{m}{\rightarrowtail} N$ and there exist maps a, h, f such that

$$\begin{array}{ccccc}
1 & \xrightarrow{o} & N & \xrightarrow{s} & N \\
\downarrow & & \downarrow{\scriptstyle f} & & \downarrow{\scriptstyle f} \\
1 & \xrightarrow{a} & A & \xrightarrow{h} & A \\
\downarrow & & \downarrow{\scriptstyle m} & & \downarrow{\scriptstyle m} \\
1 & \xrightarrow{o} & N & \xrightarrow{s} & N
\end{array}$$ commutes.

By the uniqueness property, $mf = id_N$ making m iso and hence $M = true$. □

<u>6.12 Corollary</u> Peano objects = natural number objects . □

Again we point out, that 6.12 and 6.7 provide an <u>internal</u> characterization of natural number objects. Furthermore 6.8 gives a sufficient (and clearly necessary) condition for the existence of natural number objects (already obtained in Freyd [4] Prop. 5.44 by other methods).

This concludes our selected applications which are not included for completeness, but only to illustrate the set theoretical method in pratice.

Bibliography

1. J. Bénabou, Catégories et logiques faibles, Tagungsbericht 30/1973 Oberwolfach

2. M. C. Bunge, Boolean topoi and the independence of Suslin's hypothesis, Preprint No. 25, Aarhus Universitet 1972/73

3. J. C. Cole, Categories of sets and models of set theory, Ph.D. Thesis, University of Sussex 1971

4. P. Freyd, Aspects of topoi, Bull. Austral. Math. Soc. 7 (1972). 1-76

5. J. Gray, The meeting of the Midwest Category Seminar in Zürich August 24-30, Springer Lecture Notes 195 (1971), 248-255

6. A. Kock - Ch. J. Mikkelsen, Non-standard extensions in the theory of toposes, Preprint No. 25, Aarhus Universitet 1971/72

7. A. Kock - Ch. J. Mikkelsen, Topos-theoretic factorization of non-standard extensions, Preprint Aarhus Universitet 1972

8. A. Kock - P. Lécouturier - Ch. J. Mikkelsen, Some topos theoretic concepts of finiteness, to appear in Springer Lecture Notes

9. A. Kock - G. C. Wraith, Elementary toposes, Lecture Notes No. 30, Aarhus Universitet 1971

10. F. W. Lawvere, An elementary theory of the category of sets, Proc. Nat. Acad. Sc. 51 (1964), 1506-1511

11. F. W. Lawvere, Quantifiers and sheaves, Actes Congrès Int. (Nice 1970) 1 (1971), 329-334

12. F. W. Lawvere, Toposes, algebraic geometry and logic, Springer Lecture Notes 274 (1972), 1-12

13. F. W. Lawvere, Continuously variable sets; Algebraic geometry = Geometric logic, to appear in Proc. Logic Coll. Bristol 1973

14. F. W. Lawvere - M. Tierney, Elementary topos, Lectures at the Midwest Category Seminar, Zürich 1970, summarized in [5]

15. P. Lécouturier, Quantificateur dans le topos élémentaires, Preprint, Université Zaïre, Kinshasa 1971/72

16. Ch. Maurer, Universen als interne Topoi, Dissertation, Universität Bremen 1974

17. Ch. Mikkelsen, Characterization of an elementary topos, Tagungsbericht Oberwolfach 30/1972

18. Ch. Mikkelsen, On the internal completeness of elementary topoi Tagungsbericht Oberwolfach 30/1973

19. Ch. Mikkelsen, Thesis, to appear

20. W. Mitchell, Boolean topoi and the theory of sets, J. Pure Appl. Alg. 2 (1972), 261-274

21. G. Osius, Kategorielle Mengenlehre: Eine Charakterisierung der Kategorie der Klassen und Abbildungen, to appear in Math. Annalen

22. G. Osius, Categorical set theory: A characterization of the category of sets, to appear in J. Pure Appl. Alg.

23. G. Osius, The internal and external aspect of logic and set theory in elementary topoi, to appear in Cah. Top. Géom. Diff.

24. R. Paré, Colimits in topoi, Preprint, Dalhousie University 1973

25. H. Rasiowa - R. Sikorski, The mathematics of metamathematics, PWN - Polish Scientific Publishers, Warzawa 1962

26. M. Tierney, Sheaf theory and the continuum hypothesis, Springer Lecture Notes 274 (1972), 13-42

27. M. Tierney, Foundations of analysis in topos, Tagungsbericht Oberwolfach 30/1972

28. G. Van de Wauw - De Kinder, Some properties concerning the natural number object in a topos, Tagungsbericht Oberwolfach 30/1973

29. G. C. Wraith, Lectures on elementary topoi, to appear in Springer Lecture Notes

Fachsektion Mathematik, Universität Bremen, Germany (BRD) .

ADDENDUM

(received by the editors in October 1974)

G. Osius : A Note on Kripke-Joyal Semantics for the Internal Language of Topoi

A NOTE ON KRIPKE-JOYAL SEMANTICS FOR THE INTERNAL LANGUAGE OF TOPOI

GERHARD OSIUS

The purpose of this paper is to give the important connection beetween the Kripke-Joyal-semantics and the internal interpretation of the set-theoretical language L(SET) of elementary topoi which is given in [3]. We assume familiarity with the basic parts of [3] and adopt the notations from there. In fact this note should be considered as an appendix to our paper [3], in particular since the material here is essentially known to the experts in this field for some time (but has not been published yet) and only our strict presentation seems to be original.

The now called Kripke-Joyal-semantics was developed by Joyal [unpublished] as a logical tool in certain categories (using ideas of Kripke's semantics) and has been used since in elementary topoi, e.g. in Kock-Lécouturier-Mikkelsen [1]. We will restrict ourselves here to elementary topoi and the following considerations will take place in the elementary theory ET of topoi (or, if the reader prefers, within a fixed elementary topos <u>E</u>). In this context the Kripke-Joyal-semantic appears as a particular interpretation of the set-theoretical language L(SET), namely the following.

<u>With respect to a fixed object</u> X of the topos we give an interpretation of the primitive operations of L(SET) :

- A-elements are interpreted as maps X ——> A which are now called <u>elements of</u> A <u>at the stage</u> (or: <u>time</u>, <u>place</u>) X .
- The constant 1-element is interpreted as X ——> 1 .

- For any map $A \xrightarrow{f} B$ the evaluation-operator $f(-)$ is interpreted through: $f(X \xrightarrow{a} A) := X \xrightarrow{a} A \xrightarrow{f} B$.
- The ordered-pair-operator is interpreted through:

$$\langle X \xrightarrow{a} A, X \xrightarrow{b} B \rangle := X \xrightarrow{(a,b)} A \times B \quad .$$

Now let $\varphi(x_1,..x_n)$ be a formula of L(SET) with free variables among $x_i \varepsilon A_i$ and let $X \xrightarrow{a_i} A_i$ be elements at stage X (i=1,..n). By induction on the length of formulas we define what it means that $\varphi(a_1,..a_n)$ <u>holds</u> <u>at</u> <u>stage</u> X under the interpretation, written $\models_X \varphi(a_1,..a_n)$:

(F) $\models_X$ False iff $X \simeq 0$.

(=) $\models_X X \xrightarrow{a} A = X \xrightarrow{b} A$ iff $a = b$.

(∧) $\models_X (\varphi(a_1,..a_n) \wedge \psi(a_1,..a_n))$ iff
$\models_X \varphi(a_1,..a_n)$ and $\models_X \psi(a_1,..a_n)$.

(∨) $\models_X (\varphi(a_1,..a_n) \vee \psi(a_1,..a_n))$ iff
there exists a <u>jointly</u> <u>epic</u> pair ($Y \xrightarrow{t} X$, $Z \xrightarrow{s} X$) such that
$\models_Y \varphi(a_1 t,..a_n t)$ <u>and</u> $\models_Z \psi(a_1 s,..a_n s)$.

(⇒) $\models_X (\varphi(a_1,..a_n) \Rightarrow \psi(a_1,..a_n))$ iff for all $Y \xrightarrow{t} X$
$\models_Y \varphi(a_1 t,..a_n t)$ implies $\models_Y \psi(a_1 t,..a_n t)$.

(∀) $\models_X (\forall y \varepsilon B)\, \varphi(a_1,..a_n,y)$ iff
for all $Y \xrightarrow{t} X$ and $Y \xrightarrow{b} B$: $\models_Y \varphi(a_1 t,..a_n t,b)$.

(∃) $\models_X (\exists y \varepsilon B)\, \varphi(a_1,..a_n,y)$ iff there exists an <u>epic</u> map $Y \overset{t}{\twoheadrightarrow} X$ and $Y \xrightarrow{b} B$ such that $\models_Y \varphi(a_1 t,..a_n t,b)$.

Since negation is defined as $(-) \Rightarrow$ False we get in particular

(¬) $\models_X \neg\varphi(a_1,..a_n)$ iff for all $Y \xrightarrow{t} X$
$\models_Y \varphi(a_1 t,..a_n t)$ implies $Y \simeq 0$.

And concerning the defined predicates $(-) \in (A \xrightarrow{M} \Omega)$ we note

$(\epsilon)\quad \models_X\ X \xrightarrow{a} A \in A \xrightarrow{M} \Omega \quad$ iff $\quad X \xrightarrow{a} A \xrightarrow{M} \Omega = \text{true}_X$.

For an intuitive understanding of the above definitions the maps $Y \xrightarrow{t} X$, $Z \xrightarrow{s} X$ beetween the stages should be viewed as passages from the "later" stages (times) Y, Z to the "present" (or "earlier") stage (time) X . 0 is the latest and 1 the earliest stage. In this terminology $(\forall)$ can be read: $(\forall y \varepsilon B)\ \varphi(a_1,..a_n,y)$ holds at stage X iff for all passages $Y \xrightarrow{t} X$ from later stages Y to X $\varphi(a_1t,..a_nt,b)$ holds at Y for all elements $Y \xrightarrow{b} B$. The other definitions can be read similiarly.

A formula $\varphi(x_1,..x_k)$ having <u>exactly</u> the free variables $x_1 \varepsilon A_1,..$ $x_k \varepsilon A_k$ is said to be <u>Kripke-Joyal-valid</u> iff for all stages X and all elements $X \xrightarrow{a_i} A_i$ $(i=1,..k)$ $\models_X \varphi(a_1,..a_k)$ holds.

The important connection beetween Kripke-Joyal-semantics and the internal interpretation of the language L(SET) is brought out by the

<u>Metatheorem</u>

For any formula $\varphi(x_1,..x_n)$ with free variables among $x_1 \varepsilon A_1,..\ x_n \varepsilon A_n$ and elements $X \xrightarrow{a_i} A_i$ $(i=1,..n)$ at a fixed stage X the following are equivalent: (1) $\models_X \varphi(a_1,..a_n)$

(2) $X \xrightarrow[\langle a_1,..a_n\rangle]{} A_1 \times .. \times A_n \xrightarrow[\{\langle x_1,..x_n\rangle \mid \varphi(x_1,..x_n)\}]{} \Omega = \text{true}_X$

i.e. $\models_X \langle a_1,..a_n\rangle \in \{\langle x_1,..x_n\rangle \mid \varphi(x_1,..x_n)\}$, by (ϵ) .

<u>Corollary</u> A formula of L(SET) is Kripke-Joyal-valid if and only if it is internally valid. □

The proof of the metatheorem is straightforward by induction on the length of formulas using the following

<u>Lemma</u> For subobjects $A \xrightarrow{M} \Omega$, $A \xrightarrow{N} \Omega$, elements $X \xrightarrow{a} A$, $X \xrightarrow{b} B$ and $A \xrightarrow{f} B$ we have:

(1) $\models_X a \in M$ iff $\text{image}(a) \subset M$.

(2) $\models_X a \in M \cap N$ iff $\models_X (a \in M \wedge a \in N)$,

and similiar for $\cup$ and $\Rightarrow$ (replace $\wedge$ above by $\vee$ and $\Rightarrow$) .

(3) $\models_X b \in \forall f(M)$ iff $\models_X \forall x \varepsilon A\,(fx = b \Rightarrow x \in M)$.

(4) $\models_X b \in \exists f(M)$ iff $\models_X \exists x \varepsilon A\,(fx = b \wedge x \in M)$.

Proof: (1) is obvious.

(2) is straightforward for $\cap$ and $\Rightarrow$ since $\models_X a \in M \Rightarrow N$ is by (1) equivalent to $im(a) \cap M \subset N$. To prove (2) for $\cup$ let $C \xrightarrow{m} A$, $D \xrightarrow{n} A$ be monic maps with character $\chi(m) = M$, $\chi(n) = N$. We note that by $(\vee)(\in)$ $\models_X (a \in M \vee a \in N)$ holds iff there exist a jointly epic pair $(Y \xrightarrow{t} X\,,\ Z \xrightarrow{s} X)$ and maps $Y \xrightarrow{c} C$, $Z \xrightarrow{d} D$ such that $mc = at$ and $nd = as$. Now given such maps t, s, c, d we have $Y+Z \xrightarrow{(t,s)} X \xrightarrow{a} A = Y+Z \xrightarrow{c+d} C+D \xrightarrow{(m,n)} A$ and hence $im(a) \subset im(m,n) = M \cup N$, which in turn gives $\models_X a \in M \cup N$ by (1). Conversly, the latter implies the existence of a' such that $X \xrightarrow{a} A = X \xrightarrow{a'} C \cup D \xrightarrow{m \cup n} A$, and pulling the jointly epic pair $(C \to C \cup D, D \to C \cup D)$ along a' yields a jointly epic pair (t,s) and maps c,d with the above properties.

(3) is again straightforward since $\models_X b \in \forall f(M)$ is equivalent to $f^{-1}(im(b)) \subset M$. And to establish (4) we note that $\models_X b \in \exists f(M)$ holds by $(\exists)(\wedge)(\in)$ iff there exist an epic $Y \xrightarrow{t} X$ and $Y \xrightarrow{c} C$ such that $Y \xrightarrow{c} C \xrightarrow{m} A \xrightarrow{f} B = Y \xrightarrow{t} X \xrightarrow{b} B$. Given such t and c we clearly have $im(b) \subset im(fm)$ and hence $\models_X b \in \exists f(M)$ by (1). Conversly, suppose the latter and let $C \xrightarrow{e} E \xrightarrow{k} B = C \xrightarrow{fm} B$ be the epi-mono-factrization of fm. By $(\in)$ there exists a map b' such that $X \xrightarrow{b'} E \xrightarrow{k} B = X \xrightarrow{b} B$, and pulling $C \xrightarrow{e} E$ along b' yields maps $Y \xrightarrow{t} X$, $Y \xrightarrow{c} C$ with the above properties.□

By the metatheorem the Kripke-Joyal-semantics and the internal interpretation of the language L(SET) provide "equivalent" logical tools in elementary topoi and since each method has some advantages over the other both should be used (according to the situation one may be more appropriate than the other). Since the internal interpretation has already been studied in detail in [3] we can immediatly conclude many properties of the Kripke-Joyal-semantics from the metatheorem. For example, we obtain from [3] Thm 4.23 the following interpretation of unique existence in Kripke-Joyal-semantics:

($\exists$!) $\models_X (\exists ! y \varepsilon B)\ \varphi(a_1,..a_n,y)$ holds iff for all $Y \xrightarrow{t} X$ there exists a unique $Y \xrightarrow{b} B$ such that $\models_Y \varphi(a_1t,..a_nt,b)$.

Let us finally observe how Kripke-Joyal-semantics can be modified if the topos is generated by a class $\underline{G}$ of objects which is closed under subobjects. In this case we restrict the above stages X, Y, Z,.. (i.e. the domains of elements) to members of the class $\underline{G}$ of generators, and all previous results hold unchanged for the restricted stages as well if we only replace the interpretation ($\exists$) for existential quantification by

$(\exists)_{\underline{G}}$ $\models_X (\exists y \varepsilon B)\ \varphi(a_1,..a_n,y)$ iff there exists a jointly epic family $(Y_i \xrightarrow{t_i} X)_{i\in I}$ and a family of elements $(Y_i \xrightarrow{b_i} B)_{i\in I}$ such that for all $i \in I$: $Y_i \in \underline{G}$ and $\models_{Y_i} \varphi(a_1t_i,..a_nt_i,b_i)$.

Examples

1. The class $\underline{G}$ of open objects is in well-opened topoi by definition a class of generators (for the plentitude of well-opened topoi see [2],4.). In this case yet another "external" interpretation of the language L(SET) which is closely related to Kripke-Joyal-semantics is given in [2]. We note that if in addition "support splits" in the topos then $(\exists)_{\underline{G}}$ can again be replace by the original ($\exists$).

2. In well-pointed topoi $\underline{G} = \{0,1\}$ is by definition a class of generators.

In both examples above the definitions (F) - $(\exists)_{\underline{G}}$ can be simplified because of the particular nature of the class $\underline{G}$ of generators.

<u>References</u>

[1] A. Kock - P. Lécouturier - Ch.J. Mikkelsen : Some topos theoretic concepts of finiteness, to appear in Springer Lecture Notes

[2] G. Osius : The internal and external aspect of logic and set theory in elementary topoi, to appear in Cahiers Top. Géom. Diff.

[3] G. Osius : Logical and set theoretical tools in elementary topoi, to appear in Springer Lecture Notes

Fachsektion Mathematik, Universität Bremen, Germany (BRD)

ɔl. 309: D. H. Sattinger, Topics in Stability and Bifurcation Theory. I, 190 pages. 1973. DM 20,–

ɔl. 310: B. Iversen, Generic Local Structure of the Morphisms in ommutative Algebra. IV, 108 pages. 1973. DM 18,–

ɔl. 311: Conference on Commutative Algebra. Edited by J. W. Brewer nd E. A. Rutter. VII, 251 pages. 1973. DM 24,–

ɔl. 312: Symposium on Ordinary Differential Equations. Edited by '. A. Harris, Jr. and Y. Sibuya. VIII, 204 pages. 1973. DM 22,–

ɔl. 313: K. Jörgens and J. Weidmann, Spectral Properties of Hamilnian Operators. III, 140 pages. 1973. DM 18,–

ɔl. 314: M. Deuring, Lectures on the Theory of Algebraic Functions One Variable. VI, 151 pages. 1973. DM 18,–

ɔl. 315: K. Bichteler, Integration Theory (with Special Attention to ector Measures). VI, 357 pages. 1973. DM 29,–

ɔl. 316: Symposium on Non-Well-Posed Problems and Logarithmic onvexity. Edited by R. J. Knops. V, 176 pages. 1973. DM 20,–

ɔl. 317: Séminaire Bourbaki – vol. 1971/72. Exposés 400–417. IV, 51 pages. 1973. DM 29,–

ɔl. 318: Recent Advances in Topological Dynamics. Edited by A. eck. VIII, 285 pages. 1973. DM 27,–

ɔl. 319: Conference on Group Theory. Edited by R. W. Gatterdam nd K. W. Weston. V, 188 pages. 1973. DM 20,–

ɔl. 320: Modular Functions of One Variable I. Edited by W. Kuyk. V, 95 pages. 1973. DM 20,–

ɔl. 321: Séminaire de Probabilités VII. Edité par P. A. Meyer. VI, 322 ages. 1973. DM 29,–

ɔl. 322: Nonlinear Problems in the Physical Sciences and Biology. dited by I. Stakgold, D. D. Joseph and D. H. Sattinger. VIII, 357 ages. 1973. DM 29,–

ɔl. 323: J. L. Lions, Perturbations Singulières dans les Problèmes ux Limites et en Contrôle Optimal. XII, 645 pages. 1973. DM 46,–

ɔl. 324: K. Kreith, Oscillation Theory. VI, 109 pages. 1973. DM 18.–

ɔl. 325: C.-C. Chou, La Transformation de Fourier Complexe et Equation de Convolution. IX, 137 pages. 1973. DM 18,–

ɔl. 326: A. Robert, Elliptic Curves. VIII, 264 pages. 1973. DM 24.–

ɔl. 327: E. Matlis, One-Dimensional Cohen-Macaulay Rings. XII, 57 pages. 1973. DM 20,–

ɔl. 328: J. R. Büchi and D. Siefkes, The Monadic Second Order heory of All Countable Ordinals. VI, 217 pages. 1973. DM 22,–

ɔl. 329: W. Trebels, Multipliers for (C, α)-Bounded Fourier Expanons in Banach Spaces and Approximation Theory. VII, 103 pages. 973. DM 18,–

ɔl. 330: Proceedings of the Second Japan-USSR Symposium on robability Theory. Edited by G. Maruyama and Yu. V. Prokhorov. I, 550 pages. 1973. DM 40,–

ɔl. 331: Summer School on Topological Vector Spaces. Edited by . Waelbroeck. VI, 226 pages. 1973. DM 22,–

ɔl. 332: Séminaire Pierre Lelong (Analyse) Année 1971-1972. V, 131 ages. 1973. DM 18,–

ɔl. 333: Numerische, insbesondere approximationstheoretische Beandlung von Funktionalgleichungen. Herausgegeben von R. Anorge und W. Törnig. VI, 296 Seiten. 1973. DM 27,–

ɔl. 334: F. Schweiger, The Metrical Theory of Jacobi-Perron Algothm. V, 111 pages. 1973. DM 18,–

ɔl. 335: H. Huck, R. Roitzsch, U. Simon, W. Vortisch, R. Walden, B. Vegner und W. Wendland, Beweismethoden der Differentialgeomeie im Großen. IX, 159 Seiten. 1973. DM 20,–

ɔl. 336: L'Analyse Harmonique dans le Domaine Complexe. Edité ar E. J. Akutowicz. VIII, 169 pages. 1973. DM 20,–

ɔl. 337: Cambridge Summer School in Mathematical Logic. Edited y A. R. D. Mathias and H. Rogers. IX, 660 pages. 1973. DM 46,–

ɔl: 338: J. Lindenstrauss and L. Tzafriri, Classical Banach Spaces. (, 243 pages. 1973. DM 24,–

ɔl. 339: G. Kempf, F. Knudsen, D. Mumford and B. Saint-Donat, oroidal Embeddings I. VIII, 209 pages. 1973. DM 22,–

ɔl. 340: Groupes de Monodromie en Géométrie Algébrique. SGA 7 II). Par P. Deligne et N. Katz. X, 438 pages. 1973. DM 44,–

ɔl. 341: Algebraic K-Theory I, Higher K-Theories. Edited by . Bass. XV, 335 pages. 1973. DM 29,–

Vol. 342: Algebraic K-Theory II, "Classical" Algebraic K-Theory, and Connections with Arithmetic. Edited by H. Bass. XV, 527 pages. 1973. DM 40,–

Vol. 343: Algebraic K-Theory III, Hermitian K-Theory and Geometric Applications. Edited by H. Bass. XV, 572 pages. 1973. DM 40,–

Vol. 344: A. S. Troelstra (Editor), Metamathematical Investigation of Intuitionistic Arithmetic and Analysis. XVII, 485 pages. 1973. DM 38,–

Vol. 345: Proceedings of a Conference on Operator Theory. Edited by P. A. Fillmore. VI, 228 pages. 1973. DM 22,–

Vol. 346: Fučik et al., Spectral Analysis of Nonlinear Operators. II, 287 pages. 1973. DM 26,–

Vol. 347: J. M. Boardman and R. M. Vogt, Homotopy Invariant Algebraic Structures on Topological Spaces. X, 257 pages. 1973. DM 24,–

Vol. 348: A. M. Mathai and R. K. Saxena, Generalized Hypergeometric Functions with Applications in Statistics and Physical Sciences. VII, 314 pages. 1973. DM 26,–

Vol. 349: Modular Functions of One Variable II. Edited by W. Kuyk and P. Deligne. V, 598 pages. 1973. DM 38,–

Vol. 350: Modular Functions of One Variable III. Edited by W. Kuyk and J.-P. Serre. V, 350 pages. 1973. DM 26,–

Vol. 351: H. Tachikawa, Quasi-Frobenius Rings and Generalizations. XI, 172 pages. 1973. DM 20,–

Vol. 352: J. D. Fay, Theta Functions on Riemann Surfaces. V, 137 pages. 1973. DM 18,–

Vol. 353: Proceedings of the Conference. on Orders, Group Rings and Related Topics. Organized by J. S. Hsia, M. L. Madan and T. G. Ralley. X, 224 pages. 1973. DM 22,–

Vol. 354: K. J. Devlin, Aspects of Constructibility. XII, 240 pages. 1973. DM 24,–

Vol. 355: M. Sion, A Theory of Semigroup Valued Measures. V, 140 pages. 1973. DM 18,–

Vol. 356: W. L. J. van der Kallen, Infinitesimally Central-Extensions of Chevalley Groups. VII, 147 pages. 1973. DM 18,–

Vol. 357: W. Borho, P. Gabriel und R. Rentschler, Primideale in Einhüllenden auflösbarer Lie-Algebren. V, 182 Seiten. 1973. DM 20,–

Vol. 358: F. L. Williams, Tensor Products of Principal Series Representations. VI, 132 pages. 1973. DM 18,–

Vol. 359: U. Stammbach, Homology in Group Theory. VIII, 183 pages. 1973. DM 20,–

Vol. 360: W. J. Padgett and R. L. Taylor, Laws of Large Numbers for Normed Linear Spaces and Certain Fréchet Spaces. VI, 111 pages. 1973. DM 18,–

Vol. 361: J. W. Schutz, Foundations of Special Relativity: Kinematic Axioms for Minkowski Space Time. XX, 314 pages. 1973. DM 26,–

Vol. 362: Proceedings of the Conference on Numerical Solution of Ordinary Differential Equations. Edited by D. Bettis. VIII, 490 pages. 1974. DM 34,–

Vol. 363: Conference on the Numerical Solution of Differential Equations. Edited by G. A. Watson. IX, 221 pages. 1974. DM 20,–

Vol. 364: Proceedings on Infinite Dimensional Holomorphy. Edited by T. L. Hayden and T. J. Suffridge. VII, 212 pages. 1974. DM 20,–

Vol. 365: R. P. Gilbert, Constructive Methods for Elliptic Equations. VII, 397 pages. 1974. DM 26,–

Vol. 366: R. Steinberg, Conjugacy Classes in Algebraic Groups (Notes by V. V. Deodhar). VI, 159 pages. 1974. DM 18,–

Vol. 367: K. Langmann und W. Lütkebohmert, Cousinverteilungen und Fortsetzungssätze. VI, 151 Seiten. 1974. DM 16,–

Vol. 368: R. J. Milgram, Unstable Homotopy from the Stable Point of View. V, 109 pages. 1974. DM 16,–

Vol. 369: Victoria Symposium on Nonstandard Analysis. Edited by A. Hurd and P. Loeb. XVIII, 339 pages. 1974. DM 26,–

Vol. 370: B. Mazur and W. Messing, Universal Extensions and One Dimensional Crystalline Cohomology. VII, 134 pages. 1974. DM 16,-

Vol. 371: V. Poenaru, Analyse Différentielle. V, 228 pages. 1974. DM 20,-

Vol. 372: Proceedings of the Second International Conference on the Theory of Groups 1973. Edited by M. F. Newman. VII, 740 pages. 1974. DM 48,-

Vol. 373: A. E. R. Woodcock and T. Poston, A Geometrical Study of the Elementary Catastrophes. V, 257 pages. 1974. DM 22,-

Vol. 374: S. Yamamuro, Differential Calculus in Topological Linear Spaces. IV, 179 pages. 1974. DM 18,-

Vol. 375: Topology Conference 1973. Edited by R. F. Dickman Jr. and P. Fletcher. X, 283 pages. 1974. DM 24,-

Vol. 376: D. B. Osteyee and I. J. Good, Information, Weight of Evidence, the Singularity between Probability Measures and Signal Detection. XI, 156 pages. 1974. DM 16.-

Vol. 377: A. M. Fink, Almost Periodic Differential Equations. VIII, 336 pages. 1974. DM 26,-

Vol. 378: TOPO 72 - General Topology and its Applications. Proceedings 1972. Edited by R. Alò, R. W. Heath and J. Nagata. XIV, 651 pages. 1974. DM 50,-

Vol. 379: A. Badrikian et S. Chevet, Mesures Cylindriques, Espaces de Wiener et Fonctions Aléatoires Gaussiennes. X, 383 pages. 1974. DM 32,-

Vol. 380: M. Petrich, Rings and Semigroups. VIII, 182 pages. 1974. DM 18,-

Vol. 381: Séminaire de Probabilités VIII. Edité par P. A. Meyer. IX, 354 pages. 1974. DM 32,-

Vol. 382: J. H. van Lint, Combinatorial Theory Seminar Eindhoven University of Technology. VI, 131 pages. 1974. DM 18,-

Vol. 383: Séminaire Bourbaki - vol. 1972/73. Exposés 418-435 IV, 334 pages. 1974. DM 30,-

Vol. 384: Functional Analysis and Applications, Proceedings 1972. Edited by L. Nachbin. V, 270 pages. 1974. DM 22,-

Vol. 385: J. Douglas Jr. and T. Dupont, Collocation Methods for Parabolic Equations in a Single Space Variable (Based on C^1-Piecewise-Polynomial Spaces). V, 147 pages. 1974. DM 16,-

Vol. 386: J. Tits, Buildings of Spherical Type and Finite BN-Pairs. IX, 299 pages. 1974. DM 24,-

Vol. 387: C. P. Bruter, Eléments de la Théorie des Matroïdes. V, 138 pages. 1974. DM 18,-

Vol. 388: R. L. Lipsman, Group Representations. X, 166 pages. 1974. DM 20,-

Vol. 389: M.-A. Knus et M. Ojanguren, Théorie de la Descente et Algèbres d' Azumaya. IV, 163 pages. 1974. DM 20,-

Vol. 390: P. A. Meyer, P. Priouret et F. Spitzer, Ecole d'Eté de Probabilités de Saint-Flour III - 1973. Edité par A. Badrikian et P.-L. Hennequin. VIII, 189 pages. 1974. DM 20,-

Vol. 391: J. Gray, Formal Category Theory: Adjointness for 2-Categories. XII, 282 pages. 1974. DM 24,-

Vol. 392: Géométrie Différentielle, Colloque, Santiago de Compostela, Espagne 1972. Edité par E. Vidal. VI, 225 pages. 1974. DM 20,-

Vol. 393: G. Wassermann, Stability of Unfoldings. IX, 164 pages. 1974. DM 20,-

Vol. 394: W. M. Patterson 3rd. Iterative Methods for the Solution of a Linear Operator Equation in Hilbert Space - A Survey. III, 183 pages. 1974. DM 20,-

Vol. 395: Numerische Behandlung nichtlinearer Integrodifferential- und Differentialgleichungen. Tagung 1973. Herausgegeben von R. Ansorge und W. Törnig. VII, 313 Seiten. 1974. DM 28,-

Vol. 396: K. H. Hofmann, M. Mislove and A. Stralka, The Pontryagin Duality of Compact O-Dimensional Semilattices and its Applications. XVI, 122 pages. 1974. DM 18,-

Vol. 397: T. Yamada, The Schur Subgroup of the Brauer Group. V, 159 pages. 1974. DM 18,-

Vol. 398: Théories de l'Information, Actes des Rencontres de Marseille-Luminy, 1973. Edité par J. Kampé de Fériet et C. Picard. XII, 201 pages. 1974. DM 23,-

Vol. 399: Functional Analysis and its Applications, Proceeding 1973. Edited by H. G. Garnir, K. R. Unni and J. H. Williamsor XVII, 569 pages. 1974. DM 44,-

Vol. 400: A Crash Course on Kleinian Groups - San Francisc 1974. Edited by L. Bers and I. Kra. VII, 130 pages. 1974. DM 18,

Vol. 401: F. Atiyah, Elliptic Operators and Compact Group V, 93 pages. 1974. DM 18,-

Vol. 402: M. Waldschmidt, Nombres Transcendants. VIII, 27 pages. 1974. DM 25,-

Vol. 403: Combinatorial Mathematics - Proceedings 1972. Edite by D. A. Holton. VIII, 148 pages. 1974. DM 18,-

Vol. 404: Théorie du Potentiel et Analyse Harmonique. Edité pa J. Faraut. V, 245 pages. 1974. DM 25,-

Vol. 405: K. Devlin and H. Johnsbråten, The Souslin Probler VIII, 132 pages. 1974. DM 18,-

Vol. 406: Graphs and Combinatorics - Proceedings 1973. Edite by R. A. Bari and F. Harary. VIII, 355 pages. 1974. DM 30,

Vol. 407: P. Berthelot, Cohomologie Cristalline des Schémas d Caracteristique p > o. VIII, 598 pages. 1974. DM 44,-

Vol. 408: J. Wermer, Potential Theory. VIII, 146 pages. 197 DM 18,-

Vol. 409: Fonctions de Plusieurs Variables Complexes, Séminai François Norguet 1970-1973. XIII, 612 pages. 1974. DM 47,

Vol. 410: Séminaire Pierre Lelong (Analyse) Année 1972-197 VI, 181 pages. 1974. DM 18,-

Vol. 411: Hypergraph Seminar. Ohio State University, 197 Edited by C. Berge and D. Ray-Chaudhuri. IX, 287 pages. 197 DM 28,-

Vol. 412: Classification of Algebraic Varieties and Compa Complex Manifolds. Proceedings 1974. Edited by H. Popp. 333 pages. 1974. DM 30,-

Vol. 413: M. Bruneau, Variation Totale d'une Fonction. XIV, 33 pages. 1974. DM 30,-

Vol. 414: T. Kambayashi, M. Miyanishi and M. Takeuchi, Un potent Algebraic Groups. VI. 165 pages. 1974. DM 20,-

Vol. 415: Ordinary and Partial Differential Equations, Proceeding of the Conference held at Dundee, 1974. XVII, 447 pages. 197 DM 37,-

Vol. 416: M. E. Taylor, Pseudo Differential Operators. IV, 15 pages. 1974. DM 18,-

Vol. 417: H. H. Keller, Differential Calculus in Locally Conve Spaces. XVI, 131 pages. 1974. DM 18,-

Vol. 418: Localization in Group Theory and Homotopy Theo and Related Topics Battelle Seattle 1974 Seminar. Edited P. J. Hilton. VI, 171 pages. 1974. DM 20,-

Vol. 419: Topics in Analysis - Proceedings 1970. Edited b O. E. Lehto, I. S. Louhivaara, and R. H. Nevanlinna. XIII, 391 page 1974. DM 35,-

Vol. 420: Category Seminar. Proceedings, Sydney Catego Theory Seminar 1972/73. Edited by G. M. Kelly. VI, 375 page 1974. DM 32,-

Vol. 421: V. Poénaru, Groupes Discrets. VI, 216 pages. 197 DM 23,-

Vol. 422: J.-M. Lemaire, Algèbres Connexes et Homologie de Espaces de Lacets. XIV, 133 pages. 1974. DM 23,-

Vol. 423: S. S. Abhyankar and A. M. Sathaye, Geometric Theor of Algebraic Space Curves. XIV, 302 pages. 1974. DM 28,

Vol. 424: L. Weiss and J. Wolfowitz, Maximum Probabili Estimators and Related Topics. V, 106 pages. 1974. DM 18,

Vol. 425: P. R. Chernoff and J. E. Marsden, Properties of Infinit Dimensional Hamiltonian Systems. IV, 160 pages. 1974. DM 20,

Vol. 426: M. L. Silverstein, Symmetric Markov Processes. I 287 pages. 1974. DM 28,-

Vol. 427: H. Omori, Infinite Dimensional Lie Transformatio Groups. XII, 149 pages. 1974. DM 18,-

Vol. 428: Algebraic and Geometrical Methods in Topology, Pr ceedings 1973. Edited by L. F. McAuley. XI, 280 pages. 197 DM 28,-